MATERIALS KINETICS
FUNDAMENTALS

MATERIALS KINETICS FUNDAMENTALS

Principles, Processes, and Applications

RYAN O'HAYRE

Cover images: Brushed metal: © vasabii | Thinkstock; Mother board: © PaulPaladin | Thinkstock; Silicon ingot: © Petr Sobolev | Thinkstock; Rust Texture: © iSailorr / Thinkstock; Grunge stone and rusty metal background: © kirstypargeter | Thinkstock
Cover design: Wiley
Interior illustrations by Michael Dale Sanders

This book is printed on acid-free paper.

Library of Congress Cataloging-in-Publication Data:

O'Hayre, Ryan P.
  Materials kinetics fundamentals : principles, processes, and applications / Ryan O'Hayre.
      pages cm
  Includes bibliographical references and index.
  ISBN 978-1-118-97289-2 (cloth) – 978-1-11897294-6 (epdf) – 978-1-118-97293-9 (epub)
1. Materials – Creep – Textbooks. 2. Mechanics, Analytic – Textbooks. I. Title.
  TA418.22.O33 2015
  620.1′1233 – dc23
                                                                              2014042177

Printed in the United States of America

10 9 8 7 6 5 4 3 2 1

To Lisa, who knows you should try everything at least twice

CONTENTS

III APPENDIXES

PREFACE

We are used to thinking about our material world in static terms. Consider the glass of water on your desk, the blue-colored sheets on your bed, the wooden door to your bathroom—these are essentially unchanging and permanent objects, right? But come back tomorrow and the level of water in that glass might be just a little bit lower. Come back in two years and your blue sheets might be fading to grey. Come back in one hundred years and your wooden door may have crumbled to dust.

As these examples illustrate, while it is tempting to think of our material world in static terms, the truth is that tiny changes are constantly taking place. If we could examine that glass of water at the atomic scale, we would see the water molecules churning and vibrating at a fantastic rate, with billions of the most energetic molecules escaping from the liquid and evaporating into the air of the room every second! At the atomic scale, we would see that the blue-colored sheets on your bed are constantly bombarded by high-energy photons (light), causing irreversible damage to the blue dye molecules coating the cotton fibers. From the moment the tree was chopped down to make your wooden door, it began its slow but inevitable decay back to the atmospheric carbon dioxide and water from which it was made. When we purposely process or manufacture materials, the changes we affect on them can be even more stunning. Consider the miraculous conversion (in a kiln) of soft clay into a strong and resilient ceramic pot or the conversion of common beach sand (via many, many steps) to a high-purity single-crystal silicon wafer.

Some changes are fast while others are much slower. In cold and rainy Seattle, it might take three weeks for a glass of water to evaporate, while in the desert of Arizona, it might only take a day. If your blue sheets are exposed to intense direct sunlight, they might photo bleach in a matter of months rather than years. If you burned that wooden door, you could convert it back into carbon dioxide and water in a few minutes instead of waiting 100 years for microbes to do the same work.

Understanding these changes is the domain of kinetics. At its core, kinetics deals with rates; in other words, kinetics tells us *how fast* something takes place—for example, how fast water can evaporate from a glass. In this textbook, you will uncover the secrets to understanding the kinetic processes described above as well as many others. This textbook is designed to provide you with an accessible and (hopefully) interesting introduction to the main concepts and principles underlying kinetic processes in materials systems. A key point here is that this textbook focuses on *materials* kinetics. While there are a large number of books on chemical kinetics, there are far fewer that focus on materials kinetics and fewer still that provide an accessible, introductory-level treatment of this subject. This textbook aims to equip you with that knowledge.

Following this mandate, the first part of this textbook, "Kinetic Principles," is devoted to a basic treatment of fundamental and universally important kinetic concepts such as diffusion and reaction rate theory. Illustrated diagrams, examples, text boxes, and homework questions are all designed to impart a unified, *intuitive* understanding of these basic kinetic concepts. Armed with these tools, the second part of the textbook, "Applications of Materials Kinetics," shows you how to apply them to qualitatively and quantitatively model common kinetic processes relevant to materials science and engineering. Since materials scientists and engineers are chiefly concerned with the solid state, the text focuses on gas–solid, liquid–solid, and solid–solid kinetic processes. A wide variety of exciting real-world examples are used to illustrate the application of kinetic principles to materials systems, including silicon processing and integrated circuit fabrication, gas transport through membranes, thin-film deposition, sintering, oxidation, carbon-14 dating, nucleation and growth, steel degassing, and kinetic aspects of energy conversion devices such as fuel cells and batteries.

ACKNOWLEDGMENTS

First and foremost, I would like to thank nearly a decades-worth of students at the Colorado School of Mines for pushing me to make the subject of materials kinetics as interesting and accessible as possible. Without their inspiration, this textbook would not have been written. In particular, I thank the 2014 Metallurgical and Materials Engineering Junior Class at CSM for their extensive critiques, comments, and enthusiasm while test piloting a draft version of this book. If there are fewer errors or run-on sentences in this textbook than you might otherwise expect, it is due, in large part, to their careful scrutiny. I also thank current and former colleagues, including Fritz Prinz at Stanford University for getting me started on the whole textbook writing thing, Dennis Readey for first sparking my love of materials kinetics, Bob Kee for trading textbook-writing war stories, and Corinne Packard for numerous discussions on how to best approach various topics in materials kinetics.

I would also like to thank Dr. Michael Sanders for his inspired artistic rendering of the many diagrams and illustrations that decorate this textbook. I dare say that it is rare indeed to find an individual who combines a Ph.D. in Materials Science with over a decades-worth of experience as a professional graphics artist. The clear and insightful illustrations in this textbook are the result of Michael's fortuitous and impressive expertise in both areas.

I would also like to thank the Grandey Fund for Energy-Related Pedagogy at the Colorado School of Mines, whose generous support, in part, made this textbook possible.

On a personal note, I am forever thankful for the encouragement, confidence, support, and love of Lisa, Kendra, Arthur, Morgan, and little Anna as well as friends, family, and colleagues around the world.

LEARNING OBJECTIVES

This textbook is not intended to be a comprehensive treatise on the kinetics of materials. Instead, it is intended to be an accessible and (hopefully) interesting introduction to the main concepts and principles that underpin kinetic processes in materials systems. The following list details some of the basic concepts and skills that you will acquire by studying this textbook. For students taking a university course in materials kinetics using this textbook, this list of learning objectives can perhaps be a helpful place to begin exam preparations.

After studying this textbook, you should be able to:

1. Define kinetics and explain the difference between kinetics and thermodynamics.

2. Give examples of both homogeneous and heterogeneous kinetic processes.

3. Convert quantities from one set of units to another quickly and accurately. For example, you should be able to define flux and be able to correctly convert between various units for flux [e.g., $mol/(cm^2 \cdot s)$ vs. $A/(cm^2 \cdot s)$ vs. $L/(cm^2 \cdot s)$ vs. $g/(cm^2 \cdot s)$ vs. $atoms/(cm^2 \cdot s)$].

4. Define mobility and write the general phenomenological equation for transport. Give examples of how this generalized equation can be applied to electrical/thermal conduction, diffusion, and convection, respectively.

5. Explain (in terms an intelligent high-school student could understand) the atomistic mechanisms of reactions. Define reaction order and give examples of first- and second-order reactions. Develop the general activated rate equation (Arrhenius relationship) that describes how reaction rate varies with temperature.

6. Apply kinetic reaction rate models to predict the progress of simple first- and second-order reactions (gas–gas reactions, radioactive decay).

7. Apply kinetic reaction rate equations to predict how reaction rates change with temperature, pressure, and concentration.

8. Compare and contrast gas, liquid, and solid-state diffusion processes. Predict and model (quantitatively) diffusion processes in all three phases of matter. Provide reasonable ballpark estimates for the approximate rates of diffusion in all three phases of matter.

9. Explain (in terms an intelligent high-school student could understand) the atomistic mechanisms of diffusion in gas and solid phases.

10. Discuss how diffusion in the gas phase depends on pressure and temperature.

11. Develop the general activated rate equation (Arrhenius relationship) that describes how solid-state diffusivity varies with temperature.

12. Give examples of kinetic processes that are reaction rate limited and processes that are diffusion limited. Write equations to quantitatively model simple coupled reaction/diffusion systems such as the passive oxidation of silicon.

13. Explain the difference between equilibrium, steady-state, and time-dependent (non-steady-state) processes. Provide concrete examples of each.

14. Explain the atomic mechanisms of solid-state nucleation and growth. Define the critical nucleation size and the critical nucleation barrier and sketch surface-mediated growth sites such as steps, kinks, and holes.

15. Discuss the kinetic and thermodynamic factors governing liquid–solid and solid–solid phase transformations. Explain and predict nucleation, growth, and time–temperature–transformation (TTT) processes in solid-state systems both qualitatively (through diagrams) and quantitatively (through equations).

16. Describe various types of solidification processes and apply information from phase diagrams to predict the type of microstructures that can arise from common isomorphous, eutectic, and peritectic solidification events.

17. Define surface energy and explain why surfaces have greater energy than the bulk. Provide examples of kinetic processes that are driven by surface energy considerations.

18. Describe (qualitatively) and mathematically model (quantitatively) morphological evolution processes in solid-state materials such as coarsening, grain growth, and sintering.

PART I

KINETIC PRINCIPLES

CHAPTER 1

INTRODUCTION TO MATERIALS KINETICS

You are about to embark on a journey into the world of materials kinetics. This chapter will act as a road map for your travels, setting the stage for the rest of the book. In broad terms, this chapter will acquaint you with an overview of kinetics, providing answers to some basic questions: What is kinetics? Why is it important? How can we classify the main types of kinetic processes? From this starting point, the subsequent chapters will lead you onward in your journey as you acquire a fundamental understanding of materials kinetics principles.

1.1 WHAT IS KINETICS?

Kinetics deals with rates; in other words, kinetics tells us "how fast" reactions take place, how rapidly phase transformations occur, or how quickly atoms move from one location to another. Additionally, kinetics describes how these rates are impacted by important system variables such as pressure, temperature, or concentration. Many kinetic phenomena can be described by basic concepts that broadly fall into one of two domains: *reaction processes* and *transport processes*. Reaction kinetics describes the rates at which reactions occur, while transport kinetics describes the rates at which matter, charge, or energy is physically transported from one place to another. In many kinetic processes, both reaction *and* transport are important. For example, consider the kinetic processes involved in the oxidation (rusting) of a metal, as illustrated in Figure 1.1. As the figure illustrates, even this seemingly simple process involves a surprisingly large number of more basic steps, and these steps can be quite different depending on exactly how the oxide grows! Don't worry about trying to understand

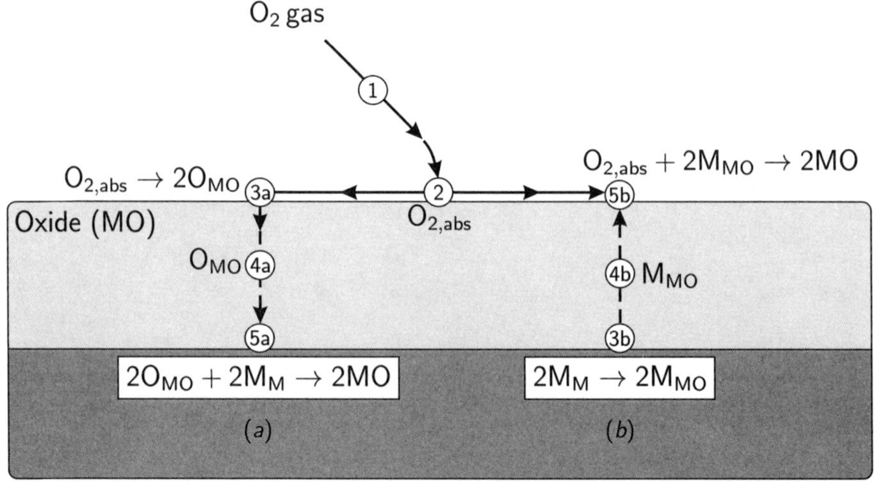

FIGURE 1.1   Oxidation of a metal (M) involves a number of steps and these steps may be different depending on whether the oxide grows from (*a*) the oxide/metal interface or (*b*) the air/oxide interface. For case (*a*), the steps might include (1) gas-phase transport of oxygen to the metal surface, (2) absorption of the oxygen gas on the surface of the material, (3a) splitting of the absorbed oxygen molecule into oxygen atoms and incorporation of these oxygen atoms into the oxide, (4a) transport (via diffusion) of the oxygen atoms through the oxide to the metal/oxide interface, and (5a) reaction of the oxygen atoms at the oxide/metal interface to create additional oxide. For case (*b*), steps such as 1 and 2 might be the same, but others might be different, such as: (3b) dissolution of metal atoms into the oxide at the metal/oxide interface, (4b) transport (via diffusion) of metal atoms through the oxide to the oxide/air interface, and (5b) splitting of the absorbed oxygen molecule into oxygen atoms and reaction with metal atoms on the surface to create additional oxide. All of these steps can be understood and modeled using the basic kinetic principles that are discussed in this textbook.

the kinetics of this oxidation process just yet. By the time you have finished this textbook, however, you will be able to understand and model all of the kinetic mechanisms illustrated in Figure 1.1.

1.2   KINETICS VERSUS THERMODYNAMICS

Although this textbook might be your first exposure to *kinetics*, you have probably already taken at least a course or two on *thermodynamics*. If you thought you could forget all that thermodynamics knowledge now that you have moved on to kinetics, think again! Kinetics and thermodynamics are closely coupled; understanding and modeling the kinetics of a process first requires a good understanding of the thermodynamic forces that are driving it. Don't worry if your thermodynamics skills are a little bit rusty. The next chapter in this textbook is designed to provide you with a brief review of the most important thermodynamic principles needed to tackle kinetics.

How can we understand the difference between thermodynamics and kinetics? Here is a simple way to think about the two subjects: Thermodynamics predicts what *should* happen, kinetics predicts *how fast* it will happen.

Figure 1.2 schematically illustrates the relationship between thermodynamics and kinetics. Thermodynamics deals with energy states and driving forces—in other words, thermodynamics can tell us the most energetically favorable state for a system under a given set of conditions (i.e., at a given temperature, pressure, and composition). For systems that are not in equilibrium (i.e., not in their lowest energy state for a given set of conditions), thermodynamics can tell us how far away from equilibrium they are and thus the magnitude of the driving forces that are acting on them in an effort to bring them to equilibrium. We can therefore use thermodynamics to predict how a system might want to change and also quantify the size of the driving forces acting on the system to help bring about those changes. While this thermodynamic information is extremely helpful in predicting what *should* happen in a system, it is insufficient for predicting if it actually *will* happen or *how fast* it will happen. In order to answer those questions, we need kinetics. Kinetics deals with the speed of changes—in other words, kinetics tells us how fast a system can change from one state to another state.

The relative thermodynamic stability of graphite versus diamond provides a classic illustration of the interplay between thermodynamics and kinetics. Graphite and diamond are both polymorphs (same composition but different phases) of carbon. At room temperature and pressure, thermodynamics tells us that diamond is less stable than graphite—in other words, there is an energetic driving force favoring the transformation of diamond into graphite. So, are diamonds forever? Thermodynamics

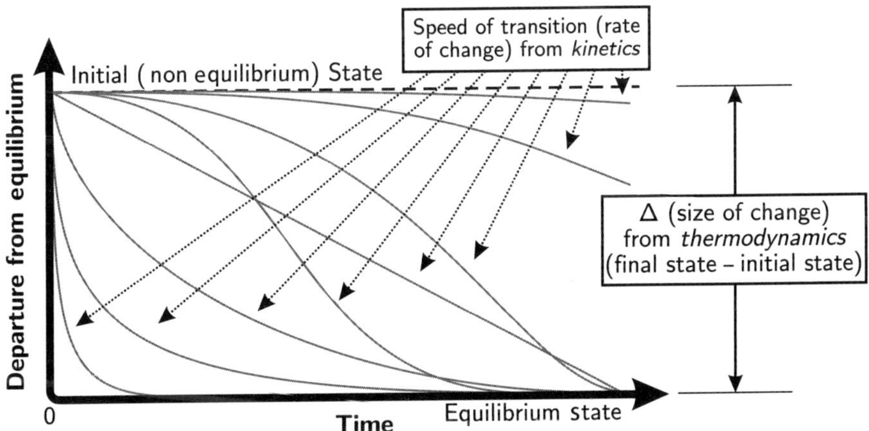

FIGURE 1.2 Schematic illustration of the interplay between thermodynamics and kinetics. Thermodynamics can be used to quantify departures from equilibrium and hence the driving forces acting on a system to return it to equilibrium. Kinetics can be used to determine the pathway and the speed at which a system returns to equilibrium. Many different scenarios are possible, from extremely fast equilibration to rates so slow that they are essentially zero.

suggests not! Fortunately for all who enjoy the sparkle of a diamond ring, however, kinetics is here to save the day. Although there is a definite thermodynamic driving force favoring the conversion of diamond into graphite, the kinetic barriers are so high that the rate of this transformation is essentially zero at room temperature and pressure. If you really wanted to get rid of a diamond ring, you would have to throw it into an extremely hot furnace (I suppose Mt. Doom might also work) in order to provide enough energy to allow kinetics to do its work. At such high temperatures, however, the diamond would likely transform (burn) into CO_2 rather than convert to graphite—unless you heated it in an oxygen-free environment. I would not suggest testing out this particular kinetic experiment, however!

1.3 HOMOGENEOUS VERSUS HETEROGENEOUS KINETICS

This textbook introduces you to the discipline of kinetics. More specifically, however, this textbook is intended to introduce you to *materials kinetics*. Why do we make a specific distinction regarding materials kinetics and what does this mean?

The field of kinetics largely evolved from chemistry and specifically from the study of chemical reactions (most often in the gas phase). The detailed kinetics knowledge acquired from the study of chemical reactions has grown and matured over the years into the well-known discipline of chemical kinetics. Today there are a number of fantastic textbooks [1–4] that cover this topic and almost every chemistry and chemical engineering department in the world teaches courses on this subject. Compared to chemical kinetics, however, the field of materials kinetics is much less well defined and much less well established. Trying to overview or even define the field of materials kinetics is a daunting task. In addition, there are very few textbooks (especially introductory textbooks) on the subject, although courses on materials kinetics are typically taught at almost every materials science and engineering department in the world. Clearly, an understanding of materials kinetics is crucial for all materials engineers and scientists, but just what is it?

There are two main aspects that distinguish materials kinetics: (1) the involvement of at least one solid phase and (2) the (frequent) presence of heterogeneity.

The first point is easy to understand. As materials scientists and engineers, we concern ourselves with the solid state. Thus, the kinetic phenomena that we encounter typically also involve the solid state or incorporate at least one solid phase. This focus on the solid state is a significant distinction from the chemical kinetics training that most chemists and chemical engineers receive, which typically focuses on gas-phase or liquid-phase reactions and processes.

The second point essentially arises directly from the first point. Because materials kinetics focuses on the solid state, the systems we encounter almost always involve heterogeneity. What is meant by heterogeneity? The simplest definition of a heterogeneous kinetic process is one in which more than one phase is involved. Thus, any kinetic process involving both a solid and a liquid or a solid and a gas is necessarily heterogeneous. The metal oxidation process discussed in Figure 1.1, for example, is a heterogeneous kinetic process that involves *three* phases: the solid metal phase, the solid oxide phase, and the gas (air) phase.

In the solid state, heterogeneity is generally manifested by the presence of boundaries or interfaces between distinct regions or phases in a system. These boundaries or interfaces often play an important, if not dominating, role in the kinetic behavior of the system. In contrast to single-phase gas or liquid systems, heterogeneity is typically present and important even for most single-phase solid-state systems as well. Consider, for example, the coarsening and grain growth that occurs during the annealing of a polycrystalline material. Understanding the properties of these grain boundaries and how quickly they can propagate through the material during the annealing process is the special domain of materials kinetics.

Essentially all of the quirks and imperfections that make solid-state systems interesting—point defects, dislocations, grain boundaries, inclusions, voids, surfaces—fall within the scope of materials kinetics. This focus on solid-state processes and heterogeneity—what many would call microstructural development—is what makes materials kinetics unique. In order to tackle this topic, we will need to borrow a lot of concepts from chemical reaction kinetics, which we will cover in Chapter 3 of this textbook, but we will also learn many other concepts that are not usually covered in traditional chemical-based treatments of kinetics. In particular, we will spend a lot of time on solid-state diffusion and transport (Chapter 4). Compared to the gas and liquid phases, transport of matter in the solid phase tends to be slower and more difficult; thus, atomic transport processes such as diffusion become much more important in determining kinetic behavior in solid-state systems.

1.4   REACTION VERSUS DIFFUSION

As discussed above, one distinguishing feature of the solid state is that the transport of matter (e.g., atoms or molecules) tends to be much slower and more difficult than in the liquid or gas state. As a result, diffusion plays a central role in many solid-state kinetic processes. As we will learn in Chapter 4, diffusion processes can be described using a set of relatively straightforward mathematical equations known as Fick's first and second laws. These equations can be applied to determine how fast a diffusion process occurs. A key quantity that appears in these equations is a parameter known as the *diffusion coefficient*, or *diffusivity*, D. This parameter quantifies the relative ease at which atoms or molecules can be transported via diffusion in a material. It changes depending on both the nature of the atoms or molecules that are moving (diffusing) as well as the nature of the material through which they are moving. Generally speaking, the higher the value of D, the faster a species can diffuse through a material.

Table 1.1 provides typical values for atomic diffusivities in the solid, liquid, and gas states. As you can see, solid-state diffusivities tend to be many orders of magnitude slower than liquid- or gas-phase diffusivities. Thus, solid-state diffusion often tends to be a *rate-limiting* step in many solid-state kinetic processes.

TABLE 1.1 Typical Atomic Diffusivities for Solid, Liquid, and Gas States

State	Diffusivity (cm^2/s)
Solid $\left(<\frac{1}{2}T_m\right)$	10^{-10}–10^{-30}
Solid (near T_m)	10^{-6}–10^{-10}
Liquid	10^{-4}–10^{-5}
Gas	10^{-1}

Note: T_m = melting temperature.

What Is a Rate-Limiting Step?

Kinetic processes often involve a number of individual steps that must be accomplished in series. Consider, for example, the metal oxidation process illustrated in Figure 1.1, which involves at least five steps. The overall rate of such processes is typically limited by the *slowest* step. This step is often described as the "rate-limiting" step. The rate-limiting step can also be thought of as the process "bottleneck." In a process where one particular step is the bottleneck, speeding up the other steps will not help the situation—we can only increase the overall rate if we speed up the rate-limiting step.

In addition to (or instead of) series steps, certain kinetic processes can have parallel pathways—for example, in the metal oxidation process illustrated in Figure 1.1, the diffusion of oxygen atoms through the oxide layer from the air/oxide interface to the oxide/metal interface (step 4a) can occur in *parallel* with the diffusion of metal atoms through the oxide layer from the oxide/metal interface to the air/oxide interface (step 4b). In such instances, the *fastest* step controls the overall kinetic behavior of the system. Thus, for metal oxidation, when oxygen diffusion through the oxide layer (4a) is faster than metal diffusion through the oxide layer (4b), kinetic pathway (*a*) dominates overall, and the oxide grows from the metal/oxide interface. In contrast, when metal diffusion through the oxide layer is faster than oxygen diffusion through the oxide layer, kinetic pathway (*b*) dominates overall, and the oxide grows from the air/oxide interface. Some metal oxidation processes (such as the oxidation of Si) proceed via pathway (*a*), while others (such as the oxidation of Ag) proceed via pathway (*b*).

In addition to diffusivity D, another kinetic parameter that we will frequently encounter is the reaction rate constant k. The reaction rate constant is used to quantify the relative ease of a chemical reaction or, in some cases, a localized reconfiguration or charge transfer process. As with diffusion processes, there are a number of mathematical expressions (rate laws) available to describe the speed of various reaction processes. Both homogeneous reactions as well as heterogeneous reactions can be quantified using rate constants, and we will see some examples of both in Chapter 3. Just as a higher value of D indicates a higher relative ease of diffusion, a higher value of k indicates a higher relative ease of reaction.

Because of the heterogeneous nature of solid-state systems, when reactions occur in these systems, they typically give rise to or proceed at interfaces (i.e., at surfaces or phase boundaries). For example, the metal oxidation process illustrated in Figure 1.1 involves several steps occurring at the air/oxide and oxide/metal interfaces which can be described using reaction kinetics (e.g., steps 2, 3a/b, 5a/b).

Because the overall rate of a kinetic process is determined by its rate-limiting step, a common theme that is frequently observed in solid-state kinetic systems is the competition between reaction and diffusion. Here, again, the metal oxidation process illustrated in Figure 1.1 provides an instructive example. During the initial stages of oxidation, when the oxide layer is very thin, it is quite common for the overall oxidation rate to be rate limited by one of the reaction processes. However, as oxidation proceeds and the oxide layer grows thicker, diffusion becomes rate limiting because the diffusing atoms face an increasingly thicker oxide through which they must diffuse.

1.5   CLASSIFYING KINETIC PROCESSES

Perhaps one of the most straightforward ways to classify kinetic processes is in terms of the phases of matter that are involved. Using this approach, kinetic processes can be grouped into six broad categories:

1. Gas–gas
2. Gas–liquid
3. Liquid–liquid
4. Gas–solid
5. Liquid–solid
6. Solid–solid

Figure 1.3 provides examples of each of these six categories. While there is certainly overlap in the interest and treatment of these categories between fields, the first three categories are a common focus of chemists and chemical engineers while materials scientists and engineers focus predominantly on the last three categories. In this textbook, you will have an opportunity to tackle examples from all six categories, although we will focus most particularly on the last three: gas–solid, liquid–solid, and solid–solid. Via the applications-oriented chapters in the second half of this textbook (Chapters 5–7) you will have the opportunity to encounter a broad range of interesting real-world examples of gas–solid, liquid–solid, and solid–solid kinetic processes.

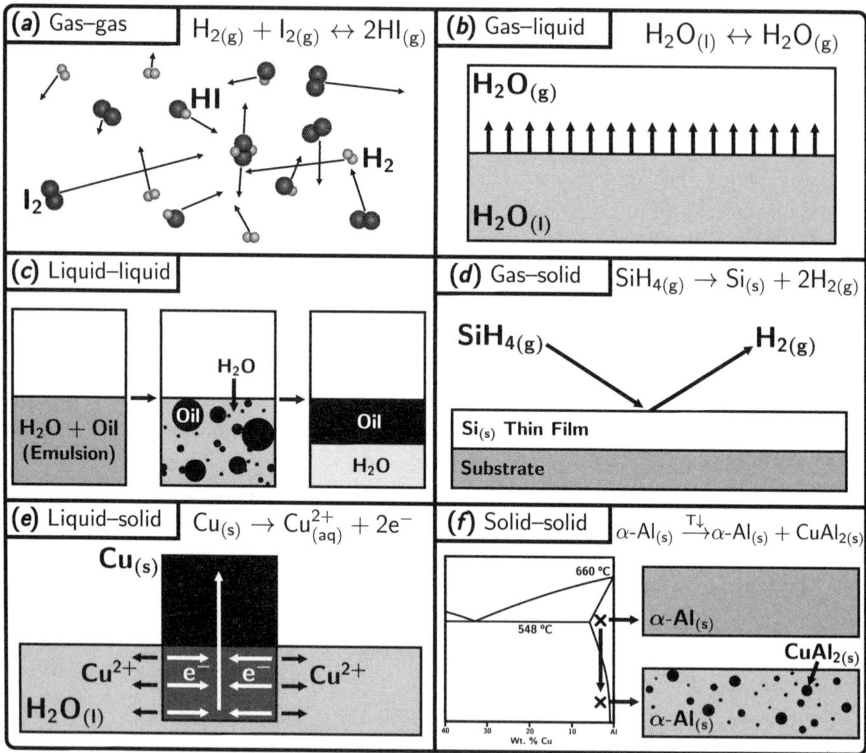

FIGURE 1.3 Examples of kinetic processes classified by types of phases involved. (*a*) Gas–gas: reaction equilibrium between hydrogen gas, iodine gas, and hydrogen iodide gas. (*b*) Gas–liquid: evaporation of liquid water from a glass. (*c*) Liquid–Liquid: gradual separation of an oil–water mixture. (*d*) Gas–solid: chemical vapor deposition of a thin Si film. (*e*) Liquid–solid: corrosion of Cu metal in seawater. (*f*) Solid–solid: precipitation of $CuAl_2$ particles from a copper–aluminum alloy during a heat treatment process.

1.6   BRIEF WORD ABOUT UNITS

Unit conversions invariably present challenges for students and professors alike. Like other technical disciplines, the field of kinetics presents many opportunities for unit confusion—in fact, it has been this author's experience over the years that unit issues result in more lost points on exams than any other single issue! In particular, instances where pressure or energy units appear in an expression are among the most common sources for unit errors. This is because non-SI units (such as atm, torr, cal, and eV) are commonly used for pressures and energies. Students are therefore highly encouraged to carefully read through Appendix A for a review on units. A number of exercises are also provided at the end of this chapter to provide unit conversion practice.

1.7 CHAPTER SUMMARY

The purpose of this chapter was to set the stage for learning about kinetics and to give a broad overview of the field of materials kinetics. The main points introduced in this chapter include:

- Thermodynamics predicts whether a process *should* happen, while kinetics predicts *how fast* it will happen. In the simplest terms, kinetics deals with rates. Additionally, kinetics describes how these rates are impacted by important system variables such as pressure, temperature, or concentration.

- When thermodynamics predicts that a process is favorable, this does not necessarily mean that it *will* happen. There are many thermodynamically favorable processes that do not occur because the kinetic barriers are too high. A "yes" from thermodynamics really means "maybe."

- Many kinetic phenomena can be described by basic concepts that broadly fall into one of two domains: *reaction processes* and *transport processes*. Reaction kinetics describes the rates at which reactions occur while transport kinetics describes the rates at which matter (e.g., atoms or molecules), or charge, or energy is physically transported from one place to another.

- An overall kinetic process can often be broken down into a set of more detailed individual kinetic steps, some of which must occur in *series*, while others can take place in *parallel*. The overall rate for a process is determined by the interaction between all of these individual steps and is often governed by one step that is much slower than the others—this is known as the rate-limiting step.

- A homogeneous kinetic process is one which occurs in a single phase, while a heterogeneous kinetic process involves several (two or more) distinct regions or phases. Almost all solid-state kinetic processes are heterogeneous because almost all solid-state systems manifest heterogeneity. Therefore, the field of materials kinetics mainly confronts heterogeneous kinetic processes.

- Compared to the gas and liquid phase, transport of matter in the solid phase tends to be much slower and more difficult—thus atomic transport processes such as diffusion become much more important in determining kinetic behavior in solid-state systems.

- Kinetic processes can be classified according to the phases of matter involved. Using this scheme, there are six main classes of kinetic processes: gas–gas, gas–liquid, liquid–liquid, gas–solid, liquid–solid, and solid–solid. The field of materials kinetics is chiefly concerned with kinetic processes involving at least one solid phase.

1.8   CHAPTER EXERCISES

Review Questions

Problem 1.1. Define kinetics. Contrast kinetics and thermodynamics.

Problem 1.2. The transformation of diamond to graphite is an example of a thermodynamically favorable but kinetically frustrated process. A kinetically frustrated process is one that will not occur over a reasonable scientifically observable time period. Come up with another example of a process that is thermodynamically favorable but kinetically frustrated.

Problem 1.3. (a) Give the definition of a homogeneous kinetic process. (b) Give an example of a homogeneous kinetic process. Provide diagrams, text, and equations to fully support your example in sufficient detail.

Problem 1.4. (a) Give the definition of a heterogeneous kinetic process. (b) Give an example of a heterogeneous kinetic process. Provide diagrams, text, and equations to fully support your example in sufficient detail.

Problem 1.5. Create a figure similar to Figure 1.3 but with different (i.e., new) examples of each of the six main types of kinetic processes.

Problem 1.6. Give an example of a kinetic process involving at least two *series* steps. Provide diagrams, text, and equations to fully support your example in sufficient detail.

Problem 1.7. Give an example of a kinetic process involving at least two *parallel* steps. Provide diagrams, text, and equations to fully support your example in sufficient detail.

Calculation Questions

Problem 1.8. The SI value of the gas constant R is $R = 8.314 \ \text{J}/(\text{mol} \cdot \text{K})$. Convert R to the following units and show all your work in each case:

(a) $\text{cal}/(\text{mol} \cdot \text{K})$

(b) $\text{L} \cdot \text{atm}/(\text{mol} \cdot \text{K})$

(c) $\text{cm}^3 \cdot \text{atm}/(\text{mol} \cdot \text{K})$

Problem 1.9. In Chapter 4, we will develop a formula based on the kinetic theory of gases to predict the self-diffusivity of a single-species ideal gas as

$$D_{\text{gas}} = \sqrt{\frac{1}{6M}} \frac{(RT)^{3/2}}{\pi d^2 P N_{\text{A}}} \tag{1.1}$$

A simplified version of this formula, using non-SI units, is often given as

$$D_{\text{gas}} = 1.61 \times 10^{-3} \sqrt{\frac{1}{M}} \frac{(T)^{3/2}}{d^2 P} \tag{1.2}$$

where T is in kelvin, P is in atm, d is in Å, and M is in grams per mole. Derive how the prefactor of 1.61×10^{-3} is obtained. Clearly show all your work.

Problem 1.10. If the energy required to move a unit electron charge across a potential difference of one volt is defined as an electron-volt, denoted eV, calculate:

(a) An electron-volt in joules

(b) The energy in joules to move 1 mol of electrons across a potential of 1 V

(c) The same as in (b) but in units of kilocalories per mole

(d) Convert from the value of R in J/(mol · K) to the value of k in eV/K.

(e) If thermal energy is given as kT, calculate the thermal energy in electron-volts at 300 K and 1000 °C.

CHAPTER 2

A SHORT DETOUR INTO THERMODYNAMICS

Thermodynamics is the study of energetics—in other words, the study of energy contents and energy flows in and between systems. By applying thermodynamic principles, we can (among other things), determine whether a system is at *equilibrium*, that is, in its lowest (most stable) energy state. Perhaps of greater relevance to our study of kinetics, however, thermodynamics can tell us when a system is not in equilibrium, and it allows us to calculate the size of energetic deviations from equilibrium. Displacements from equilibrium induce energetic driving forces that act on a system to try to return it to equilibrium. The sizes of these energetic driving forces become key inputs into many kinetic equations, and thus we will need appropriate thermodynamic tools to calculate these quantities. Although some previous exposure to thermodynamics will be assumed, in this chapter we will revisit a number of the most important thermodynamic principles and review how to calculate some of the most important thermodynamic quantities that we will need as we continue in our development of kinetics.

2.1 DYNAMIC EQUILIBRIUM

While equilibrium indicates a stable state, it is crucial to understand that equilibrium does not necessarily imply a completely static and unchanging state. We will often encounter the concept of *dynamic equilibrium* in our discussions of kinetics. In a dynamic equilibrium, backward and forward kinetic processes of equal but opposite rates occur. For example, a glass of water can be in dynamic equilibrium with 100% relative humidity air above it. In such a situation, the overall level of water in the glass will not change, even after weeks or months of waiting. However, if we could examine this system at the atomistic scale, we would see that water molecules are constantly escaping the liquid phase for the gas phase, while an equal and opposite number (stochastically averaged over time and space) of water molecules are moving from the gas phase to the liquid phase. Thus, even in this equilibrium and seemingly static system, kinetic processes (evaporation and condensation) are constantly occurring.

From the discussion above, it is clear that equilibrium in most systems should be thought of as a dynamic *balancing act*. We will return to this concept of balance frequently in this textbook. Making even small changes to a system at equilibrium can upset this balance. If we were to heat the glass of water in the example discussed above, this would increase the rate of evaporation from the glass while decreasing the rate of condensation, thereby pushing the system out of equilibrium. In most materials systems, there are three main levers that we can use to push systems into or out of equilibrium and hence induce change:

1. Changes in temperature
2. Changes in pressure
3. Changes in composition

In the sections that follow, we will review the main thermodynamic tools that we can use to calculate the equilibrium conditions for a system and also to determine the size of the thermodynamic deviations that occur when a system is subjected to changes in temperature, pressure, and/or composition that force it away from equilibrium.

2.2 ENTHALPY (*H*), ENTROPY (*S*), AND GIBBS FREE ENERGY (*G*)

The three most important thermodynamic quantities that we will encounter again and again throughout this textbook are *enthalpy* (H), *entropy* (S), and (Gibbs) *free energy* (G). Usually, we will be concerned with quantifying changes in these thermodynamic quantities (i.e., ΔH, ΔS, ΔG) during a process or reaction rather than the absolute values of H, S, or G. It is important to remember that changes in thermodynamic functions are always calculated as *final state − initial state*. Thus, $\Delta G = G_{\text{final}} - G_{\text{initial}}$.

While you have hopefully already been acquainted with enthalpy, entropy, and Gibbs free energy before, here is a brief review of these three important thermodynamic quantities:

Enthalpy H can be considered to be a measure of the heat value of a system. For a reversible thermodynamic process at constant pressure, ΔH represents the heat that is released (if ΔH is negative) or the heat that is absorbed (if ΔH is positive) during the process. An *exothermic* process is one where ΔH is negative (heat is released), while an *endothermic* process is one where ΔH is positive (heat is absorbed).

Entropy S is a measure of the disorder of a system or more precisely, a measure of the number of potential microscopic configurations available to a system. The absolute entropy (S) for any system at any temperature above absolute zero must be positive. Furthermore, the total net change in entropy (ΔS) experienced by the combination of a system and its surrounding during any thermodynamic process must always be greater than or equal to zero. (This is the second law of thermodynamics). However, a system itself can experience either a negative or positive entropy change (ΔS) during a thermodynamic process. For a system to experience a negative entropy change during a thermodynamic process, its surroundings must experience at least an equal if not greater positive change in entropy. When a system manifests a negative change in entropy, this can be interpreted as the system moving toward a more ordered state (e.g., a gas condensing to a liquid), while a positive change in entropy can be interpreted as a move toward a more disordered state.

Gibbs free energy G represents the maximum amount of energy that is *free* or available in a system to do work or to affect change. The Gibbs free energy is perhaps the most crucial of the three thermodynamic quantities we have discussed. It is the key for determining the *spontaneity* or energetic favorability of a thermodynamic process. If ΔG is zero, then this indicates that there is no free energy available to affect change in a system. Worse yet, if ΔG is greater than zero, then additional energy must be supplied to a system in order to make the process happen. Therefore, the sign of ΔG indicates whether or not a thermodynamic process is spontaneous (favorable):

- $\Delta G > 0$: nonspontaneous (energetically unfavorable)
- $\Delta G = 0$: equilibrium
- $\Delta G < 0$: spontaneous (energetically favorable)

A spontaneous process is energetically favorable; it is a "downhill" process. Although spontaneous processes are energetically favorable, spontaneity is no guarantee that a process will occur, nor does it indicate how fast a process will occur. Many spontaneous processes do not occur because they are impeded by kinetic barriers. Thus, our calculation of ΔG only provides us the first step in our quest to understand the rate of processes. Once we have determined ΔG for a process, we will then need to apply kinetic laws to determine *how quickly* (if at all) the process will happen!

2.2.1   Relationship between ΔG, ΔH, and ΔS

For a closed thermodynamic system at constant temperature and pressure, the following equation can be used to relate changes experienced by the three thermodynamic quantities discussed above during a thermodynamic process:

$$\Delta G = \Delta H - T\,\Delta S \tag{2.1}$$

You will find this equation extremely useful to help you understand whether a process is thermodynamically favorable or not and to gain insight into how a process's favorability is affected by temperature.

ΔG will necessarily be negative, and hence a process will be thermodynamically favorable if ΔH is negative and ΔS is positive. Exothermic processes (i.e., $\Delta H < 0$) tend to be thermodynamically favorable, unless they are also associated with large negative entropy changes (i.e., $\Delta S < 0$), in which case the magnitudes of ΔH and $T \Delta S$ must be compared to determine thermodynamic favorability.

For processes where ΔS is positive, increasing the temperature will tend to increase the thermodynamic favorability. In other words, increasing temperature favors processes where system disorder (entropy) is increased. Thus, both melting (going from the ordered solid state to the less ordered liquid state) and evaporation (going from the liquid state to the even less ordered gas state) are increasingly favored at higher temperatures. This is a principle that we are intuitively familiar with from our personal experiences; you can apply it to many other situations as well.

When ΔH is negative and ΔS is also negative, thermodynamic favorability again depends on the temperature, but in this case, lower temperatures lead to greater favorability. This is the situation for condensation or solidification reactions, which release heat (and hence are exothermic) but lead to an increase in system order (and hence have negative ΔS). As we already understand intuitively, condensation and solidification are favored as we decrease the temperature!

If ΔH is positive and ΔS is negative, then ΔG will necessarily be positive, and hence the process will be thermodynamically unfavorable regardless of the temperature.

These concepts are summarized in Table 2.1.

TABLE 2.1 Relationship between ΔH, ΔS, and ΔG

ΔH	ΔS	ΔG	
−	+	−	Favorable
−	−	?	Less favorable with increasing T
+	−	+	Unfavorable
+	+	?	More favorable with increasing T

Energy, Entropy, and Heat Death of Universe

Energy is conserved. This is the first law of thermodynamics. Consider the chain of energy conversions involved in the movement of a car. The energy contained in the car's fuel is converted by the engine to thermal energy (heat) as well as kinetic energy to propel the car forward. The kinetic energy moving the car forward is itself subsequently converted into additional heat energy as the car's body pushes through the air and the car's tires overcome the frictional resistance of the road. Finally, the car's brakes dissipate any remaining kinetic energy to further heat as the car comes to a stop. In terms of the energy of the universe—nothing has changed in this process—there is just as much energy in the universe after the car has been driven as before it was driven. What has changed, however, is the *availability* of that energy. What was before highly concentrated energy stored in the form of chemical bonds in the car's fuel has now been spread out into an essentially unrecoverable form as a slight increase in the thermal energy of the car, the road, and the air.

If you track the flows and conversions of energy associated with essentially any process, the tendency is that some, if not all, of the energy involved is eventually dissipated as heat. The more technical way of stating this is that $\Delta G < 0$ for any spontaneous process, while $\Delta G/T = -\Delta S_{universe}$. Thus the entropy of the universe always increases during any spontaneous process (essentially every change that occurs). The early scientists who developed thermodynamics recognized the profound implications of this conclusion when taken to its logical extreme: Eventually, all the energy of the universe must be dissipated into a dilute, uniform thermal background, at which point nothing more can ever change. This concept is known as the "heat death" of the universe.

2.3 MOLAR QUANTITIES

Typical notation distinguishes between intensive and extensive variables. Intensive quantities such as temperature and pressure do not scale with the system size; extensive quantities such as internal energy and entropy do scale with system size. For example, if the size of a box of gas molecules is doubled and the number of molecules in the box doubles, then the internal energy and entropy double while the temperature and pressure are constant. Frequently, certain conventions are used to denote intensive versus extensive properties, such as using lowercase for intensive variables (e.g., p for pressure) and uppercase for extensive variables (e.g., S for entropy). Unfortunately, such nomenclature is not standardized and is frequently inconsistent. For example, temperature is almost always represented with an uppercase T, although it is an intensive quantity.

Many extensive properties can be normalized by the size of the system to give *specific* properties, that is, properties per unit mass, or per mole, or per volume. For example, consider the extensive Gibbs free-energy change (ΔG_{rxn}) involved in a chemical reaction, which can be normalized to an intensive quantity ($\Delta \hat{G}_{rxn}$) via

$$\Delta \hat{G}_{rxn} = \frac{\Delta G_{rxn}}{n} \tag{2.2}$$

where n is the number of moles of substance involved in the reaction and $\Delta \hat{G}_{rxn}$ is the *molar* Gibbs free-energy change for the reaction (kJ/mol). An occasional practice is to denote *specific* properties with a particular convention, such as the "hat" (^) symbol used above or by underlining them (ΔG). Again, however, such nomenclature is not standardized and is frequently inconsistent within or between texts.

In this text, no attempt will be made to use nomenclature to distinguish between extensive and intensive properties or between an extensive property and its mass or mole-normalized specific counterpart. You should be aware, then, that quantities appearing in certain equations may represent an extensive property or its specific (e.g., per-mole) counterpart depending on the situation. For example, ΔG could have units of kilojoules or kilojoules per mole depending on the context. In general, this should not cause confusion, as the context and units involved will almost always be spelled out explicitly.

Extensive thermodynamic quantities such as ΔH, ΔS, and ΔG are almost always normalized per mole of substance involved to produce intensive, molar-based values for these quantities. This is because it is often useful to quantify energy changes due to a reaction on a per-mole basis. Thus, when you encounter these quantities in this textbook, they almost always refer to the specific (per-mole) value.

Calculating Extensive versus Specific Thermodynamic Quantities

Remember, the Δ symbol denotes a change during a thermodynamic process (such as a reaction), calculated as *final state − initial state*. Therefore, a negative energy change means energy is released during a process; a negative volume change means the volume decreases during a process. For example, the combustion of H_2 and O_2 to produce water,

$$H_2 + \frac{1}{2}O_2 \rightleftharpoons H_2O \tag{2.3}$$

has $\Delta G_{rxn} = -237 \text{ kJ/mol } H_2$ at room temperature and pressure. For every 1 mol H_2 gas consumed (or every $\frac{1}{2}$ mol O_2 gas consumed or 1 mol H_2O produced), the Gibbs free-energy change is -237 kJ. If 5 mol of O_2 gas is reacted, the *extensive* Gibbs free-energy change (ΔG_{rxn}) would be

$$5 \text{ mol } O_2 \times \left(\frac{1 \text{ mol } H_2}{1/2 \text{ mol } O_2} \right) \times \left(\frac{-237 \text{ kJ}}{1 \text{ mol } H_2} \right) = -2370 \text{ kJ} \tag{2.4}$$

Of course the *specific* (per-mole) Gibbs free energy of this reaction is still $\Delta G_{rxn} = -237 \text{ kJ/mol } H_2$. In both cases, a quick inspection of the units makes it clear whether the ΔG involved is an extensive or specific (molar) quantity.

2.4 STANDARD STATE

Because most thermodynamic quantities depend on temperature and pressure, it is convenient to reference everything to a standard set of conditions. There are two types of standard conditions:

The *thermodynamic standard state* describes the standard set of conditions under which reference values of thermodynamic quantities are typically given. Standard-state conditions specify that all reactant and product species are present in their pure, most stable forms at unit activity. For gases, this implies 1 atm partial pressure. For liquids, this implies a pure liquid under 1 atm hydrostatic pressure. For a solute, it implies an ideal solution at 1 M concentration (1 mol/L). For solids, it implies a pure solid under 1 atm pressure. Standard-state conditions are designated by a degree symbol. For example, $\Delta H°$ represents an enthalpy change under standard-state thermodynamic conditions. Importantly, there is no "standard temperature" in the definition of thermodynamic standard-state conditions. However, since most tables list standard-state thermodynamic quantities at 25 °C (298.15 K), this temperature is usually implied. At temperatures other than 25 °C, it is sometimes necessary to apply temperature corrections to $\Delta H°$ and $\Delta S°$ values obtained at 25 °C, although it is frequently approximated that these values change only slightly with temperature, and hence this issue can be ignored. For temperatures far from 25 °C, however, this approximation should not be made. Furthermore, it should be noted that $\Delta G°$ changes much more strongly with temperature (as shown in Equation 2.1) and therefore $\Delta G°$ values should always be adjusted by temperature using at least the linear dependence predicted by Equation 2.1.

Standard temperature and pressure, or STP, is the standard condition most typically associated with gas law calculations. STP conditions are taken as room temperature (298.15 K) and atmospheric pressure. (Standard-state pressure is actually defined as 1 bar = 100 kPa. Atmospheric pressure is taken as 1 atm = 101.325 kPa. These slight differences are usually ignored.)

2.5 CALCULATING THERMODYNAMIC QUANTITIES

When a chemical reaction occurs, the corresponding changes in the system's thermodynamic functions can be calculated by computing the differences in the thermodynamic values between the reactants and products. For a general reaction

$$a\text{A} + b\text{B} \rightarrow m\text{M} + n\text{N} \tag{2.5}$$

where A and B are reactants, M and N are products, and a, b, m, n represent the number of moles of A, B, M, and N, respectively, $\Delta H°_{\text{rxn}}$ may be calculated as

$$\Delta H°_{\text{rxn}} = [m\,\Delta H°_{\text{f,M}} + n\,\Delta H°_{\text{f,N}}] - [a\,\Delta H°_{\text{f,A}} + b\,\Delta H°_{\text{f,B}}] \tag{2.6}$$

where the molar standard-state formation enthalpy $\Delta H°_{\text{f},i}$ tells how much enthalpy is required to form 1 mol of chemical species i under standard-state conditions from the reference species.

Thus, the molar enthalpy of reaction ($\Delta H°_{\text{rxn}}$) is computed from the difference between the *molar weighted* reactant and product formation enthalpies. Note that enthalpy changes (like all energy changes) are computed in the form of *final state − initial state*, or in other words, *products − reactants*.

An expression analogous to Equation 2.6 may be written for the molar standard-state entropy of a reaction, $\Delta S^\circ_{\text{rxn}}$, using *standard molar entropy* values S° for the species taking part in the reaction. See Example 2.6 for details. The standard-state Gibbs free energy of reaction can likewise be calculated from the Gibbs free energies of the constituent species, or alternatively, once $\Delta H^\circ_{\text{rxn}}$ and $\Delta S^\circ_{\text{rxn}}$ are obtained, $\Delta G^\circ_{\text{rxn}}$ can be obtained from the relationship $\Delta G = \Delta H - T\Delta S$.

Example 2.1

Question: Calculate $\Delta H^\circ_{\text{rxn}}$, $\Delta S^\circ_{\text{rxn}}$, and $\Delta G^\circ_{\text{rxn}}$ (at $T = 298$ K) for the decomposition of hydrogen iodide:

$$2\text{HI}(g) \rightarrow \text{H}_2(g) + \text{I}_2(g) \tag{2.7}$$

Solution: By consulting a thermodynamic database, we can obtain the $\Delta H^\circ_{\text{f}}$ and S° values for HI, H$_2$, and I$_2$ at $T = 298$ K as follows:

Chemical Species	$\Delta H^\circ_{\text{f},i}$ (kJ/mol)	S°_i [J/(mol $\cdot$ K)]
HI(g)	26.5	206.6
H$_2$(g)	—	130.7
I$_2$(g)	62.43	260.7

Following Equation 2.6, the $\Delta H^\circ_{\text{rxn}}$ for HI decomposition is calculated as

$$\Delta H^\circ_{\text{rxn}} = [\Delta H^\circ_{\text{f},\text{H}_2} + \Delta H^\circ_{\text{f},\text{I}_2}] - [2\Delta H^\circ_{\text{f},\text{HI}}] \tag{2.8}$$

$$= [0 + 62.43] - [2 \cdot 26.5]$$

$$= 9.43 \text{ kJ/mol}$$

Similarly, $\Delta S^\circ_{\text{rxn}}$ is calculated as

$$\Delta S^\circ_{\text{rxn}} = [S^\circ_{\text{H}_2} + S^\circ_{\text{I}_2}] - [2S^\circ_{\text{HI}}] \tag{2.9}$$

$$= [130.7 + 260.7] - [2 \cdot 206.6]$$

$$= -21.8 \text{ J/(mol} \cdot \text{K)}$$

Finally, $\Delta G^\circ_{\text{rxn}}$ may be calculated from the relationship $\Delta G = \Delta H - T\Delta S$ as

$$\Delta G^\circ_{\text{rxn}} = \Delta H^\circ_{\text{rxn}} - T\Delta S^\circ_{\text{rxn}} \tag{2.10}$$

$$= 9.43 \text{ kJ/mol} - (298 \text{ K})[-21.8 \text{ J/(mol} \cdot \text{K)}]\left(\frac{1 \text{ kJ}}{1000 \text{ J}}\right)$$

$$= 15.93 \text{ kJ/mol}$$

2.6   REACTION QUOTIENT Q AND EQUILIBRIUM CONSTANT K

Most thermodynamic quantities are given at the standard state. As previously discussed, the standard state assumes unit activity for all species participating in the process. However, we often want to know the ΔG for a process under non-standard-state conditions. For example, in the hydrogen/iodine/hydrogen iodide system, maybe we would like to calculate ΔG for specific amounts (partial pressures) of the three gases involved in the reaction. In order to deal with non-standard-state conditions, it is necessary to introduce the concept of *chemical potential*. Chemical potential measures how the Gibbs free energy of a system changes as the chemistry of the system changes. Each chemical species in a system is assigned a chemical potential. Formally

$$\mu_i^{\alpha} = \left(\frac{\partial G}{\partial n_i} \right)_{T,P,n_{j \neq i}} \tag{2.11}$$

where μ_i^{α} is the chemical potential of species i in phase α and $(\partial G/\partial n_i)_{T,P,n_{j \neq i}}$ expresses how much the Gibbs free energy of the system changes for an infinitesimal increase in the quantity of species i (while temperature, pressure, and the quantities of all other species in the system are held constant). When we change the amounts (concentrations) of chemical species in a system, we are changing the free energy of the system. This change in free energy in turn changes the equilibrium point for the system. Understanding chemical potential is therefore key to understanding how changes in concentration can tip a system toward (or away from) equilibrium.

Chemical potential is related to concentration through *activity a*:

$$\mu_i = \mu_i^{\circ} + RT \ \ln a_i \tag{2.12}$$

where μ_i° is the reference chemical potential of species i at standard-state conditions and a_i is the activity of species i. The activity of a species depends on its chemical nature and is described in more detail in the dialog box below.

The activity of a species depends on its chemical nature. Here are a few simple guidelines to help you calculate the activity for various species depending on their state:

For an *ideal gas*, $a_i = p_i/p^{\circ}$, where p_i is the partial pressure of the gas and p° is the standard-state pressure (1 atm). For example, the activity of oxygen in air at 1 atm is approximately 0.21. The activity of oxygen in air pressurized to 2 atm would be 0.42. Since we accept $p^{\circ} = 1$ atm, we are often lazy and write $a_i = p_i$, recognizing that p_i is a *unitless* gas partial pressure.

For a *nonideal gas*, $a_i = \gamma_i(p_i/p^{\circ})$, where γ_i is an activity coefficient describing the departure from ideality ($0 < \gamma_i < 1$).

For a *dilute (ideal) solution*, $a_i = c_i/c^\circ$, where c_i is the molar concentration of the species and c° is the standard-state concentration (1 M = 1 mol/L). For example, the activity of Na^+ ions in 0.1 M NaCl is 0.10.

For *nonideal solutions*, $a_i = \gamma_i(c_i/c^\circ)$. Again, we use γ_i to describe departures from ideality ($0 < \gamma_i < 1$).

For *pure components*, $a_i = 1$. For example, the activity of gold in a chunk of pure gold is 1. The activity of platinum in a platinum electrode is 1. The activity of liquid water is usually taken as 1.

Combining Equations 2.11 and 2.12, it is possible to calculate changes in the Gibbs free energy for a system of i chemical species by

$$dG = \sum_i \mu_i \, dn_i = \sum_i (\mu_i^\circ + RT \, \ln a_i) \, dn_i \qquad (2.13)$$

Consider an arbitrary chemical reaction placed on a molar basis for species A in the form

$$1A + bB \rightleftharpoons mM + nN \qquad (2.14)$$

where A and B are reactants, M and N are products, and 1, b, m, and n represent the number of moles of A, B, M, and N, respectively. On a molar basis for species A, ΔG for this reaction may be calculated from the chemical potentials of the various species participating in the reaction (assuming a single phase):

$$\Delta G = (m\mu_M^\circ + n\mu_N^\circ) - (\mu_A^\circ + b\mu_B^\circ) + RT \, \ln \frac{a_M^m a_N^n}{a_A^1 a_B^b} \qquad (2.15)$$

Recognizing that the lumped standard-state chemical potential terms represent the standard-state molar free-energy change for the reaction, ΔG°, the equation can be simplified to a final form:

$$\Delta G = \Delta G^\circ + RT \, \ln \frac{a_M^m a_N^n}{a_A^1 a_B^b} \qquad (2.16)$$

$$\Delta G = \Delta G^\circ + RT \, \ln Q \qquad (2.17)$$

This equation, called the van't Hoff isotherm, tells how the Gibbs free energy of a system changes as a function of the activities (read concentrations or gas pressures) of the reactant and product species.

Example 2.2

Question: Calculate ΔG_{rxn} for the hydrogen iodide decomposition reaction under the following conditions:

1. $P_{HI} = 10.0$ atm, $P_{I_2} = 1.5$ atm, $P_{H_2} = 2.0$ atm, $T = 298$ K
2. $P_{HI} = 1.0$ atm, $P_{I_2} = 2.0$ atm, $P_{H_2} = 1.0$ atm, $T = 400$ K

Solution: From Example 2.1 we know $\Delta G_{rxn}^{\circ} = 15.93$ kJ/mol at $T = 298$ K. Using Equation 2.16 to adjust for pressure and Equation 2.1 to adjust for temperature, we can therefore determine ΔG_{rxn} under any arbitrary set of non-standard-state conditions.

For the conditions given in 1, only the pressure adjustment is needed, since the temperature given (298 K) is the same as was used to calculate ΔG_{rxn}° in Example 2.1. Applying the conditions given in 1 to Equation 2.16 yields

$$\Delta G_{rxn} = \Delta G_{rxn}^{\circ} + RT \ \ln \frac{a_{I_2(g)} a_{H_2(g)}}{a_{HI(g)}^2} \tag{2.18}$$

$$= \Delta G_{rxn}^{\circ} + RT \ \ln \frac{(P_{I_2}/P_{I_2}^{\circ})(P_{H_2}/P_{H_2}^{\circ})}{(P_{HI}/P_{HI}^{\circ})^2}$$

$$= 15{,}930 \ \text{J/mol} + [8.314 \ \text{J/(mol} \cdot \text{K}) \cdot 298 \ \text{K}]$$

$$\times \ln \frac{(1.5/1.0)(2.0/1.0)}{(10.0/1.0)^2}$$

$$= 7240 \ \text{J/mol} = 7.2 \ \text{kJ/mol}$$

For the second set of conditions, the temperature given in 2 (400 K) is different from the temperature that was used to calculate ΔG_{rxn}° in Example 2.1 (298 K). Thus, we must first determine ΔG_{rxn}° at this new temperature. This would best be done by consulting a thermodynamic database to get values for ΔH_f° and S° for HI(g), I_2(g), and H_2(g) at $T = 400$ K and then using those values to calculate ΔG_{rxn}° at 400 K in a manner analogous to what was done in Example 2.1 at 298 K. Alternatively, values for ΔH_f° and S° at 400 K could be estimated from the values at 298 K if we knew the heat capacities of HI(g), I_2(g), and H_2(g). However, we can also make the assumption (reasonable for modest changes in temperature) that these heat capacity effects are negligible. In this case, we can approximate ΔG_{rxn}° at 400 K with Equation 2.1 using the previously calculated values for ΔH_{rxn}° and ΔS_{rxn}° at $T = 298$ K:

$$\Delta G_{rxn}^{\circ} = \Delta H_{rxn}^{\circ} - T\Delta S_{rxn}^{\circ} \tag{2.19}$$

$$\Delta G_{rxn}^{\circ}|_{T=400K} \approx \Delta H_{rxn}^{\circ}|_{T=298K} - (400 \ \text{K})\Delta S_{rxn}^{\circ}|_{T=298K}$$

$$\approx 9430 \ \text{J/mol} - [400 \ \text{K} \cdot -21.8 \ \text{J/(mol} \cdot \text{K})]$$

$$\approx 18150 \ \text{J/mol} = 18.2 \ \text{kJ/mol}$$

Applying this estimated value for $\Delta G^{\circ}_{\mathrm{rxn}}$ at 400 K together with the other conditions given in 2 to Equation 2.16 then yields

$$\Delta G_{\mathrm{rxn}} = \Delta G^{\circ}_{\mathrm{rxn}} + RT \ \ln \frac{a_{\mathrm{I_2(g)}} a_{\mathrm{H_2(g)}}}{a^2_{\mathrm{HI(g)}}} \tag{2.20}$$

$$= \Delta G^{\circ}_{\mathrm{rxn}} + RT \ \ln \frac{(P_{\mathrm{I_2}}/P^{\circ}_{\mathrm{I_2}})(P_{\mathrm{H_2}}/P^{\circ}_{\mathrm{H_2}})}{(P_{\mathrm{HI}}/P^{\circ}_{\mathrm{HI}})^2}$$

$$= 18150 \ \mathrm{J/mol} + [8.314 \ \mathrm{J/(mol \cdot K)} \cdot 400 \ \mathrm{K}] \ln \frac{(2.0/1.0)(1.0/1.0)}{(1.0/1.0)^2}$$

$$= 20450 \ \mathrm{J/mol} = 20 \ \mathrm{kJ/mol}$$

Difference between ΔG and ΔG°

It is crucial to recognize the distinction between ΔG and ΔG°. At a given temperature, a reaction can have only ONE value for the standard-state free-energy change ΔG°. This corresponds to the free-energy change for an idealized reaction process that never really happens—the complete conversion of pure reactants at unit activity into pure products at unit activity under standard-state conditions. In contrast, a reaction can have an infinite number of values for ΔG, which describes the free energy for an actual system containing any arbitrary mixture of reactant and product species under any arbitrary temperature and pressure conditions.

The term that appears in the natural logarithm of Equation 2.16, which is a ratio of product and reactant activities raised to their appropriate stochiometric coefficients, has a special name. It is known as the *reaction quotient Q*. As we discussed in the section on dynamic equilibrium, it is important to think of all processes, including chemical reactions, as dynamic processes that can occur in both the forward and backward directions. Thus, you should think of a chemical reaction as being like a balance between the reactant and product species—and the reaction quotient is essentially a quantitative indicator of that balance. The reaction quotient indicates whether the current balance of a reaction under any arbitrary set of conditions has been skewed more toward the reactant or product side as compared to the standard-state conditions. At the standard-state condition, all of the reactants and product species are at unit activity, and thus the reaction quotient is 1. In this case, $\Delta G = \Delta G^{\circ}$, which makes sense, since ΔG° is the free energy under standard-state conditions.

Recall that for a system at equilibrium, $\Delta G = 0$. This is the definition of thermodynamic equilibrium. Applying this definition to Equation 2.16 enables us to determine the precise ratio of reactant and product activities that lead to a perfect balance (equilibrium) between the reactant and product states in a chemical system. This specific value of the reaction quotient has a special name. It is known as the *equilibrium*

TABLE 2.2   Relationship between Q, K, and Reaction Equilibria

Q vs. K	Reaction Balance
$Q < K$	Reaction will proceed in forward direction (reactants $\rightarrow$ products)
$Q = K$	Reaction is at equilibrium, forward and backward rates equal
$Q > K$	Reaction will proceed in backward direction (reactants $\leftarrow$ products)

constant K: At equilibrium $\Delta G = 0$; thus $0 = \Delta G° + RT \ln K$, and therefore

$$K = e^{-\Delta G°/(RT)} \tag{2.21}$$

When the reaction quotient Q differs from the equilibrium constant K, this indicates a system that is not in equilibrium. For example, given a system that is initially at equilibrium ($Q = K, \Delta G = 0$), if we increase the product activities relative to the reactant activities, the reaction quotient increases ($Q > K$). In this case, $\Delta G > 0$. In other words, increasing the product activities has decreased the favorability of the forward reaction (or in the other sense, increased the favorability of the backward reaction). In order for the system to return to equilibrium, some of those excess product species must be converted back into reactant species. A mechanical analogy is helpful here: You can think of the increase in the product activities (and hence Q) as an increase in the *chemical pressure* on the right-hand side of the reaction equation, which forces the reaction balance back to the left. On the other extreme, an increase in the reactant species activities would decrease the reaction quotient ($Q < K$). In this case, the forward-reaction direction becomes favored ($\Delta G < 0$), and in order to restore equilibrium, some of those excess reactant species must be converted into product species. Using the mechanical analogy again, by increasing the reactant species activities, we have increased the chemical pressure on the left-hand side of the reaction equation, which forces the reaction to the right.

Thus, we can use the reaction quotient (Q) and the equilibrium constant (K) as guides to help us understand reaction equilibria. For reactions that are not at equilibrium, we can compare Q versus K to determine the direction a reaction should proceed in order to restore its equilibrium. The key points of our discussion of Q and K are summarized in Table 2.2.

Le Châtelier's Principle

The analogy we developed in thinking about how chemical activities can act like pressures to force a reaction to shift its equilibrium to the right or to the left is an example of Le Châtelier's principle:

If a chemical system at equilibrium experiences a change in concentration, temperature, volume, or partial pressure, then the equilibrium shifts to counteract the imposed change and a new equilibrium is established.

This principle is quite general and quite versatile. It is worth keeping in mind to help you predict shifts in the direction of thermodynamic processes when various disturbances such as changes in temperature, pressure, or concentration are applied.

Example 2.3

Question: A reaction vessel is filled with the following gas partial pressures at a temperature of 298 K: $P_{HI} = 20.0$ atm, $P_{I_2} = 3.0$ atm, and $P_{H_2} = 4.0$ atm. The gases are then allowed to react (via the hydrogen iodide decomposition reaction) until equilibrium is established. Calculate the resulting equilibrium partial pressures for all three gases.

Solution: From Equation 2.21 we can calculate the equilibrium constant K, for the hydrogen iodide decomposition reaction at $T = 298$ K:

$$K = \exp\left(-\frac{\Delta G^\circ}{RT}\right) \tag{2.22}$$

$$K|_{T=298} = \exp\left(-\frac{15930 \text{ J/mol}}{8.314 \text{ J/(mol} \cdot \text{K)} \cdot 298 \text{ K}}\right) = 1.6 \times 10^{-3}$$

We can then calculate Q based on the initial gas partial pressures in the reaction vessel prior to equilibration and compare this to K in order to determine the direction of the reaction:

$$Q = \frac{a_{I_2(g)}^{\text{init}} a_{H_2(g)}^{\text{init}}}{(a_{HI(g)}^{\text{init}})^2} = \frac{(P_{I_2}^{\text{init}}/P_{I_2}^\circ)(P_{H_2}^{\text{init}}/P_{H_2}^\circ)}{(P_{HI}^{\text{init}}/P_{HI}^\circ)^2} = \frac{(3.0/1.0)(4.0/1.0)}{(20.0/1.0)^2} = 3.0 \times 10^{-2} \tag{2.23}$$

In this example, $Q > K$, so the hydrogen iodide decomposition reaction will actually proceed backward to the direction written in Equation 2.7! In other words, the I_2 and H_2 gases will react to form additional HI in this scenario. This result may not be obvious based on how much larger the initial HI gas pressure is compared to the initial I_2 and H_2 gas pressures. This illustrates why it is important to evaluate both Q and K to determine the direction of the reaction. In the case of the hydrogen iodide decomposition reaction, ΔG° is large and positive, and therefore the reverse reaction is strongly favored.

Now that we have determined the direction that the reaction will proceed in moving toward equilibrium, we can use this fact coupled with the reaction stoichiometry, the initial partial pressures of the gas species, and the definition of K to determine the equilibrium partial pressure values. In order to do this, we use an unknown variable, x, to express the amount of I_2 gas that reacts in order to reach equilibrium. Since equal amounts of I_2 and H_2 are consumed by the reaction, x also expresses the amount of H_2 gas that reacts in order to reach equilibrium. During this process, the amount of HI gas that is formed is $2x$.

Therefore the equilibrium pressures of I_2, H_2, and HI can be given by

$$P_{I_2}^{eq} = P_{I_2}^{init} - x \tag{2.24}$$

$$P_{H_2}^{eq} = P_{H_2}^{init} - x \tag{2.25}$$

$$P_{HI}^{eq} = P_{HI}^{init} + 2x \tag{2.26}$$

Applying these relationships to the definition of the equilibrium constant (evaluated in terms of pressures) gives

$$K = \frac{(P_{I_2}^{eq}/P_{I_2}^{\circ})(P_{H_2}^{eq}/P_{H_2}^{\circ})}{\left(\dfrac{P_{HI}^{eq}}{P_{HI}^{\circ}}\right)^2} = \frac{[(P_{I_2}^{init} - x)/P_{I_2}^{\circ}][(P_{H_2}^{init} - x)/P_{H_2}^{\circ}]}{[(P_{HI}^{init} + 2x)/P_{HI}^{\circ}]^2} \tag{2.27}$$

$$= \frac{(P_{I_2}^{init} - x)(P_{H_2}^{init} - x)}{(P_{HI}^{init} + 2x)^2} \qquad (P_{I_2}^{\circ} = P_{H_2}^{\circ} = P_{HI}^{\circ} = 1) \tag{2.28}$$

This can then be written in terms of a quadratic formula for x:

$$(4K - 1)x^2 + (4KP_{HI}^{init} + P_{I_2}^{init} + P_{H_2}^{init})x + [K(P_{HI}^{init})^2 - P_{I_2}^{init}P_{H_2}^{init}] = 0 \tag{2.29}$$

For the specific initial HI, I_2, and H_2 gas partial pressures given in this example, this quadratic equation has two real roots for x:

$$x = 2.39 \text{ atm} \qquad x = 4.79 \text{ atm}$$

Only the first root is relevant, however, since the second value of x would lead to a negative equilibrium partial pressure of I_2 and H_2, which is impossible. Thus, taking the first root as the valid solution, we can finally provide the equilibrium pressures for the three gases:

$$P_{I_2}^{eq} = P_{I_2}^{init} - x = 3.0 \text{ atm} - 2.39 \text{ atm} = 0.6 \text{ atm}$$

$$P_{H_2}^{eq} = P_{H_2}^{init} - x = 4.0 \text{ atm} - 2.39 \text{ atm} = 1.6 \text{ atm}$$

$$P_{HI}^{eq} = P_{HI}^{init} + 2x = 20.0 \text{ atm} + 2 \cdot 2.39 \text{ atm} = 24.8 \text{ atm}$$

2.7 TEMPERATURE DEPENDENCE OF K

In addition to being affected by reactant and product activities, the equilibrium balance of a chemical process can also be affected by temperature. This effect shows up in the temperature dependence of K. This temperature effect can best be seen by combining Equations 2.1 and 2.21:

$$K = \exp\left(\frac{-(\Delta H° - T\Delta S°)}{RT}\right)$$

$$= \exp\left(\frac{-\Delta H°}{RT}\right)\exp\left(\frac{\Delta S°}{R}\right) \tag{2.30}$$

As indicated by Equation 2.30, the sign of $\Delta H°$ determines the effect that temperature has on the equilibrium constant for a reaction. For an *exothermic reaction* ($\Delta H° < 0$), decreasing the temperature will increase K. This will cause the equilibrium balance to shift to the right, in favor of the products. As with our previous discussion on activity effects, this temperature effect is another excellent example of Le Châtelier's principle. Recall that heat is released in an exothermic process, that is,

$$\text{reactants} \rightleftharpoons \text{products} + \text{heat} \tag{2.31}$$

When an exothermic process is viewed in this way, it can be seen that a decrease in temperature acts as a disturbing force which will cause the equilibrium to shift to the right (in favor of the reactants) so as to produce heat and thereby minimize the effect of the disturbance (the temperature decrease). In contrast, an increase in temperature would cause the equilibrium to shift left: Heat would be consumed and K would decrease. For an endothermic reaction, the exact opposite temperature behavior would be observed.

If both $\Delta H°$ and K are known at one temperature T_1, then the value of K can be estimated at a new temperature T_2 by setting up a ratio:

$$K_1 = \exp\left(\frac{-(\Delta H° - T_1\,\Delta S°)}{RT_1}\right)$$

$$K_2 = \exp\left(\frac{-(\Delta H° - T_2\,\Delta S°)}{RT_2}\right) \tag{2.32}$$

$$\frac{K_1}{K_2} = \frac{\exp\left(\frac{-\Delta H°}{RT_1}\right)}{\exp\left(\frac{-\Delta H°}{RT_2}\right)} = \exp\left[-\frac{\Delta H°}{R}\left(\frac{1}{T_1} - \frac{1}{T_2}\right)\right]$$

Please note that in developing this equation it is assumed that $\Delta H°$ and $\Delta S°$ do not change with temperature (this is the same assumption we made in Example 2.2). As long as there are no intermediate phase changes and the two temperatures T_1 and T_2 are not too far apart (within say a few hundred degrees), this is generally a reasonable approximation.

Example 2.4

Question: For the hydrogen iodide decomposition reaction, you have determined that $K = 1.6 \times 10^{-3}$ at $T = 298$ K while $K = 8.0 \times 10^{-3}$ at $T = 516$ K. Based on this information, estimate $\Delta H°$ and $\Delta S°$ for this reaction under the assumption that these quantities do not change with temperature.

Solution: Using Equation 2.32, we can first calculate $\Delta H°$:

$$\frac{K_1}{K_2} = \exp\left[-\frac{\Delta H°}{R}\left(\frac{1}{T_1} - \frac{1}{T_2}\right)\right]$$

$$\Delta H° = -R\left(\frac{T_1 T_2}{T_2 - T_1}\right)\ln\left(\frac{K_1}{K_2}\right)$$

$$= -8.314\ \text{J/(mol} \cdot \text{K)}\left(\frac{298\ \text{K} \cdot 516\ \text{K}}{516\ \text{K} - 298\ \text{K}}\right)\ln\left(\frac{1.6 \times 10^{-3}}{8.0 \times 10^{-3}}\right)$$

$$= 9473\ \text{J/mol} = 9.5\ \text{kJ/mol}$$

We can then use either K_1 (and T_1) or K_2 (and T_2) to determine $\Delta S°$. We should get the same answer either way (this is a good check to make sure you have done things correctly!):

$$K_1 = \exp\left(\frac{-(\Delta H° - T_1 \Delta S°)}{RT_1}\right)$$

$$\Delta S° = \frac{\Delta H°}{T_1} + R\ln K_1$$

$$\Delta S° = \frac{9473\ \text{J/mol}}{298\ \text{K}} + [8.314\ \text{J/(mol} \cdot \text{K)}\ln(1.6 \times 10^{-3})]$$

$$= -21.73\ \text{J/(mol} \cdot \text{K)} = -22\ \text{J/(mol} \cdot \text{K)}$$

$$K_2 = \exp\left(\frac{-(\Delta H° - T_2\ \Delta S°)}{RT_2}\right)$$

$$\Delta S° = \frac{\Delta H°}{T_2} + R\ln K_2$$

$$= \frac{9473\ \text{J/mol}}{516\ \text{K}} + [8.314\ \text{J/(mol} \cdot \text{K)}\ln(8.0 \times 10^{-3})]$$

$$= -21.78\ \text{J/(mol} \cdot \text{K)} = -22\ \text{J/(mol} \cdot \text{K)}$$

Reassuringly, these values for $\Delta H°$ and $\Delta S°$ are very close to what we calculated in Example 2.1 using the thermodynamic tables.

2.8 THERMODYNAMICS OF PHASE TRANSFORMATIONS

A phase transformation is a special type of reaction that involves a change in the structure (i.e., atomic arrangement, electronic/magnetic configuration) of a system and sometimes (but not always) a change in its composition as well. As a result of these changes, the new phase will posses physical properties (e.g., density, compressibility, magnetic induction) that are distinct from the previous phase. Well-known phase changes include solid-to-liquid (melting) and liquid-to-vapor (vaporization) and their corresponding reverse processes (solidification and condensation). Solid–solid phase transformations are also quite common; the ability to understand and regulate such processes is often central to materials engineering.

Consider a simple phase change involving the melting of solid ice to liquid water:

$$H_2O_{(s)} \rightleftharpoons H_2O_{(l)} \qquad (2.33)$$

The thermodynamic favorability of this phase change is determined by ΔG for the transformation process, which is itself dependent on the temperature, pressure, and composition (purity) of the phases. As with any other thermodynamic process, ΔG for a phase transformation can be determined from the G values for the final versus initial states:

$$\Delta G_{(s \to l)} = G_{H_2O(l)} - G_{H_2O(s)} \qquad (2.34)$$

where

$$G_{H_2O(l)} = H_{H_2O(l)} - TS_{H_2O(l)}$$

$$G_{H_2O(s)} = H_{H_2O(s)} - TS_{H_2O(s)}$$

As shown in the plot of G versus T in Figure 2.1, the Gibbs free energy for both the solid and liquid phases can be represented as straight lines (assuming H and S are independent of temperature) with slopes determined by $S_{H_2O(s)}$ and $S_{H_2O(l)}$, respectively. Because liquid water has greater entropy than solid ice (the less rigidly bonded liquid molecules have a greater number of accessible microstates), the liquid-phase curve has a more negative slope. Thus, at higher temperatures the liquid phase is more stable (lower free energy), while at lower temperatures the solid phase is more stable. The two curves intersect at $T = 0\,°C$. At this temperature, both the solid and liquid phases can coexist in thermodynamic equilibrium.

When examining the kinetics of a phase transformation, a typical first step is to calculate the thermodynamic driving force for the process. Consider again the water/ice example above. At equilibrium ($T = T_{eq} = 0\,°C$), the thermodynamic driving force for transformation is zero. However, if the temperature is displaced from this equilibrium temperature, the balance will be upset and a thermodynamic driving force for phase transformation will ensue. How can we quantify this driving force?

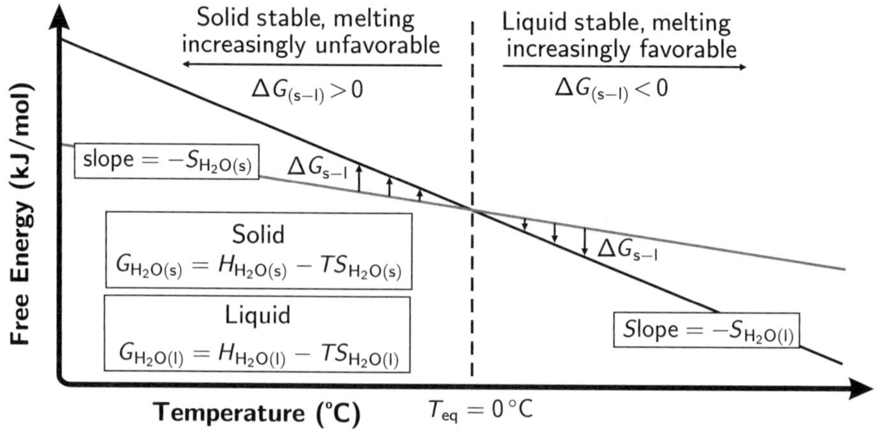

FIGURE 2.1   Free energy (G)–temperature (T) diagram illustrating the thermodynamics of the $H_2O(s) \rightarrow H_2O(l)$ phase transformation. Assuming pure ice and pure water at atmospheric pressure, equilibrium between the solid (ice) and liquid (water) phases is achieved at $0\,°C$ (the two curves cross at $T = 0\,°C$). When $T < 0\,°C$, the solid phase has a lower free energy and hence there is a thermodynamic driving force for solidification. When $T > 0\,°C$, the liquid phase has a lower free energy and hence there is a thermodynamic driving force for melting.

Starting from Equation 2.34, we can express $\Delta G_{(s \rightarrow l)}$ in terms of $\Delta H_{(s \rightarrow l)}$ and $\Delta S_{(s \rightarrow l)}$:

$$\begin{aligned}
\Delta G_{(s \rightarrow l)} &= G_{H_2O(l)} - G_{H_2O(s)} \\
&= (H_{H_2O(l)} - H_{H_2O(s)}) - T(S_{H_2O(l)} - S_{H_2O(s)}) \\
&= \Delta H_{(s \rightarrow l)} - T\,\Delta S_{(s \rightarrow l)}
\end{aligned} \tag{2.35}$$

Recognize that when $T = T_{eq}$, $\Delta G_{(s \rightarrow l)} = 0$. We thus have

$$\Delta G_{(s \rightarrow l)}|_{T=T_{eq}} = 0 = \Delta H_{(s \rightarrow l)} - T_{eq}\,\Delta S_{(s \rightarrow l)}$$

Thus

$$\Delta H^{\circ}_{(s \rightarrow l)} = T_{eq}\,\Delta S_{(s \rightarrow l)} \tag{2.36}$$

We can then express the thermodynamic driving force for phase change at any arbitrary temperature T in terms of $\Delta S_{(s \rightarrow l)}$ and ΔT ($\Delta T = T_{eq} - T$):

$$\begin{aligned}
\Delta G_{(s \rightarrow l)} &= \Delta H_{(s \rightarrow l)} - T\,\Delta S_{(s \rightarrow l)} \\
&= T_{eq}\,\Delta S_{(s \rightarrow l)} - T\,\Delta S_{(s \rightarrow l)} \\
&= \Delta T\,\Delta S_{(s \rightarrow l)} \quad \text{where } \Delta T = T_{eq} - T
\end{aligned} \tag{2.37}$$

An example of this type of calculation is provided in the exercise below.

Example 2.5

Question: You are provided with the following thermodynamic data for the solidification of iron ($Fe_{(l)} \rightleftharpoons Fe_{(s)}$):

Heat of solidification: $\Delta H_{(l \to s)} = -13.81$ kJ/mol

Entropy of solidification: $\Delta S_{(l \to s)} = -7.62$ J/(mol · K)

(a) What is the melting point (T_M) of iron?

(b) If liquid iron is cooled 100 °C below its melting temperature, what is the driving force ($\Delta G_{(l \to s)}$) for solidification?

(c) If liquid iron is heated 50 °C above its melting temperature, what is the driving force ($\Delta G_{(l \to s)}$) for solidification?

(d) As liquid iron is cooled further and further below its melting point (i.e., increasing "undercooling"), does the thermodynamic driving force for solidification increase, decrease, or remain constant?

Solution:

(a) We can use Equation 2.36 to calculate the melting point of iron based on $\Delta H_{(l \to s)}$ and $\Delta S_{(l \to s)}$:

$$\Delta G_{(l \to s)}|_{T=T_{eq}} = 0 = \Delta H_{(l \to s)} - T_{eq}\,\Delta S_{(l \to s)}$$

Thus

$$T_{eq} = T_M = \frac{\Delta H_{(s \to l)}}{\Delta S_{(s \to l)}}$$

$$T_M = \frac{-13810 \text{ J/mol}}{-7.62 \text{ J/(mol · K)}}$$

$$= 1810 \text{ K}$$

(b) We can use Equation 2.37 to determine the thermodynamic driving force for solidification ($\Delta G_{(l \to s)}$) for $\Delta T = 100\,°C$ (100 °C undercooling):

$$\Delta G_{(l \to s)} = \Delta T\,\Delta S_{(l \to s)} \qquad \text{where } \Delta T = T_M - T$$

$$= (100 \text{ K})[-7.62 \text{ J/(mol · K)}]$$

$$= -762 \text{ J/mol} \tag{2.38}$$

The fact that $\Delta G_{(l \to s)}$ is negative here indicates that solidification is thermodynamically favorable when iron is cooled below its melting temperature. This makes sense!

(c) The thermodynamic driving force for solidification when iron is heated $50\,°C$ above its melting point can be calculated in a similar fashion:

$$\Delta G_{(l \to s)} = \Delta T\ \Delta S_{(l \to s)} \qquad \text{where}\ \ \Delta T = T_M - T$$

$$= (-50\ \text{K})[-7.62\ \text{J}/(\text{mol} \cdot \text{K})]$$

$$= 380\ \text{J}/\text{mol}$$

In this case, $\Delta G_{(l \to s)}$ is positive, which indicates that solidification is thermo-dynamically unfavorable when iron is heated above its melting temperature. Again, this makes sense!

(d) As can be seen in Equation 2.38, for increasing ΔT (increasing undercooling), $\Delta G_{(l \to s)}$ will become increasingly negative (since $\Delta S_{(l \to s)}$ is negative). Thus, the thermodynamic driving force for solidification increases with increasing undercooling.

Note that this example considers a solidification process $l \to s$ whereas the text discussion on the ice/water example considered a melting process $s \to l$. Converting thermodynamic quantities for solidification to the corresponding quantities for melting is easy. It only requires flipping the sign. In other words:

$$\Delta H_{(l \to s)} = -\Delta H_{(s \to l)} \tag{2.39}$$

$$\Delta S_{(l \to s)} = -\Delta S_{(s \to l)} \tag{2.40}$$

$$\Delta G_{(l \to s)} = -\Delta G_{(s \to l)} \tag{2.41}$$

Generally, solidification processes are exothermic while melting processes are endothermic. (Quiz: Can you explain why this is so?)

2.9 IDEAL GAS LAW

Throughout this textbook, the application of the ideal gas law will come up again and again in one form or another. While you have almost certainly encountered the ideal gas law many times already, it is worth briefly reviewing it here. The ideal gas law is most commonly expressed as

$$PV = nRT \tag{2.42}$$

where P is the gas pressure, V is the gas volume, n is the number of moles of gas, R is the gas constant, and T is the temperature. The ideal gas law is derived from basic thermodynamic principles. Although detailed derivation of the ideal gas law is outside the scope of this textbook, it is important that you are able to apply it to calculate various properties of gases using a variety of different units.

In our study of kinetics, one of the most common uses of the ideal gas law will be to convert information about the partial pressures of various gas species (e.g., in units of atmospheres) into information about the molar concentrations of those gases (e.g., in units of mol/m^3). Gas concentrations can be calculated from gas partial pressures using a relatively straightforward manipulation of the ideal gas law:

$$C_{gas} = \frac{n}{V} = \frac{P}{RT} \tag{2.43}$$

where C_{gas} represents the concentration (in units of mol/volume) of the gas. *While this expression is indeed quite straightforward, students (and even professors) often make mistakes when evaluating this expression using actual numbers.* Units are a source of nearly unending confusion here. Beware!

A good recommendation is to always use SI units when evaluating this expression. If different units are eventually needed, such a conversion can be done *after* the gas concentration has been calculated using SI values. Using this approach, you will always use the exact same set of units when evaluating Equation 2.43 and you will always use the exact same value for the gas constant R. Thus, there will be fewer opportunities for error or confusion. See Example 2.6 for an illustration of this approach.

Example 2.6

Question: The air pressure at the top of Mount Everest is only about 33% of the value at sea level. Assuming the composition of air at the top of Mount Everest is the same as that at sea level, calculate the concentration (in units of mol/cm^3) of oxygen in the air at the top of Mount Everest. Assume an air temperature of $-20\,^\circ C$ at the top of Mount Everest.

Solution: In order to calculate the concentration of oxygen in the air at the top of Mount Everest, we first need to determine the partial pressure of oxygen in air at the top of Mount Everest. The composition of air is 21% O_2, 78% N_2, and 1% other species, mostly Ar. The problem statement tells us that we can assume that this composition still holds true at the top of Mount Everest. On the other hand, the *total atmospheric pressure* is much lower at the top of Mount Everest than it is at sea level, and hence the *partial pressure* of oxygen gas will be much lower at the top of Mount Everest as well. Since the problem statement tells us that the air pressure at the top of Mount Everest is 33% of the sea-level value, we can calculate the partial pressure of oxygen gas at the top of Mount Everest as

$$P_{O_2,\,\text{Everest}} = 0.21 \cdot P_{\text{Total, Everest}}$$
$$= 0.21 \cdot (0.33 \cdot P_{\text{Total, Sea Level}})$$
$$= 0.21 \cdot (0.33 \cdot 1 \text{ atm}) = 0.0693 \text{ atm}$$

It is strongly recommended to *always* use SI units when working with the ideal gas law. Indeed, it is generally a good policy to use SI units for *all* calculations and only convert solutions to other units at the very end if/when needed.

In this case, we must therefore first convert the oxygen partial pressure to SI units (Pa) before we apply the ideal gas law:

$$P_{O_2, \text{ Everest}} = 0.0693 \text{ atm} \cdot \frac{101{,}325 \text{ Pa}}{1 \text{ atm}} = 7021.8 \text{ Pa}$$

Then, we can evaluate the ideal gas law using SI units:

$$C_{O_2, \text{ Everest}} = \frac{P_{O_2, \text{ Everest}}}{RT}$$

$$= \frac{7021.8 \text{ Pa}}{8.314 \text{ J/(mol} \cdot \text{K)} \cdot 253.15 \text{ K}} = 3.37 \text{ mol/m}^3$$

If SI units are used, the resulting concentration will also be in SI units: mol/m^3. Since the problem statement asks for the answer in units of mol/cm^3, a final unit conversion is required:

$$C_{O_2, \text{ Everest}} = 3.37 \text{ mol/m}^3 \cdot \frac{1 \text{ m}^3}{(100 \text{ cm})^3} = 3.4 \times 10^{-6} \text{ mol/cm}^3$$

2.10 CALCULATING CONCENTRATIONS FOR LIQUIDS OR SOLIDS

Another frequent mistake among students is to try to apply the ideal gas law to calculate the concentrations of species in condensed-matter phases (e.g., liquid or solid phases). *Do not make this mistake!* The ideal gas law only applies to gases. To calculate concentrations for liquid or solid species, information about the *density* (ρ_i) of the liquid or solid phase is required. Both mass densities and molar densities (concentrations) as well as molar and atomic volumes may be of interest. The complexity of calculating these quantities tends to increase with the complexity of the material under consideration. In this section, we will consider three levels of increasing complexity: pure materials, simple compounds or dilute solutions, and more complex materials involving mixtures of multiple phases/compounds.

2.10.1 Calculating Densities/Concentrations in Pure Materials

Mass densities (g/cm^3) are readily available for most pure materials. For example, the density of Si is $\rho_{\text{Si}} = 2.33 \text{ g/cm}^3$ while the density of SiO_2 is $\rho_{SiO_2} = 2.65 \text{ g/cm}^3$. Calculating the molar densities (molar concentrations) and molar volumes of pure Si and pure SiO_2 (or any other pure material) from their mass densities and their molecular weights is quite straightforward:

$$c_i = \frac{\rho_i}{M_i} \tag{2.44}$$

$$V_{\text{m},i} = \frac{1}{c_i} = \frac{M_i}{\rho_i} \tag{2.45}$$

where c_i, $V_{m,i}$, and M_i are respectively the molar concentration (typical units are mol/cm^3), molar volume (typical units are cm^3/mol), and molecular weight (typical units are g/mol) of a pure material i (e.g., pure Si or SiO_2). Thus, the molar concentration and volume of pure SiO_2 can be calculated as

$$c_{SiO_2} = \frac{\rho_{SiO_2}}{M_{SiO_2}} = \frac{2.65 \text{ g/cm}^3}{60.1 \text{ g/mol}} = 4.41 \times 10^{-2} \text{ mol SiO}_2/\text{cm}^3$$

$$V_{m,SiO_2} = \frac{1}{c_{SiO_2}} = 22.7 \text{ cm}^3/\text{mol } SiO_2$$

Molar concentrations and molar volumes can be converted into number densities (n_i) and atomic volumes Ω_I by multiplying or dividing by Avogadro's number, respectively:

$$n_i = c_i N_A \tag{2.46}$$

$$\Omega_i = \frac{V_{m,i}}{N_A} \tag{2.47}$$

These quantities correspond to the number of atoms or molecules of species i per unit volume (e.g., no./cm^3), and the volume associated with a single atom or molecule of a species i (e.g., cm^3), respectively. Thus, the number density and atomic volume of SiO_2 are:

$$n_{SiO_2} = c_{SiO_2} N_A = 4.41 \times 10^{-2} \text{ mol SiO}_2/\text{cm}^3 \cdot 6.022 \times 10^{23} \text{ no./mol}$$

$$= 2.66 \times 10^{22} \text{ SiO}_2 \text{ atoms/cm}^3$$

$$\Omega_{SiO_2} = \frac{V_{m,SiO_2}}{N_A} = \frac{22.7 \text{ cm}^3/\text{mol } SiO_2}{6.022 \times 10^{23} \text{ no./mol}} = 3.77 \times 10^{-23} \text{ cm}^3$$

2.10.2 Calculating Densities/Concentrations in Stoichiometric Compounds or Dilute Solutions

When calculating the mass density, molar concentration, or molar volume of a specific individual species that is present in combination with other species (e.g., in a compound or solution), further work is needed. If the material's composition can be expressed in terms of a single stoichiometric compound or formula unit, the approach is still fairly straightforward—it just requires application of the compound stoichiometry. Similarly, dilute solutions, where the solute species is present in very low concentrations relative to the host solvent, can be handled in a relatively straightforward manner by assuming that the host material's density is not affected by the presence of the solute species.

As an example of the first case (dealing with one species in a stoichiometric compound), consider how to calculate the molar concentration of *oxygen atoms* in SiO_2:

$$c_{O,SiO_2} = \frac{n_O \rho_{SiO_2}}{M_{SiO_2}}$$

$$= \frac{2 \cdot 2.65 \text{ g/cm}^3}{60.1 \text{ g/mol}} = 8.82 \times 10^{-2} \text{ mol O/cm}^3$$

Compared to our previous example, the only additional information required is the number of moles of oxygen per formula unit of the compound, n_O, which can be obtained by examination of the compound's stoichiometry.

Knowledge of the compound stoichiometry can similarly be applied to calculate the mass density of oxygen atoms in SiO_2 as

$$\rho_{O,SiO_2} = \left(\frac{2M_O}{2M_O + M_{Si}} \right) \rho_{SiO_2} = W_{O,SiO_2} \; \rho_{SiO_2}$$

$$= \frac{2 \cdot 16.0 \text{ g/mol}}{2 \cdot 16.0 \text{ g/mol} + 28.1 \text{ g/mol}} \cdot 2.65 \text{ g/cm}^3 = 1.41 \text{ g/cm}^3$$

where W_{O,SiO_2} is the oxygen *mass fraction* in SiO_2 or, in other words, the fraction of the total mass of SiO_2 that is due to the oxygen. The factors of 2 appearing in this expression again reflect the fact that there are *two* moles of oxygen atoms in every mole of SiO_2.

The molar concentration and mass density of Si in SiO_2 can be calculated from the compound stoichiometry in a similar manner:

$$c_{Si,SiO_2} = \frac{n_{Si} \rho_{SiO_2}}{M_{SiO_2}}$$

$$= \frac{1 \cdot 2.65 \text{ g/cm}^3}{60.1 \text{ g/mol}} = 4.41 \times 10^{-2} \text{ mol Si/cm}^3$$

$$\rho_{Si,SiO_2} = \left(\frac{M_{Si}}{2M_O + M_{Si}} \right) \rho_{SiO_2} = W_{Si,SiO_2} \; \rho_{SiO_2}$$

$$= \frac{1 \cdot 28.1 \text{ g/mol}}{2 \cdot 16.0 \text{ g/mol} + 28.1 \text{ g/mol}} \cdot 2.65 \text{ g/cm}^3 = 1.24 \text{ g/cm}^3$$

Calculating concentration of a species that is present at low levels (say, <1%) as a solute in a solid solution is also fairly straightforward. The typical approach is to assume that the density of the host material (ρ_M) is not changed by the presence of the solute species. In this case, the mass density ($\rho_{i,M}$) and molar concentration ($c_{i,M}$) of the solute species i in the host material, M, can be calculated as

$$\rho_{i,M} = W_{i,M} \; \rho_M \tag{2.48}$$

$$c_{i,M} = X_{i,M} \frac{\rho_M}{M_M} \tag{2.49}$$

where $W_{i,\mathrm{M}}$ is the mass fraction of species i in material M, $X_{i,\mathrm{M}}$ is the mole fraction of species i in material M, and M_{M} is the molecular weight of material M.

In order to illustrate this approach, consider the calculation of the molar concentration (density) of arsenic (As) in a piece of Si that has been doped with 100 ppm (parts per million) As atoms. Since 100 out of every 10^6 Si atoms have been replaced by As in this material, the mole fraction of As in Si can be calculated as

$$
\begin{aligned}
X_{\mathrm{As,Si}} &= \frac{Z_{\mathrm{As,Si}}}{Z_{\mathrm{As,Si}} + Z_{\mathrm{Si,Si}}} \\
&= \frac{100}{100 + (10^6 - 100)} = 10^{-4}
\end{aligned}
$$

where $Z_{\mathrm{As,Si}}$ and $Z_{\mathrm{Si,Si}}$ are the parts-per-million values (which are the same as the relative molar amounts) of the As and Si atoms in the As-doped Si solid solution, respectively. Assuming that the As substitution does not appreciably affect the density of Si, the molar concentration of As in Si can then be estimated as

$$
\begin{aligned}
c_{\mathrm{As,Si}} &= X_{\mathrm{As,Si}} \frac{\rho_{\mathrm{Si}}}{M_{\mathrm{Si}}} \\
&= 10^{-4} \cdot \frac{2.33 \ \mathrm{g/cm}^3}{28.1 \ \mathrm{g/mol}} = 8.29 \times 10^{-6} \ \mathrm{mol \ As/cm}^3
\end{aligned}
$$

In problem 2.9, you will have the opportunity to continue this example by calculating the mass density of As in this As-doped Si solid solution.

2.10.3 Calculating Densities/Concentrations for Mixtures of Multiple Phases/Compounds

In materials that are composed of two or more distinct compounds or phases, expressions are sometimes required for the "average" density of the mixture, or the "average" mass density or molar concentration of a particular species i in the mixture. Such systems represent an additional level of complexity, and a more general approach is required in order to calculate such "material-averaged" quantities. For example, consider a material M that is made up of N distinct compounds or phases, each of which has its own well-defined density and composition. In such a situation, the overall average density of the material, ρ_{M}, can be determined from the densities (ρ_N) and weight fractions (W_N) of the individual N phases from which the material is composed:

$$
\frac{1}{\rho_{\mathrm{M}}} = \sum_N \frac{W_N}{\rho_N} \tag{2.50}
$$

Once the overall density of the material has been calculated, the average mass density for any species i in the material ($\rho_{i,\mathrm{M}}$) can be calculated from the knowledge of the mass fraction of species i in each component N in the material ($W_{i,N}$)

and the mass fraction of each component N in the overall composition of the material (W_N):

$$\rho_{i,M} = W_{i,M}\ \rho_M = \left(\sum_N W_{i,N} W_N\right)\rho_M \tag{2.51}$$

Similarly, the average molar concentration for any species i in the material ($c_{i,M}$) can be calculated from the knowledge of the stoichiometry of species i in each component N in the material in combination with the densities (ρ_N) and weight fractions (W_N) of each component N:

$$c_{i,M} = \left(\sum_N W_N \frac{n_{i,N}}{M_N}\right)\rho_M \tag{2.52}$$

where $n_{i,N}$ is the number of moles of species i per formula unit of compound N, which can be obtained by examination of the compound's stoichiometry.

In order to illustrate this approach, consider the calculation of the average density of a two-phase composite mixture consisting of 70% SiO_2 and 30% Si (by weight):

$$\frac{1}{\rho_M} = \frac{W_{SiO_2}}{\rho_{SiO_2}} + \frac{W_{Si}}{\rho_{Si}}$$

$$= \frac{0.7}{2.65\ \text{g/cm}^3} + \frac{0.3}{2.33\ \text{g/cm}^3} = 0.3929\ \text{cm}^3/\text{g}$$

$$\rho_M = \frac{1}{0.3929\ \text{cm}^3/\text{g}} = 2.55\ \text{g/cm}^3$$

The average mass density and molar concentration of oxygen atoms in this two-phase composite may then be calculated as

$$\rho_{O,M} = (W_{O,SiO_2} W_{SiO_2} + W_{O,Si} W_{Si})\rho_M$$

$$= \left(\frac{2\cdot 16.0\ \text{g/mol}}{2\cdot 16.0\ \text{g/mol} + 28.1\ \text{g/mol}}\cdot 0.7 + 0\cdot 0.3\right) 2.55\ \text{g/cm}^3$$

$$= 0.950\ \text{g/cm}^3$$

$$c_{O,M} = \left(W_{SiO_2}\frac{n_{O,SiO_2}}{M_{SiO_2}} + W_{Si}\frac{n_{O,Si}}{M_{Si}}\right)\rho_M$$

$$= \left(0.7\cdot\frac{2}{60.1\ \text{g/mol}} + 0.3\cdot\frac{0}{28.1\ \text{g/mol}}\right) 2.54\ \text{g/cm}^3$$

$$= 5.94\times 10^{-2}\ \text{mol O/cm}^3$$

A similar exercise can be applied to calculate the average mass density and molar concentration of silicon atoms in this two-phase composite (see problem 2.10 at the end of this chapter).

2.10.4 Calculating Densities/Concentrations from Crystallographic Information

For crystalline solid compounds, species densities and concentrations can also be calculated from information on the crystal structure and unit cell size as well as the lattice occupancy of the species in question. It is sometimes more convenient to calculate species concentrations this way, particularly when structural information is more readily available than mass density information. The basic idea is to calculate density/concentration from the weight and volume of the crystallographic unit cell. Example 2.7 illustrates this approach.

Example 2.7

Question: The crystal structure of lanthanum gallate ($LaGaO_3$) can be approximated as cubic, with lattice parameters $a \approx b \approx c \approx 3.9$ Å (see Figure 2.2). Based on this information, calculate (1) the density (g/cm^3) of $LaGaO_3$, (2) the molar volume (cm^3/mol) of $LaGaO_3$, and (3) the concentration of oxygen atoms (ions) in pure $LaGaO_3$ (mol/cm^3).

Solution:

1. Assuming a perfect cubic unit cell, the unit cell volume of $LaGaO_3$ can be calculated from the lattice parameter as

$$V_{cell} = a^3 = (3.9 \text{ Å})^3 = 59.3 \text{ Å}^3 = 5.93 \times 10^{-29} \text{ m}^3$$

A single unit cell of $LaGaO_3$ contains one La ion, one Ga ion, and three oxygen ions. The total weight associated with a single $LaGaO_3$ unit cell can therefore be calculated from the molecular weights of La, Ga, and O as

$$W_{cell} = \frac{M_{La} + M_{Ga} + 3M_{O}}{N_A}$$

$$= \frac{138.9 \text{ g/mol} + 69.7 \text{ g/mol} + 3 \cdot 16.0 \text{ g/mol}}{6.022 \times 10^{23} \text{ no./mol}} = 4.26 \times 10^{-22} \text{ g}$$

Then the density can be calculated from the cell volume and weight as

$$\rho_{LaGaO_3} = \frac{W_{cell}}{V_{cell}} = \frac{4.26 \times 10^{-22} \text{ g}}{5.93 \times 10^{-29} \text{ m}^3} = 7.19 \times 10^6 \text{ g/m}^3 = 7.19 \text{ g/cm}^3$$

2. To calculate the molar volume of $LaGaO_3$, we need to first calculate the molar density (concentration) of $LaGaO_3$. This can be calculated from the mass density of $LaGaO_3$ and the molecular weight:

$$c_{LaGaO_3} = \frac{\rho_{LaGaO_3}}{M_{LaGaO_3}} = \frac{\rho_{LaGaO_3}}{M_{La} + M_{Ga} + 3M_{O}}$$

$$= \frac{7.19 \text{ g/cm}^3}{138.9 \text{ g/mol} + 69.7 \text{ g/mol} + 3 \cdot 16.0 \text{ g/mol}}$$

$$= 2.80 \times 10^{-2} \text{ mol/cm}^3$$

Then, the molar volume is simply the inverse of the molar concentration:

$$V_{\text{LaGaO}_3} = \frac{1}{c_{\text{LaGaO}_3}} = \frac{1}{2.80 \times 10^{-2} \text{ mol/cm}^3} = 35.7 \text{ cm}^3/\text{mol}$$

3. There are three moles of oxygen atoms (ions) per molar formula unit of LaGaO_3, so the concentration of oxygen ions can be determined directly from the concentration of LaGaO_3 and the stoichiometry:

$$c_{\text{O,LaGaO}_3} = 3c_{\text{LaGaO}_3} = 3 \cdot 2.80 \times 10^{-2} \text{ mol/cm}^3 = 8.40 \times 10^{-2} \text{ mol/cm}^3$$

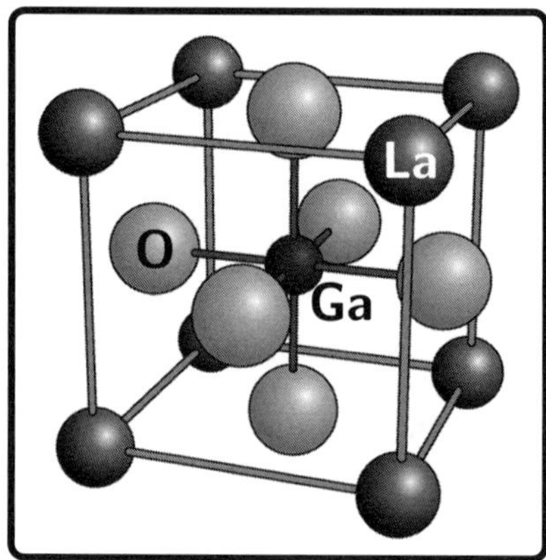

FIGURE 2.2 LaGaO_3 forms in the cubic perovskite crystal structure with La atoms at the cube corners, O atoms at the cube faces, and Ga at the cube center.

2.10.5 Calculating Site Fractions

A final concentration-like unit that is sometimes needed is the site fraction, which is essentially a unitless "occupancy" factor that can typically be obtained from crystallographic information and defect models of a solid. A site fraction gives the fraction of sites (e.g., crystalline lattice sites) of a particular type in a material which are occupied by a particular species i. Thus, it is a ratio of the number of sites of a particular type (j) that are occupied by a certain species i divided by the total number of sites of that particular type in the material:

$$X_i^j = \frac{N_i^j}{N_{\text{total}}^j} \tag{2.53}$$

In the As-doped Si example discussed earlier, the site fraction of Si sites that are occupied by As dopants (assuming perfect substitution) would be

$$X_{As}^{Si} = \frac{100}{10^6} = 10^{-4}$$

Example 2.8 explores a slightly more complex scenario.

Example 2.8

Question: $LaGaO_3$ is often doped with Sr to create oxygen vacancies, giving it an extremely high oxygen-ion conductivity and making it attractive for high-temperature fuel cell and gas sensor applications. In $LaGaO_3$, La is in the 3+ oxidation state, while Sr takes the 2+ oxidation state, so for every two La^{3+} ions substituted by two Sr^{2+} ions, one oxygen ion vacancy is created to maintain charge neutrality. Assume $LaGaO_3$ is doped with 20 mol % Sr to give $La_{0.8}Sr_{0.2}GaO_{3-\delta}$: (1) Determine the value of δ in the formula $La_{0.8}Sr_{0.2}GaO_{3-\delta}$, (2) calculate the oxygen vacancy site fraction in this material, and (3) calculate the concentration of oxygen vacancies (mol/cm^3) in this material, assuming it has the same lattice constant and density as pure $LaGaO_3$.

Solution:

1. For every two Sr ions substituted into $LaGaO_3$, one oxygen vacancy is created. Therefore for the $Sr_{0.2}$ subsitition, $V_{0.1}$ oxygen vacancies will be created. This results in a formula of $La_{0.8}Sr_{0.2}GaO_{3-0.1}$, or $La_{0.8}Sr_{0.2}GaO_{2.9}V_{O,0.1}$, which more explicitly accounts for the vacant oxygens (V_O). Thus, the value of δ is 0.1.

2. For each molar formula unit of $La_{0.8}Sr_{0.2}GaO_{2.9}V_{O,0.1}$, there are 3 mol of oxygen ion sites, of which 2.9 mol are occupied, and 0.1 mol are unoccupied. Therefore, the oxygen vacancy site fraction in this material is

$$X_V^O = \frac{N_V^O}{N_{total}^O} = \frac{0.1}{3} = 0.0333$$

In other words, 1 in 30, or about 3.33%, of the oxygen sites are vacant. It is this large number of vacant oxygen ion sites which permits significant oxygen ion motion (and hence oxygen ion conductivity) in this material.

3. Assuming that the Sr-doped $LaGaO_3$ material has the same lattice constant and density as pure $LaGaO_3$, we can approximate the oxygen vacancy concentration in this material by multiplying the oxygen ion concentration previously determined for pure $LaGaO_3$ in Example 2.2 by the oxygen vacancy site fraction, X_V^O:

$$c_{V_O} = X_V^O \cdot c_{O,LaGaO_3} = 0.0333 \cdot 8.40 \times 10^{-2} \ mol/cm^3$$

$$= 2.80 \times 10^{-3} \ mol/cm^3$$

2.11 CHAPTER SUMMARY

The purpose of this chapter was to briefly review the main principles underlying chemical thermodynamics. Thermodynamics is important because we can use it to tell us when a system is in equilibrium. For systems that are not in equilibrium, we can use thermodynamics to calculate the size of the driving forces acting on a system to return it to equilibrium. These driving forces will become a key input into later kinetic equations. The main points introduced in this chapter include the following

- Equilibrium is *dynamic*. Equilibrium represents the most stable, lowest energy state for a system. However, in most systems this equilibrium point represents a dynamic balance between forward and backward processes that continue to occur with equal rates. Even small changes can upset this balance. The most common ways that we can shift the equilibrium point of a system are by changing temperature, pressure, or composition.

- The most common thermodynamic quantities that we will encounter in our exploration of materials kinetics are enthalpy (H), entropy (S), and Gibbs free energy (G). Usually we are concerned with quantifying the *changes* in these thermodynamic functions (i.e., ΔH, ΔS, ΔG) during a process of reaction rather than the absolute values. Changes in thermodynamic functions are always calculated as *final state − initial state*.

- Gibbs free energy is perhaps the most important of the main thermodynamic functions. The sign of the Gibbs free-energy change (ΔG) during a thermodynamic process indicates whether or not the process is spontaneous and hence the direction in which it can proceed. If $\Delta G > 0$, the process is nonspontaneous (or will occur in the reverse direction). If $\Delta G = 0$, the process is at equilibrium. If $\Delta G < 0$, the process is spontaneous and will occur in the forward direction. While ΔG determines the thermodynamically predicted direction of a process, kinetic laws (discussed starting in the next chapter) are required to calculate *how quickly* a process will occur.

- The quantities ΔH, ΔS, and ΔG are related by the following important equation: $\Delta G = \Delta H - T\Delta S$. This equation can be used to determine one of the three quantities when the other two are known. It is also particularly helpful for estimating how ΔG for a process changes as a function of temperature.

- For a reaction under an arbitrary set of non-standard-state conditions, ΔG can be determined from the standard-state value of $\Delta G°$ by taking into account the activities of the reactant and product species via the reaction quotient Q:

$$\Delta G = \Delta G° + RT \ln Q$$

  where

$$Q = \frac{a_{\text{M}}^{m} a_{\text{N}}^{n}}{a_{\text{A}}^{a} a_{\text{B}}^{b}}$$

- The equilibrium constant K is a unique value for Q that is obtained when (and only when) a system is at equilibrium. At equilibrium,

$$\Delta G = 0 = \Delta G° + RT \ln K$$

where

$$K = \frac{(a_M^{eq})^m (a_N^{eq})^n}{(a_A^{eq})^a (a_B^{eq})^b}$$

and a_M^{eq}, a_N^{eq}, a_A^{eq}, and a_B^{eq} are the activities of the product and reactant species at equilibrium.

By comparing Q versus K, the direction of a reaction can be determined. If $Q < K$, the reaction will proceed in the forward direction; if $Q = K$, the reaction is at equilibrium; if $Q > K$, the reaction will proceed in the backward direction.

- Le Châtelier's principle is a useful concept that can help you to predict which way a thermodynamic process will shift when various disturbances, such as changes in temperature, pressure, or concentration, are applied. The principle states that if a chemical system at equilibrium experiences a change in concentration, temperature, volume, or partial pressure, then the equilibrium shifts to counteract the imposed change and a new equilibrium is established.

- Because $\Delta G°$ is (approximately) linearly dependent on temperature as expressed by the relationship $\Delta G = \Delta H - T\,\Delta S$, the equilibrium constant K is exponentially temperature dependent:

$$K = \exp\left(-\frac{\Delta G°}{RT}\right)$$

$$= \exp\left(\frac{-(\Delta H° - T\Delta S°)}{RT}\right)$$

$$= \exp\left(\frac{-\Delta H°}{RT}\right)\exp\left(\frac{\Delta S°}{R}\right)$$

- The thermodynamic driving force for a phase change is determined by the entropy change (ΔS) of the phase transformation and the magnitude of the departure from the equilibrium temperature (ΔT) of the phase transformation. For example, the driving force for a solidification process can be estimated as

$$\Delta G_{(l \to s)} = \Delta T\,\Delta S_{(l \to s)} \qquad \text{where } \Delta T = T_{eq} - T$$

In this case, $T_{eq} = T_m$ (i.e., the melting temperature), which is the temperature at which both phases involved in the transformation (solid and liquid) can coexist in equilibrium.

- The ideal gas law $PV = nRT$ is extremely useful in converting partial pressures of gas species (e.g., in units of atmospheres) into concentrations (e.g., in units of moles/vol, mass/vol, atoms/vol). While the ideal gas law is straightforward, units can cause a significant challenge! A good recommendation is to always use SI units when evaluating the ideal gas law. Thus, pressures must be evaluated in units of Pa (Pascals), where 1 atm = 101300 Pa.

- When calculating concentrations for liquid or solid species (which will be frequently encountered in materials kinetics problems), the ideal gas law DOES NOT APPLY! Instead, information about the density (or atomic structure and packing) is needed. These calculations can become increasingly complicated depending on the number of phases/components involved. The last section of the chapter provides detailed examples of such calculations. Mastering these concepts will be extremely useful as we move forward in our exploration of materials kinetics, as most kinetic equations involve species concentration.

2.12 CHAPTER EXERCISES

Review Questions

Problem 2.1. If an isothermal reaction involving gases exhibits a large negative volume change, will the entropy change for the same reaction likely be negative or positive? Why?

Problem 2.2. (a) If ΔH for a reaction is negative and ΔS is positive, can you say anything about the spontaneity of the reaction? (b) What if ΔH is negative and ΔS is negative? (c) What if ΔH is positive and ΔS is negative? (d) What if ΔH is positive and ΔS is positive?

Problem 2.3. For the following reactions, indicate whether the reaction is thermodynamically favorable, unfavorable, or in equilibrium as written:

(a) $CH_3OH_{(l)} + \frac{3}{2}O_{2(g)} \rightarrow CO_{2(g)} + 2H_2O_{(l)}$ $(\Delta G_{rxn} = -702 \text{ kJ/mol})$

(b) $H_2O_{(l)} \rightarrow H_2O_{(s)}$ (at $P = 1$ atm, $T = 273.15$ K, pure components)

(c) $H_2O_{(s)} \rightarrow H_2O_{(l)}$ (at $P = 1$ atm, $T = 253.15$ K, pure components)

Problem 2.4. Reaction A has $\Delta G_{rxn} = -100$ kJ/mol. Reaction B has $\Delta G_{rxn} = -200$ kJ/mol. Can you say anything about the relative speeds (reaction rates) for these two reactions?

Calculation Questions

Problem 2.5. Iron is undergoing active gas corrosion at atmospheric pressure and $T = 1027\,^{\circ}\text{C}$ by the following reaction:

$$Fe_{(s)} + 2HCl_{(g)} \rightleftharpoons FeCl_{2(g)} + H_{2(g)}$$

If the equilibrium constant for this reaction is $K_{eq} = 0.384$ at $T = 1027\,^{\circ}\text{C}$, determine the equilibrium pressures for the three gases: P_{HCl}, P_{FeCl_2}, and P_{H_2} at $T = 1027\,^{\circ}\text{C}$ (in units of atmospheres). Assume that the only H_2 and $FeCl_2$ present are due to reaction.

Problem 2.6. You are provided with the following thermodynamic data for the solidification of silver ($Ag_{(l)} \rightleftharpoons Ag_{(s)}$):

Melting temperature: $T_M = 1235$ K

Heat of solidification: $\Delta H_{rxn} = -11.28$ kJ/mol

(a) At $T = T_M$, what is the driving force (ΔG_{rxn}) for solidification?

(b) What is the entropy of solidification (ΔS_{rxn}) for this reaction?

(c) If liquid silver is cooled $T = 100\,°C$ below its melting temperature, what is the driving force (ΔG_{rxn}) for solidification?

Problem 2.7. A reaction vessel is initially filled with pure HI gas at 1 atm pressure at $T = 298$ K. The gas is then allowed to react (via the hydrogen iodide decomposition reaction), generating I_2 and H_2 gases until equilibrium is established. Calculate the resulting equilibrium partial pressures for all three gases.

Problem 2.8. A reaction vessel contains an equilibrium mixture of I_2, H_2, and HI gases with a total pressure of 10 atm at 298 K. Assume that the I_2 and H_2 gas pressures are equal. The temperature in the reaction vessel is then increased to 500 K and the gases are allowed to react until equilibrium is reestablished. Calculate the resulting equilibrium pressures of all three gases. Assume $K = 1.6 \times 10^{-3}$ for this reaction at $T = 298$ K and $\Delta H° = 9.43$ kJ/mol. As usual, assume that $\Delta H°$ and $\Delta S°$ do not change with temperature. No other information should be needed to solve this problem.

Problem 2.9. In Section 2.10.2, we calculated the molar concentration of As in an 100-ppm As-doped Si solid solution. Building on that exercise, calculate the mass density (g/cm^3) of As in this 100-ppm As-doped Si solid solution.

Problem 2.10. Calculate the average mass density (g/cm^3) and molar concentration (mol/cm^3) of silicon atoms in a two-phase composite consisting of 70% SiO_2 and 30% Si (by weight).

Problem 2.11. Metal hydrides are metal alloys that can store significant quantities of hydrogen in their atomic lattice. For this reason, they are considered potentially attractive for the on-board storage of hydrogen for hydrogen fuel cell vehicles. A certain metal hydride alloy can store 2 wt% hydrogen and has a density (including the hydrogen) of 12 g/cm^3. Treating hydrogen as an ideal gas, what pressure would be needed to compress gaseous hydrogen to the same equivalent concentration that it attains when absorbed in the metal hydride?

CHAPTER 3

CHEMICAL REACTION KINETICS

At the most basic level, a chemical reaction involves the breaking, forming, and/or rearrangement of atomic bonds. Consider a simple chemical reaction: the combustion of hydrogen and oxygen to make water. For this reaction to occur, hydrogen–hydrogen and oxygen–oxygen bonds must be broken, while hydrogen–oxygen bonds must be formed. These bond-breaking and bond-making processes occur at the subatomic scale and involve the redistribution of bonding electron clouds (or "orbitals") between the atoms as they either pull apart or come together. The process is probabilistic in nature, and this means that the bond-breaking/forming process can occur in either direction. However, the direction that leads to a lower free-energy state is heavily favored and hence occurs much more frequently. For the hydrogen–oxygen reaction example above, under most temperature and pressure conditions the final free energy of the product species (water) is much lower than the combined free energy of the reactant species (hydrogen + oxygen), and thus the reaction is biased heavily toward the formation of water.

Despite the strong biasing of reactions toward the energetically "downhill" direction, they cannot occur at infinite speed. Reaction rates are finite even if they are energetically "downhill" because:

1. We depend on random thermal motions/vibrations to bring reactants together so that the bond-breaking and bond-making processes can occur.

2. An energy barrier (called an *activation energy*) impedes the conversion of reactants into products. As illustrated in Figure 3.1, in order for reactants to be converted into products, they must first make it over this activation "hill." The *probability* that reactant species can make it over this barrier helps determine the rate at which the reaction occurs.

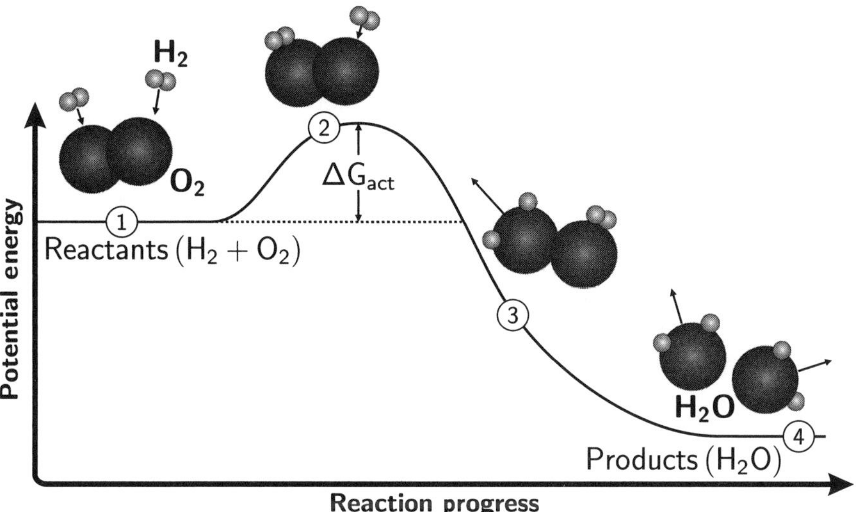

FIGURE 3.1 Schematic of the H_2–O_2 combustion reaction. (Arrows indicate relative motions of the molecules participating in the reaction.) Starting with the reactant H_2 and O_2 gases (1), hydrogen–hydrogen and oxygen–oxygen bonds must first be broken, requiring energy input (2) before hydrogen–oxygen bonds are created, releasing energy and forming H_2O product (3,4). The activation energy (ΔG_{act}) quantifies the minimum amount of input energy required to begin breaking the reactant bonds. Because of this barrier, the rate at which reactants are converted into products (the reaction rate) is limited.

These factors limit the rate at which reactions can occur and furthermore help explain two fundamental principles of reactions kinetics: (1) that reaction rates tend to be sensitive to the concentration of the reacting species and (2) that reaction rates tend to be highly sensitive to temperature. As the concentration of the reactants increases, the frequency with which they encounter one another increases proportionally. Likewise, as temperature increases, the random thermal motion and vibration of atoms increase, thereby increasing both the frequency with which the reactants can interact and the probability that the reactants can make it over the activation barrier to form products. As we will see in this chapter, these factors generally lead to a direct relationship between reaction rate and reactant concentration and an *exponential* increase in reaction rate with increasing temperature.

While increasing temperature tends to increase reaction rates, increasing reaction complexity tends to decrease reaction rates. This third fundamental principle of reaction kinetics can also be understood from an atomic-level perspective of reaction processes. In general, as the number of reactant atoms or product atoms involved in a reaction increases, the likelihood that they can all converge at the same time and place in order to react decreases. This effect is usually captured in a term known as the *order of the reaction*. The order of a reaction has a direct impact on how the reaction kinetics are treated mathematically, with "zero-order reactions" being the simplest (both mechanistically and mathematically), followed by first-order reactions, second-order reactions, and so on.

In the sections that follow, we will delve deeply into the atomistic world of reaction kinetics and learn how to predict the rates of a number of fairly simple zero, first, and second-order reaction processes. While this chapter will focus mostly on simple gas-phase chemical reaction processes, the principles learned here will apply just as well to the solid-state materials kinetic examples that we will confront later in the textbook. This is because bond-breaking and bond-forming processes are remarkably similar at the atomistic level whether they happen between molecules in the gas phase or between atoms in a solid. Thus, most reaction processes can be described using a common set of approaches. Toward the end of the chapter, in preparation for later solid-state applications of reaction kinetic principles, we will examine how reaction rates can be affected by a catalyst or a surface, and we will learn how to model several gas–solid surface reaction processes relevant to materials science and engineering.

3.1 HOMOGENEOUS VERSUS HETEROGENEOUS CHEMICAL REACTIONS

In the sections that follow, we will learn about both homogeneous and heterogeneous chemical reactions. It is important to understand the difference between them.

A *homogeneous chemical reaction* is a reaction where the reactants and products are all in one phase and the reaction can proceed anywhere throughout the volume of the considered phase/system. The hydrogen–oxygen combustion reaction is an example of a homogeneous chemical reaction (assuming that the water is produced in the gas phase). Many gas–gas and nuclear decay reactions are homogeneous.

A *heterogeneous chemical reaction* is a reaction where more than one phase is involved. Heterogeneous reactions can generally only proceed at the interface between the involved phases. Solid–liquid and solid–gas reactions are necessarily heterogeneous since at least two phases are involved in these processes. The chemical vapor deposition of a Si thin film from silane gas ($SiH_{4(g)}$), as was illustrated in Figure 1.3*d*, is a good example of a heterogeneous reaction. The reaction only proceeds at the surface of the solid where it is in contact with the gas phase. Even gas–gas reactions can occur via a heterogeneous route. For example, solid catalysts are employed to accelerate certain gas-phase reactions by providing a heterogeneous surface reaction pathway that is faster than the alternative homogeneous gas-phase reaction pathway would be. Interestingly, many solid–solid reaction processes exhibit aspects of both homogeneous and heterogeneous reaction kinetics. For example, recall the nucleation and growth of second-phase precipitates that was illustrated in Figure 1.3*f*. In such a process, the initial formation of the second-phase nuclei may proceed homogeneously throughout the entire volume of the solid. However, the subsequent growth of these nuclei is a heterogeneous process that can only occur along the surface of the precipitates. This solid–solid transformation process will be discussed in great detail in Chapter 6.

3.2 HOMOGENEOUS CHEMICAL REACTIONS

3.2.1 Reaction Rate Equation and *k*

Chemical reaction rates are modeled using rate equations. The goal of a rate equation is to describe the rate at which reactants are transformed into products during a reaction. Reaction rates are generally quantified by how quickly the concentration of one of the species involved in the reaction changes as a function of time (dc/dt). Consider an arbitrary chemical reaction in the form

$$a\mathrm{A} + b\mathrm{B} + c\mathrm{C} + \cdots \rightleftharpoons m\mathrm{M} + n\mathrm{N} + \cdots \tag{3.1}$$

where A, B, C, … are reactants, and M, N, … are products, and $a, b, c, m,$ and n represent the number of moles of A, B, C, M, and N, respectively. The rate of this reaction can be calculated (e.g., on the basis of species A) as

$$\frac{dc_\mathrm{A}}{dt} = -kc_\mathrm{A}^{\alpha} c_\mathrm{B}^{\beta} c_\mathrm{C}^{\kappa} \cdots \tag{3.2}$$

where $c_\mathrm{A}, c_\mathrm{B}, c_\mathrm{C}, \ldots$ are the concentrations of the reactant species, k is the rate constant (which is strongly temperature dependent), and $\alpha, \beta, \kappa, \ldots$ are the orders of the reaction with respect to reactants A, B, C, …. In this expression and all subsequent reaction rate expressions discussed in this chapter, it is assumed that the reaction proceeds in the forward direction as written, so that reactants (A, B, C, …) are consumed and products (M, N, …) are produced. The negative sign in front of the rate expression reflects this situation.

3.2.2 Order of Reaction

It is extremely important to note that the reaction orders $\alpha, \beta, \kappa, \ldots$ are not the same as the stoichiometric coefficients a, b, c. Reaction order cannot be simply determined from inspection of the balanced chemical equation. It requires detailed information about the kinetic mechanisms underlying the reaction and can generally only be determined experimentally or through careful kinetic studies (see the dialog box on reaction mechanism for details). The overall order of the reaction is given by the sum of the reaction orders with respect to the various reactants. In other words,

$$\text{Overall reaction order} = \alpha + \beta + \kappa + \cdots \tag{3.3}$$

In general, first- and second-order reactions are most commonly seen, but reactions of other orders are also important. Direct analytical solutions are easily acquired for zero-order, first-order, and second-order reactions. Reactions of third-order or higher generally require numerical methods for solution. In the sections that follow, we will cover several examples of zero-order, first-order, and second-order homogeneous chemical reactions.

Reaction Order Depends on Reaction Mechanism

The concentration of ozone in Earth's upper atmosphere is regulated by the following reaction:

$$2O_3(g) \rightleftharpoons 3O_2(g) \tag{3.4}$$

The experimentally determined rate law for this reaction is

$$\frac{dc_{O_3}}{dt} = -k\frac{c_{O_3}^2}{c_{O_2}} \tag{3.5}$$

This rate law is an example of a *mixed-order* reaction. The reaction is second order with respect to O_3 and inverse first order (-1 order) with respect to O_2. It is clear that this rate law cannot be obtained from a simple inspection of the chemical reaction! So from where does this seemingly strange rate law come?

Many chemical reactions, including the ozone–oxygen reaction above, do not occur in a single step as written. Instead, they proceed by a number of smaller intermediate reaction steps. These are known as the *elementary* reaction steps, and it is these individual elementary reaction steps which determine the rate law for the overall chemical reaction.

For the ozone–oxygen reaction, the overall reaction mechanism involves two elementary reaction steps that occur in series:

1. $O_3(g) \rightleftharpoons O_2(g) + O(g)$ (rapid, equilibrium described by K)

2. $O(g) + O_3(g) \rightarrow 2O_2(g)$ (slow, rate constant $= k_2$)

The first step is very rapid and thus achieves an equilibrium condition where the reactant and product concentrations can be described by an equilibrium constant, K. Meanwhile the second step is slow with a reaction rate described by a rate constant k_2. Because the second step is much slower, this step determines the overall reaction rate. The rate law for the second step (and thus for the overall reaction) can be expressed as

$$\frac{dc_{O_3}}{dt} = -k_2 c_{O_3} c_O \tag{3.6}$$

Note that the species O does not appear in the overall chemical reaction—it is an *intermediate species*. However, we can use the first (equilibrium) reaction step to express the concentration of this intermediate species in terms of the oxygen and ozone concentrations:

$$K = \frac{c_{O_2} c_O}{c_{O_3}} \tag{3.7}$$

$$c_O = K\frac{c_{O_3}}{c_{O_2}} \tag{3.8}$$

This result can then be inserted into Equation 3.6 to obtain the final rate law in terms of ozone and oxygen:

$$\frac{dc_{O_3}}{dt} = -k_2 c_{O_3} \left(K \frac{c_{O_3}}{c_{O_2}} \right)$$

$$= -k \frac{c_{O_3}^2}{c_{O_2}} \quad \text{(where } k = k_2 K) \tag{3.9}$$

This example is intended to reinforce the fact that the rate law for an overall chemical reaction usually cannot be determined directly from the reaction stoichiometry. Instead, the overall rate law for a reaction is determined by the sequence of elementary steps, or in other words the detailed *reaction mechanism*, by which the reactants are converted into products. The overall rate law for a reaction is dominated by the rate law for the slowest step in the reaction.

3.2.3 Zero-Order Reactions

Zero-order reactions are not very common. However, they provide an excellent place to begin because the mathematics are quite straightforward. In a zero-order reaction, the reaction rate is independent of the concentration of the reactant(s). Mathematically, this means

$$\frac{dc_A}{dt} = -k(c_A)^0 = -k \tag{3.10}$$

The reaction is called a zero-order reaction because the concentration of the reactants can be included in the rate expression with zero-order exponents to explicitly convey that the rate does not depend on their concentration (as shown in the equation above). In a zero-order reaction, increasing the concentration of the reactant(s) will not speed up the rate of the reaction. The most common zero-order reactions are endothermic high-temperature thermal decomposition reactions, where a large amount of thermal energy is required to break a chemical species apart. In such cases, as long as the temperature is not too high, the available thermal energy controls the rate rather than the concentration of the reactants.

An example of a zero-order reaction is the reverse Haber process:

$$2NH_{3(g)} \rightarrow N_{2(g)} + 3H_{2(g)} \tag{3.11}$$

where the rate at which ammonia decomposes to nitrogen and hydrogen is given by

$$\frac{dc_{NH_3}}{dt} = -k(c_{NH_3})^0 = -k \tag{3.12}$$

Note that this example further illustrates how the reaction order (and hence the rate equation) cannot be determined from a simple inspection of the reaction stoichiometry!

It is often desired to have an equation that directly expresses how the concentration of a reacting species of interest varies as a function of time during a reaction process. This can be done by integrating the rate expression (which is a differential equation) to obtain an *integrated rate law*. For a zero-order reaction, this integration is extremely straightforward. Assuming a zero-order reaction where reactant species A is being consumed,

$$\frac{dc_A}{dt} = -k$$

$$\int dc_A = -k \int dt$$

$$c_A - c_{A_0} = -kt$$

$$c_A = c_{A_0} - kt \tag{3.13}$$

where c_A is the concentration of species A at time t and c_{A_0} is the initial concentration of species A at time $t = 0$. For a zero-order reaction, then, the reactant(s) concentration decreases linearly with time, as shown in Figure 3.2. As should be easily seen by inspection of Equation 3.13, the rate constant k for a zero-order reaction has units of concentration/time. As we will soon see, the rate constants for first- and second-order reactions will have different units.

3.2.4 First-Order Reactions

Reactions displaying first-order reaction kinetics are extremely common. Fortunately, the mathematics needed to describe first-order reactions are also quite straightforward. In a first-order reaction, the reaction rate is directly proportional to the concentration of one of the reactant concentrations. Thus, increasing the concentration of this reactant will speed up the rate of the reaction proportionally. This behavior reflects the fundamental kinetic principle that the speed of most

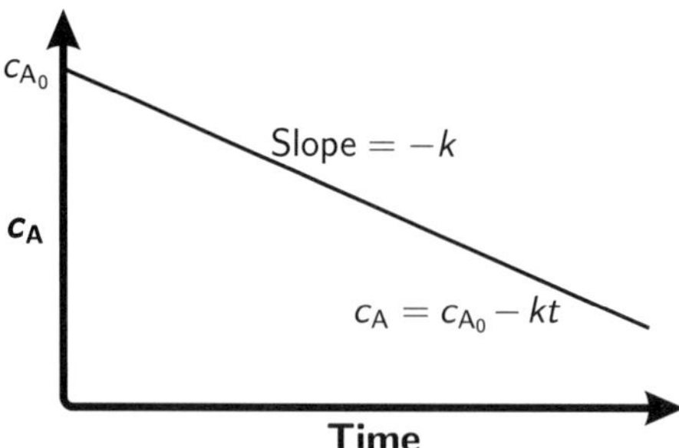

FIGURE 3.2 Variation in reactant concentration as a function of time for a zero-order reaction that consumes species A. The concentration of species A decreases linearly with time from an initial concentration c_{A_0}. The slope is given by the rate constant k.

reactions depends on the frequency with which the reactants encounter one another. Mathematically,

$$\frac{dc_A}{dt} = -kc_A^1 = -kc_A \tag{3.14}$$

It is important to note that a first-order reaction depends on the concentration *of only one reactant*. Other reactants can be present, but they will have zero order.

As we did with the zero-order rate law, this differential equation can be integrated to obtain an equation that directly expresses how the concentration of the reactant varies as a function of time during the reaction process. For a general first-order reaction involving the consumption of a reacting species A, this integration yields,

$$\frac{dc_A}{dt} = -kc_A$$

$$\int \frac{dc_A}{c_A} = -k \int dt$$

$$\ln \frac{c_A}{c_{A_0}} = -kt$$

$$c_A = c_{A_0} e^{-kt} \tag{3.15}$$

where c_A is the concentration of species A at time t and c_{A_0} is the initial concentration of species A at time $t = 0$. For a first-order reaction, then, the reactant concentration decreases exponentially with time, as shown in Figure 3.3. The rate constant k has units of time^{-1}.

Most nuclear decay processes obey first-order reaction kinetics. An example is the radioactive decay of carbon-14 (an unstable radioactive isotope of carbon) to nitrogen-14 (which is the stable isotope of nitrogen) via the emission of an electron and an antineutrino ($\overline{v}_e$):

$$^{14}_{6}C \rightarrow {}^{14}_{7}N + e^- + \overline{v}_e \tag{3.16}$$

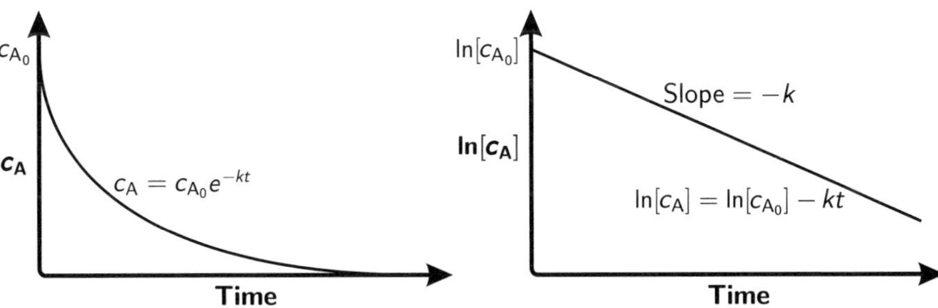

FIGURE 3.3   Variation in reactant concentration as a function of time for a first-order reaction that consumes species A. (*a*) The concentration of species A decreases exponentially with time from an initial concentration c_{A_0}. (*b*) A logarithmic plot of concentration versus time yields a straight line with a slope that is given by the rate constant k.

The carbon-14 decay rate can be expressed by a first-order rate law of the form

$$\frac{dc_{14C}}{dt} = -kc_{14C} \tag{3.17}$$

First-order reactions are frequently characterized in terms of their *half-life*. The half-life, $t_{1/2}$, is the amount of time required for a reaction to consume one-half of the initial reactant concentration. In other words, $c_A = \frac{1}{2}c_{A_0}$ when $t = t_{1/2}$. Using this definition, the half-life can easily be evaluated from Equation 3.15 as

$$c_A = \frac{1}{2}c_{A_0} = c_{A_0}e^{-kt_{1/2}}$$

$$\ln\frac{1}{2} = -kt_{1/2}$$

$$t_{1/2} = -\frac{\ln\frac{1}{2}}{k}$$

$$t_{1/2} = \frac{0.693}{k} \tag{3.18}$$

Since k (for a first order reaction) has units of time^{-1}, the half-life will have units of time. Examples 3.1–3.2 provide practice dealing with the mathematics of first-order reaction kinetics.

Example 3.1

Question: The half-life for the radioactive decay of carbon-14 is 5730 years. Calculate the rate constant for this decay reaction.

Solution: Based on the derivation of the half-life expression given in Equation 3.18, if we know $t_{1/2}$, we can solve for k:

$$t_{1/2} = \frac{0.693}{k}$$

$$k = \frac{0.693}{t_{1/2}} = \frac{0.693}{5730 \text{ yr}}$$

$$k = 1.21 \times 10^{-4}/\text{yr} = 2.30 \times 10^{-10}/\text{min} = 3.83 \times 10^{-12}/\text{s}$$

Example 3.2

Question: Living organic matter maintains an equilibrium carbon-14 concentration via constant exchange with the atmosphere. Once organic matter dies, this atmospheric exchange ceases and the carbon-14 concentration begins to decrease by the radioactive decay process. This can be used by archeologists to date carbon-containing artifacts. The initial rate of decay for organic matter, N_0, is 15.3 decays/(g min). An ancient wooden door beam recovered from an archeological site in Mesa Verde in the desert southwest of Colorado is measured to have a present decay rate $N = 14.0$ decays/(g min). How old is the door?

Solution: The chief challenge of this problem is to relate the decay rate to one of the mathematical quantities that we have discussed in terms of reaction kinetics. As illustrated by Equation 3.16, each time a decay occurs, this corresponds to one atom of carbon-14 converting to an atom of nitrogen-14. Thus, the decay rate N essentially quantifies the change in the carbon-14 concentration (atoms per unit mass) per unit time. In other words,

$$N = -\frac{dc_{14C}}{dt} \tag{3.19}$$

Inserting this expression into the first-order rate law for the carbon-14 decay reaction yields

$$\frac{dc_{14C}}{dt} = -N = -kc_{14C} \tag{3.20}$$

$$N = kc_{14C} \tag{3.21}$$

$$c_{14C} = \frac{N}{k} = \frac{14.0 \text{ decays/(g min)}}{2.30 \times 10^{-10}/\text{min}} = 6.09 \times 10^{-10} \text{ atoms/g} \tag{3.22}$$

Similarly,

$$N_0 = kc_{14C_0} \tag{3.23}$$

$$c_{14C_0} = \frac{N_0}{k} = \frac{15.3 \text{ decays/(g min)}}{2.30 \times 10^{-10}/\text{min}} = 6.65 \times 10^{-10} \text{ atoms/g} \tag{3.24}$$

Now that both the initial and present concentrations of carbon-14 in the door have been determined, these quantities can be inserted into the integrated expression of the rate law to determine the time since the tree used for the door beam was cut down (hence triggering the start of the decay process):

$$c_{14C} = c_{14C_0} e^{-kt} \tag{3.25}$$

$$\frac{c_{14C}}{c_{14C_0}} = e^{-kt} \tag{3.26}$$

$$t = -\frac{\ln(c_{14C}/c_{14C_0})}{k} \tag{3.27}$$

$$= -\frac{\ln\left(\frac{6.09\times10^{-10} \text{ atoms/g}}{6.65\times10^{-10} \text{ atoms/g}}\right)}{2.30 \times 10^{-10}/\text{min}} \tag{3.28}$$

$$= 3.82 \times 10^8 \text{ min} = 728 \text{ yr} \tag{3.29}$$

Alternative Solution: A common strategy that can dramatically simplify solution for many types of problems is to see if you can set up a ratio of equations. In this particular example, we can recognize that for a first-order reaction the reaction rate is directly proportional to the reactant (carbon-14) concentration, and thus we can set up a ratio of reaction rates that enables a much easier approach to the solution:

$$\frac{dc_{14_C}}{dt} = -N = -kc_{14_C} \tag{3.30}$$

$$\frac{dc_{14_{C_0}}}{dt} = -N_0 = -kc_{14_{C_0}} \tag{3.31}$$

$$\frac{dc_{14_C}/dt}{dc_{14_{C_0}}/dt} = \frac{N}{N_0} = \frac{kc_{14_C}}{kc_{14_{C_0}}} = \frac{c_{14_C}}{c_{14_{C_0}}} \tag{3.32}$$

From the integrated form of the rate law, we also have

$$c_{14_C} = c_{14_{C_0}} e^{-kt} \tag{3.33}$$

$$\frac{c_{14_C}}{c_{14_{C_0}}} = e^{-kt} \tag{3.34}$$

Combining these two results yields

$$\frac{N}{N_0} = \frac{c_{14_C}}{c_{14_{C_0}}} = e^{-kt} \tag{3.35}$$

from which the time can easily be evaluated in terms of the ratio of the initial and present decay rates:

$$t = -\frac{\ln(N/N_0)}{k} \tag{3.36}$$

$$= -\frac{\ln\left(\frac{14.0 \text{ decays/(g min)}}{15.3 \text{ decays/(g min)}}\right)}{2.3 \times 10^{-10}/\text{min}} \tag{3.37}$$

$$= 3.86 \times 10^8 \text{ min} = 735 \text{ yr} \tag{3.38}$$

3.2.5 Second-Order Reactions

As we move from first-order to second-order reactions, complexity increases significantly. To begin, there are several different types of second-order reactions. A second-order reaction can either be second order with respect to a single reactant or first order with respect to two distinct reactants (and therefore second order overall). In order to illustrate these possibilities, consider the following general reaction:

$$a\text{A} + b\text{B} \rightarrow m\text{M} \tag{3.39}$$

Assuming that the reaction is overall second order, there are three different potential second-order rate expressions that could govern the kinetics:

$$\frac{dc_A}{dt} = -kc_A^2 c_B^0 = -kc_A^2 \tag{3.40}$$

$$\frac{dc_A}{dt} = -kc_A^1 c_B^1 = -kc_A c_B \tag{3.41}$$

$$\frac{dc_A}{dt} = -kc_A^0 c_B^2 = -kc_B^2 \tag{3.42}$$

Remember, you cannot simply determine which of the three second-order rate laws applies from inspection of the reaction—this must be determined experimentally.

If the reaction is second order with respect to a single reactant, the mathematic treatment remains fairly straightforward. Integration of the rate law (Equation 3.40 above) yields

$$\frac{dc_A}{dt} = -kc_A^2 \tag{3.43}$$

$$\int \frac{dc_A}{c_A^2} = -k \int dt \tag{3.44}$$

$$\frac{1}{c_A} = \frac{1}{c_{A_0}} + kt \tag{3.45}$$

where c_A is the concentration of species A at time t and c_{A_0} is the initial concentration of species A at time $t = 0$.

Alternatively, if the reaction is second order with respect to B but it is still desired to express the reaction rate in terms of A, then an expression is needed to relate the concentrations of reactants A and B during reaction. This can be done by applying the stoichiometry of the chemical reaction:

$$c_B = c_{B_0} + \frac{b}{a}(c_A - c_{A_0}) \tag{3.46}$$

where a and b are the stoichiometric coefficients for reactants A and B appearing in the chemical reaction formula (Equation 3.39) and c_{A_0} and c_{B_0} are the initial concentrations of A and B. This relationship can then be substituted into the rate law and integrated:

$$\frac{dc_A}{dt} = -kc_B^2 = -k\left[c_{B_0} + \frac{b}{a}(c_A - c_{A_0})\right]^2 \tag{3.47}$$

$$\int \frac{dc_A}{[c_{B_0} + (b/a)(c_A - c_{A_0})]^2} = -k \int dt \tag{3.48}$$

$$c_A = c_{A_0} - \frac{a}{b}c_{B_0} + \frac{a^2 c_{B_0}}{ab + b^2 c_{B_0} kt} \tag{3.49}$$

While messy, the resulting expression is nonetheless mathematically straightforward. By retaining c_A as our integration variable in this exercise, we have implicitly assumed that there is a stoichiometric excess of A relative to B (so that when the reaction approaches completion there will be A leftover).

If the reaction is first order with respect to two reactants, the same substitution process can be used to obtain an integrated expression:

$$\frac{dc_A}{dt} = -kc_A c_B = -kc_A \left[c_{B_0} + \frac{b}{a}(c_A - c_{A_0}) \right] \tag{3.50}$$

$$\int \frac{dc_A}{c_A[c_{B_0} + (b/a)(c_A - c_{A_0})]} = -k \int dt \tag{3.51}$$

$$c_A = \frac{c_{B_0} - \frac{b}{a}c_{A_0}}{(c_{B_0}/c_{A_0}) \exp\left[-kt\left(\frac{b}{a}c_{A_0} - c_{B_0}\right)\right] - b/a} \tag{3.52}$$

A simpler but equivalent expression for this integrated rate law can be obtained by retaining the time-dependent concentrations of both A and B in the equation:

$$\frac{c_A}{c_B} = \frac{c_{A_0}}{c_{B_0}} \exp\left[kt\left(\frac{b}{a}c_{A_0} - c_{B_0}\right)\right] \tag{3.53}$$

For second-order reactions, the precise shape of the reactant decay–time curve depends on whether the reaction is second order with respect to a single reactant or first order with respect to two reactants, as shown in Figure 3.4, and also on the relative starting concentrations of A and B. Depending on the relative starting reactant concentrations, a reaction that is second order with respect to a single species A can

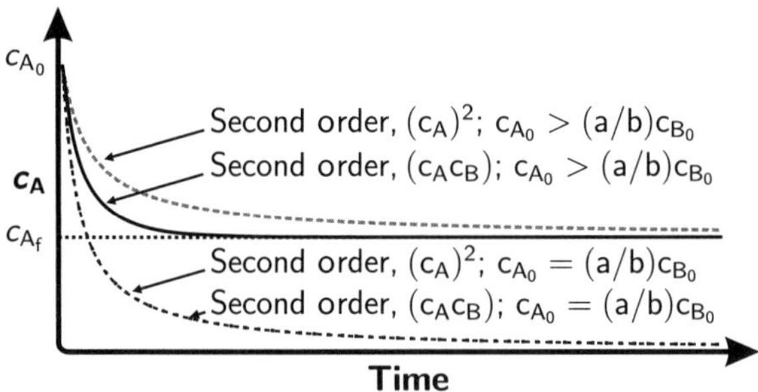

FIGURE 3.4 Variation in reactant concentration as a function of time for a reaction that is second order with respect to a single reactant A (dashed lines) or first order with respect to two reactants A and B (solid lines). In both situations, curves are shown for cases where species A and B are present in stoichiometric equal amounts (so that all the reactants are depleted, overlapped light-gray curves) or for cases where reactant A is present in stoichiometric excess (dark black curves) and so there will still be A left over when the reaction is complete ($c_{A_f} \neq 0$). In all cases, the concentration of species A decreases non linearly with time from an initial concentration c_{A_0}.

proceed more quickly, less quickly, or at the same rate as a reaction that is first order with respect to two species. Note that for cases where one of the reactants is present in stoichiometric excess (A is assumed to be in excess here), there will still be A left over when the reaction is complete. The final amount of A left over (c_{A_f}) can be determined from the initial amounts of the reactants and the reaction stoichiometry using Equation 3.46 by setting $c_B = 0$ and solving for c_A. For all types of second order reactions, the rate constant k has units of (concentration·time)$^{-1}$.

A practice problem involving a reaction which is second order with respect to a single reactant is provided in Example 3.3.

Example 3.3

Question: The decomposition of NO_2 to NO and O_2,

$$2NO_2(g) \rightarrow 2NO(g) + O_2(g) \tag{3.54}$$

has been experimentally determined to be second order with respect to the NO_2 concentration. A reaction begins at time $t = 0$ with an unknown initial concentration of pure $NO_2(g)$. After 2 h, the rate of O_2 gas production is measured as 1.0×10^{-10} mol/(cm^3·s). If the reaction rate constant for this reaction is $k = 10$ (mol/cm$^3 \cdot$ s)$^{-1}$, determine what the initial concentration of $NO_2(g)$ must have been when the reaction began.

Solution: The problem informs us that this decomposition reaction has been experimentally determined to be second order with respect to the NO_2 concentration. Thus, we can write the rate law for the reaction as

$$\frac{dc_{NO_2}}{dt} = -k(c_{NO_2})^2 \tag{3.55}$$

and the integrated form of the rate law (assuming initially only NO_2 reactant and no products at time $t = 0$):

$$\frac{1}{c_{NO_2}} = \frac{1}{c_{NO_2 0}} + kt \tag{3.56}$$

We have been asked to determine the initial (unknown) starting concentration of NO_2 reactant (i.e., $c_{NO_2 0}$) given that the O_2 gas production rate after 2 h of reaction is $dc_{O_2}/dt = 1.00 \times 10^{-10}$ mol/(cm$^3 \cdot$ s). Based on the reaction stoichiometry, this measured oxygen production rate after 2 h of reaction can be converted to a corresponding NO_2 consumption rate:

$$\left.\frac{dc_{NO_2}}{dt}\right|_{t=2\,h} = -2\left.\frac{dc_{O_2}}{dt}\right|_{t=2\,h}$$

where the factor of -2 reflects the fact that for every molecule of O_2 gas that is produced, two molecules of NO_2 are consumed. From this instantaneous NO_2

consumption rate at $t = 2$ h we can calculate the instantaneous NO_2 concentration at $t = 2$ h using Equation 3.55:

$$\left.\frac{dc_{NO_2}}{dt}\right|_{t=2h} = -2\left.\frac{dc_{O_2}}{dt}\right|_{t=2h} = -k\left(c_{NO_2}|_{t=2h}\right)^2$$

$$c_{NO_2}|_{t=2h} = \sqrt{\left(\frac{2}{k}\right)\left.\frac{dc_{NO_2}}{dt}\right|_{t=2h}}$$

$$= \sqrt{\left(\frac{2}{10\ (mol/cm^3 \cdot s)^{-1}}\right) 1.00 \times 10^{-10}\ mol/(cm^3 \cdot s)}$$

$$= 4.47 \times 10^{-6}\ mol/cm^3$$

Then, having obtained the NO_2 concentration at the specific time $t = 2$ h, we can use Equation 3.56 to calculate the initial NO_2 concentration:

$$\frac{1}{c_{NO_2}|_{t=2h}} = \frac{1}{c_{NO_{20}}|_{t=2h}} + kt$$

$$c_{NO_{20}} = \frac{c_{NO_2}|_{t=2h}}{1 - c_{NO_2}|_{t=2h}kt}$$

$$= \frac{4.47 \times 10^{-6}\ mol/cm^3}{1 - 4.47 \times 10^{-6}\ mol/cm^3 \cdot 10\ (mol/cm^3 \cdot s)^{-1} \cdot 7200\ s}$$

$$= 6.59 \times 10^{-6}\ mol/cm^3$$

Pseudo-First-Order Reactions   Under certain circumstances, second-order reactions can sometimes be approximated as first-order reactions. For example, consider a second-order reaction that depends on the concentrations of two different reactants (each to the first order). If one of the reactant concentrations is much larger than the other reactant concentration, then it will remain essentially constant (only slightly depleted) during the reaction process while the concentration of the other reactant is fully consumed. In this situation, the second-order rate law can be rewritten as a pseudo-first-order rate law. As an example, consider a second-order reaction that is first order with respect to two reactants A and B. The rate law for this reaction is

$$\frac{dc_A}{dt} = -kc_A c_B$$

However, if $c_B \gg c_A$, we can assume that $c_B \approx$ constant and the rate law can be rewritten as

$$\frac{dc_A}{dt} = -k'c_A \qquad \text{where } k' = kc_B \tag{3.57}$$

k' is the pseudo-first-order rate constant for the reaction and has units of time^{-1} whereas k, the true rate constant for the reaction, has units of (concentration $\cdot$ time)$^{-1}$.

An example problem utilizing the pseudo-first-order reaction approximation is treated in Example 3.4.

Example 3.4

Question: Under certain conditions, the chemical reaction

$$CO(g) + NO_2(g) \rightarrow CO_2(g) + NO(g) \tag{3.58}$$

is experimentally determined to be first order with respect to CO and first order with respect to NO_2 (second order overall). In order to simplify kinetic measurements, the system is reduced to pseudo first order by operating with a very large excess of CO(g) so that the kinetics are controlled only by the $NO_2(g)$ concentration. When a CO(g) concentration of 5.0×10^{-5} mol/cm^3 is used [which is well in excess of the $NO_2(g)$ concentration], the apparent half-life of this pseudo-first-order reaction is measured to be 2.0 h. Determine both the apparent rate constant k' (units in s^{-1}) and the true rate constant k [units in (mol/cm$^3 \cdot$ s)$^{-1}$] for this reaction. If the CO(g) concentration was doubled, how would this affect the apparent rate constant k'? What about the true rate constant k?

Solution: Because this reaction has been experimentally determined to be first order with respect to CO and first order with respect to NO_2 (second order overall), the rate law for the reaction can be written as

$$\frac{dc_{NO_2}}{dt} = -kc_{NO_2}c_{CO} \tag{3.59}$$

However, since the problem states that $c_{CO} \gg c_{NO_2}$, we can assume that $c_{NO_2} \approx$ constant and so the rate law can be rewritten as

$$\frac{dc_{NO_2}}{dt} = -kc'_{NO_2} \qquad \text{where } k' = kc_{CO} \tag{3.60}$$

This pseudo-first-order rate law can be integrated in exactly the same manner as a normal-first-order rate law, yielding an exponential function

$$c_{NO_2} = c_{NO_20}e^{-k't}$$

with the apparent half-life given by

$$t'_{1/2} = \frac{0.693}{k'}$$

The apparent half-life of this pseudo-first-order reaction is stated to be $t'_{1/2} =$ 2.0 h when the CO(g) concentration is 5.0×10^{-5} mol/cm^3. Thus, we can determine k' as

$$k' = \frac{0.693}{t'_{1/2}} = \frac{0.693}{7200 \text{ s}} = 9.6 \times 10^{-5}/\text{s}$$

As expected, k' has units of s^{-1}.

The true rate constant k can be determined from the pseudo-first-order rate constant and the CO(g) concentration:

$$k = \frac{k'}{c_{CO}} = \frac{9.6 \times 10^{-5}/\text{s}}{5.0 \times 10^{-5} \text{ mol/cm}^3} = 1.9 \ (\text{mol/cm}^3 \cdot \text{s})^{-1}$$

The final part of the question asks what would happen to the apparent rate constant k' and the true rate constant k if the CO(g) concentration was doubled. *Be careful!* This can be tricky. As shown in Equation 3.59, the true rate constant is independent of concentration. Thus, it would not change if the CO(g) concentration was doubled. However, the apparent rate constant k' is proportional to the CO(g) concentration, as indicated in Equation 3.60. Thus, doubling the CO(g) concentration would double k'. The apparent half-life for the pseudo-first-order reaction would commensurately decrease by a factor of 2. This concentration dependence is a common property of apparent (pseudo) first-order rate constants and half-lives. True rate constants (k) do not depend on concentration. Herein lies the reason that it is very important to distinguish between them.

3.2.6   Incomplete Reactions/Equilibrium Reactions

In all of the discussions so far, it has been implicitly assumed that the reactions go to completion. What this means is that the reactions are assumed to continue in the forward direction as written until one of the reactants is completely depleted. For many reactions, this assumption is reasonable. However, there are many other reactions that do not go to completion. Instead, the reaction only proceeds partway and an equilibrium state is reached where considerable concentrations of both the reactants and product species remain in coexistence. This connects back to the concept of *dynamic equilibrium* that we discussed in Chapter 2, when the forward and backward reaction rates reach a balancing point.

Zero-order reaction kinetics are not consistent with incomplete reactions as they predict reactions that will continue to progress until the reactant is completely depleted. However, for first- and second-order reactions, incomplete reactions can be approximately dealt with using the same mathematical framework that we have already developed for complete reactions by simply incorporating an offset value to the equations to account for the fact that when the reaction reaches equilibrium, a certain concentration of reactant species will still be present. For first- and single-species second-order reactions, the resulting modified equations are

$$\text{First order:} \qquad c_A - c_{A_{eq}} = (c_{A_0} - c_{A_{eq}})e^{-kt} \qquad (3.61)$$

$$\text{Second order:} \qquad \frac{1}{c_A - c_{A_{eq}}} = \frac{1}{c_{A_0} - c_{A_{eq}}} + kt \qquad (3.62)$$

where $c_{A_{eq}}$ is the final (equilibrium) concentration of A when the reaction reaches its equilibrium state. In these equations, it is assumed that $c_{A_0} > c_{A_{eq}}$ so that the reaction proceeds in the forward direction and A is consumed until the equilibrium concentration of A is attained. If $c_{A_0} < c_{A_{eq}}$, the reaction would instead proceed in the reverse direction and A would be produced until the equilibrium concentration was attained.

The equilibrium concentration of the reactants (e.g., $c_{A_{eq}}$) can be determined from thermodynamic equilibrium calculations (recall Chapter 2). Example 3.5 demonstrates this approach for an equilibrium reaction involving $H_2(g)$, $I_2(g)$, and $HI(g)$.

Example 3.5

Question: As we learned in Chapter 2, the hydrogen iodide decomposition reaction

$$2HI(g) \rightleftharpoons H_2(g) + I_2(g) \tag{3.63}$$

does not go to completion at most temperatures. Under most conditions, the $HI(g)$ only partially decomposes. Consider a reaction vessel initially filled with pure $HI(g)$ at room temperature and 1 atm pressure. At time $t = 0$ the $HI(g)$ begins to decompose. Assuming that the reaction is second order with respect to the $HI(g)$ concentration and that the rate constant $k = 1 \times 10^4$ $(mol/cm^3 \cdot s)^{-1}$, determine the time it takes for the HI, H_2, and I_2 gases to attain 90% of their final (equilibrium) values. Recall that the equilibrium constant for the hydrogen iodide decomposition reaction is $K = 1.6 \times 10^{-3}$ at room temperature ($T = 300$ K).

Solution: In order to calculate the time it takes for the reaction to progress 90% of the way to equilibrium, we first need to establish the equilibrium values of the HI, H_2, and I_2 gases. We learned how to do this in Exercise 2.3 (Chapter 2) and we can apply the same approach here, although in this case the calculation is simplified by the fact that the initial partial pressures of the product species are zero. The reaction begins with $P_{HI}^{init} = 1$ atm and $P_{I_2}^{init} = P_{H_2}^{init} = 0$. After reaction, we can represent the resulting equilibrium partial pressures of the three gases as $P_{HI}^{eq} = 1 - 2x$ and $P_{I_2}^{eq} = P_{H_2}^{eq} = x$. Following the same approach used in Exercise 2.3 (except in this case H_2 and I_2 are being produced while HI is being consumed), we can calculate x from the equilibrium constant:

$$K = \frac{(P_{I_2}^{eq}/P_{I_2}^\circ)(P_{H_2}^{eq}/P_{H_2}^\circ)}{(P_{HI}^{eq}/P_{HI}^\circ)^2}$$

$$= \frac{x^2}{(1 - 2x)^2} \qquad (P_{I_2}^\circ = P_{H_2}^\circ = P_{HI}^\circ = 1)$$

The solution for x in this case is quite straightforward:

$$x = \frac{\sqrt{K}}{1 + 2\sqrt{K}} = 0.037 \text{ atm}$$

Thus the equilibrium pressures for the three gases are

$$P_{I_2}^{eq} = P_{I_2}^{init} + x = 0 \text{ atm} + 0.037 \text{ atm} = 0.037 \text{ atm}$$

$$P_{H_2}^{eq} = P_{H_2}^{init} + x = 0 \text{ atm} + 0.037 \text{ atm} = 0.037 \text{ atm}$$

$$P_{HI}^{eq} = P_{HI}^{init} - 2x = 1 \text{ atm} - 2 \cdot 0.037 \text{ atm} = 0.926 \text{ atm}$$

The problem asks to determine the time it takes for the HI, H_2, and I_2 gases to attain 90% of their final (equilibrium) values, in other words, when

$$P_{I_2}|_{90\%} = 0.9 \cdot P_{I_2}^{eq} = 0.9 \cdot 0.037 \text{ atm} = 0.033 \text{ atm}$$

$$P_{H_2}|_{90\%} = 0.9 \cdot P_{H_2}^{eq} = 0.9 \cdot 0.037 \text{ atm} = 0.033 \text{ atm}$$

$$P_{HI}|_{90\%} = 0.1 \cdot (P_{HI}^{init} - P_{HI}^{eq}) + P_{HI}^{eq} = 0.933 \text{ atm}$$

The modified (approximate) second-order rate law for this reaction taking into account the fact that the reaction does not go to completion but instead reaches an equilibrium is given by Equation 3.62. For the HI reaction, this becomes

$$\frac{1}{c_{HI} - c_{HI_{eq}}} = \frac{1}{c_{HI_0} - c_{HI_{eq}}} + kt \tag{3.64}$$

Since this rate law involves gas concentrations instead of gas pressures, the relevant HI partial pressures must be converted to concentrations (mol/cm^3) using the ideal gas law:

$$c_{HI}|_{90\%} = \frac{P_{HI}|_{90\%}}{RT}$$

$$= \frac{0.933 \text{ atm} \cdot 101{,}300 \text{ Pa/atm}}{8.314 \text{ J/(mol} \cdot \text{K)} \cdot 300 \text{ K}} \frac{1 \text{ m}^3}{(100 \text{ cm})^3} = 3.79 \times 10^{-5} \text{ mol/cm}^3$$

$$c_{HI_0} = \frac{P_{HI}^{init}}{RT}$$

$$= \frac{1 \text{ atm} \cdot 101{,}300 \text{ Pa/atm}}{8.314 \text{ J/(mol} \cdot \text{K)} \cdot 300 \text{ K}} \frac{1 \text{ m}^3}{(100 \text{ cm})^3} = 4.06 \times 10^{-5} \text{ mol/cm}^3$$

$$c_{HI_{eq}} = \frac{P_{HI}^{eq}}{RT}$$

$$= \frac{0.926 \text{ atm} \cdot 101{,}300 \text{ Pa/atm}}{8.314 \text{ J/(mol} \cdot \text{K)} \cdot 300 \text{ K}} \frac{1 \text{ m}^3}{(100 \text{ cm})^3} = 3.76 \times 10^{-5} \text{ mol/cm}^3$$

Finally, inserting these concentrations into Equation 3.64 and solving for time yields

$$t|_{90\%} = \frac{1}{k}\left(\frac{1}{c_{HI}|_{90\%} - c_{HI_{eq}}} - \frac{1}{c_{HI_0} - c_{HI_{eq}}}\right)$$

$$= \frac{1}{1 \times 10^4 \ (\text{mol/cm}^3 \cdot \text{s})^{-1}}\left(\frac{1}{3 \times 10^{-7} \ \text{mol/cm}^3} - \frac{1}{3.0 \times 10^{-6} \ \text{mol/cm}^3}\right)$$

$$= 300 \ \text{s}$$

Exact Treatment of Equilibrium Reaction Kinetics

The treatment of incomplete reaction kinetics via Equations 3.61 and 3.62 is a rough approximation. A more exact treatment of incomplete/equilibrium reaction kinetics requires simultaneous consideration of the rates for both the forward and reverse reaction processes. For a simple first-order reaction of the form

$$A \rightleftharpoons B \tag{3.65}$$

the speed of the forward reaction process can be described with a forward rate constant k_f while the speed of the reverse reaction process can be described with a reverse rate constant k_r. These rate constants are not necessarily equal. When the reaction reaches *equilibrium*, however, the forward and reverse reaction rates must be equal, and so

$$\text{At equilibrium:} \quad \frac{dc_A}{dt} = -k_f c_{A_{eq}} = -\frac{dc_B}{dt} = -k_r c_{B_{eq}}$$

Therefore

$$c_{A_{eq}} k_f = c_{B_{eq}} k_r$$

$$\frac{c_{A_{eq}}}{c_{B_{eq}}} = \frac{k_r}{k_f} \tag{3.66}$$

As you may recall from Chapter 2, the equilibrium constant for a reaction, K, is given by the ratio of the equilibrium product versus reactant activities (concentrations) raised to their corresponding stoichiometric coefficients (which are all 1 in this elementary reaction). This leads to an extremely important relationship between the equilibrium constant and the ratio of the forward and reverse reaction rates for an elementary reaction:

$$K \equiv \frac{c_{A_{eq}}}{c_{B_{eq}}} = \frac{k_r}{k_f} \tag{3.67}$$

The solution of the integrated rate laws for even this very simple equilibrium reaction between A and B is lengthy as it involves a system of differential equations. However, integrated rate law expressions for the reactant (A) and product (B) concentrations as a function of time can eventually be obtained:

$$c_A = c_{A_0} \frac{1}{k_f + k_r}(k_r + k_f e^{-(k_f+k_r)t}) + c_{B_0} \frac{k_r}{k_f + k_r}(1 - e^{-(k_f+k_r)t}) \tag{3.68}$$

$$c_B = c_{A_0} \frac{k_f}{k_f + k_r}(1 - e^{-(k_f+k_r)t}) + c_{B_0} \frac{1}{k_f + k_r}(k_f + k_r e^{-(k_f+k_r)t}) \tag{3.69}$$

where c_{A_0} and c_{B_0} are the initial concentrations of A and B, respectively.

3.2.7 Summary of Homogeneous Reaction Kinetics

Table 3.1 summarizes the main kinetic equations associated with zero-, first-, and second-order reactions (assuming complete reaction).

3.3 TEMPERATURE DEPENDENCE OF REACTION KINETICS: ACTIVATION THEORY

At the outset of this chapter, we noted three fundamental principles of reaction kinetics:

1. Reaction rates depend (usually) on reactant concentration.
2. Reaction rates depend on reaction complexity.
3. Reaction rates depend exponentially on temperature.

TABLE 3.1 Summary of Zero-, First-, and Second-Order Reactions

	Zero Order	First Order	Second Order c_A^2	Second Order $c_A c_B$
Rate law	$\dfrac{dc_A}{dt} = -k$	$\dfrac{dc_A}{dt} = -kc_A$	$\dfrac{dc_A}{dt} = -kc_A^2$	$\dfrac{dc_A}{dt} = -kc_A c_B$
Integrated	$c_A = c_{A_0} - kt$	$c_A = c_{A_0} e^{-kt}$	$\dfrac{1}{c_A} = \dfrac{1}{c_{A_0}} + kt$	$\dfrac{c_A}{c_B} = \dfrac{c_{A_0}}{c_{B_0}} \exp\left[kt\left(\dfrac{b}{a}c_{A_0} - c_{B_0}\right)\right]$
Units for k	$\dfrac{\text{concentration}}{\text{time}}$	$\dfrac{1}{\text{time}}$	$\dfrac{1}{\text{concentration} \cdot \text{time}}$	$\dfrac{1}{\text{concentration} \cdot \text{time}}$
Half-life	$\dfrac{c_{A_0}}{2k}$	$\dfrac{\ln 2}{k}$	$\dfrac{1}{kc_{A_0}}$	$-\dfrac{\ln\left(2 - \dfrac{b}{a}\dfrac{c_{A_0}}{c_{B_0}}\right)}{k\left(\dfrac{b}{a}c_{A_0} - c_{B_0}\right)}$

Note: For second-order reaction involving both A and B, A is assumed to be in stoichiometric excess relative to B, i.e., $c_{A_0} > (a/b)c_{B_0}$, where a and b are the stoichiometric coefficients for reactants A and B appearing in the balanced chemical reaction equation.

The first two principles are captured directly by the reaction rate laws and the concepts of reaction order that we have discussed in the foregoing sections. However, where does the temperature dependence of the reaction rate enter into these equations?

The answer is that the temperature dependence is embedded inside of k, the rate constant. As we will see in this section, the rate constant depends on temperature because it captures the probability that reactants will have sufficient energy to undergo reaction, and this probability generally increases exponentially with increasing temperature.

In order to understand k and its temperature dependence, it is helpful to examine a typical chemical reaction process at the atomistic scale, such as the hydrogen iodide decomposition reaction shown in Figure 3.5. As this figure illustrates, in order for the hydrogen iodide decomposition reaction to occur, two HI molecules must first collide (Figure 3.5a). However, not all collisions lead to successful reaction. In order to successfully react, molecules must collide at just the right angle and must possess sufficient energy to overcome the energetic barrier to the reaction. This minimum required energy is known as the *activation energy* ΔG_{act} (or E_{act} if expressed in units of eV). The activation energy reflects the fact that a certain amount of energy must be supplied in order to begin to break the reactant bonds (in this case the H–I bonds),

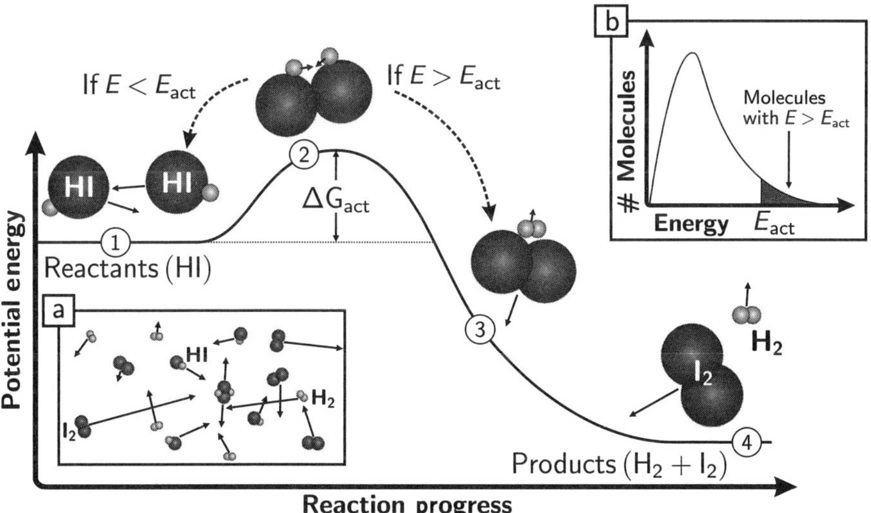

FIGURE 3.5 (*a*) Schematic illustration of the hydrogen iodide decomposition reaction at the atomic scale. The reaction rate depends on the frequency at which HI molecules collide with one another and the probability that these molecules possess sufficient energy to react when they do collide. Only molecules that have sufficient energy to overcome the activation barrier (ΔG_{act}) can react. Because the energetic distribution of gas molecules follows a Maxwell–Boltzmann distribution as shown in (*b*), only the highest energy gas molecules in the exponential tail of the distribution possess sufficient energy to successfully react.

thereby leading to an "activated state" where it then becomes energetically downhill to form new product bonds (in this case H–H and I–I bonds) in their place (Figure 3.5).

In general, only a small fraction of reactant molecules will have this minimum sufficient energy to react. In fact, because the energetic distribution of gas molecules follows a Maxwell–Boltzmann distribution as shown in Figure 3.5b, only the highest energy gas molecules in the exponential tail of the distribution possess sufficient energy ($E > E_{act}$) to successfully react.

Summarizing this atomic-level picture of the reaction process, we can understand that the overall reaction rate should be given by frequency at which the reactants collide multiplied by the probability that they have sufficient energy to react when they do collide:

$$\frac{\text{Reaction}}{\text{rate}} = \frac{\text{Collision}}{\text{frequency}} \times \frac{\text{probability that colliding molecules}}{\text{possess necessary energy to react}}$$

Because collision requires two HI molecules to come together, the collision frequency scales with the square of the HI concentration:

$$\text{Collision frequency} = f \cdot (c_{HI})^2$$

where f is a collision frequency factor that depends (among other things) on steric considerations, the temperature, and the masses and collision diameters of the reactants. Typical collision frequency factors for gas-phase reactions may be on the order of 10^9/s–10^{12}/s.[1]

Because only the highest energy gas molecules in the exponential tail of the distribution possess sufficient energy to successfully react, the probability that colliding molecules possess sufficient energy to react is given by an exponential function that depends on the activation energy and the temperature:

$$\frac{\text{Probability that colliding molecules}}{\text{possess necessary energy to react}} = e^{-\Delta G_{act}/RT}$$

Combining these terms leads to an overall expression for the reaction rate:

$$\text{Reaction rate} = \frac{dc_{HI}}{dt} = -f \cdot (c_{HI})^2 \cdot e^{-\Delta G_{act}/RT}$$

Thus

$$\frac{dc_{HI}}{dt} = -k(c_{HI})^2 \qquad (\text{where } k = fe^{-\Delta G_{act}/RT})$$

where we have made this expression equivalent to our previous expression for second-order reaction kinetics by grouping the non-concentration-dependent terms into the rate constant k. The rate constant can thus be identified as an exponentially

---

[1] For a second-order reaction, f will have units of $1/(s \cdot mol)$.

temperature-sensitive term that depends on the reaction activation energy and a frequency factor. Oftentimes you will see the temperature dependence of k expressed as

$$k = k_0 e^{-\Delta G_{act}/RT} \tag{3.70}$$

where k_0 essentially takes on the same meaning as f and incorporates the collision rate, steric/geometric factors, and other small corrections into account (although it is most often determined empirically).

Example 3.6

Question: A reaction has $\Delta G_{act} = 50$ kJ/mol. What temperature would be required to double k for this reaction compared to its value at $T = 300$ K?

Solution: As we have done for a number of problems, the easiest way to solve this problem is to set up a ratio. At temperature T_1, the rate constant is k_1, while at temperature T_2, the rate constant is k_2. From Equation 3.70 we can write

$$\frac{k_1}{k_2} = \frac{k_0 e^{-\Delta G_{act}/RT_1}}{k_0 e^{-\Delta G_{act}/RT_2}}$$

$$= \exp\left[-\frac{\Delta G_{act}}{R}\left(\frac{1}{T_1} - \frac{1}{T_2}\right)\right] \tag{3.71}$$

Solving this expression for T_2 and noting that if the rate constant has doubled at T_2, then $k_2 = 2k_1$ allows us to determine T_2:

$$T_2 = \left(\frac{1}{T_1} + \frac{R}{\Delta G_{act}}\ln\frac{k_1}{k_2}\right)^{-1}$$

$$= \left(\frac{1}{T_1} + \frac{R}{\Delta G_{act}}\ln\frac{k_1}{2k_1}\right)^{-1}$$

$$= \left(\frac{1}{300\ \text{K}} + \frac{8.314\ \text{J/(mol}\cdot\text{K)}}{50{,}000\ \text{J/mol}}\ln\frac{1}{2}\right)^{-1}$$

$$= 310\ \text{K} \tag{3.72}$$

Thus, a 10 K increase in temperature is sufficient to double the rate constant for this reaction! Many biological rate processes that occur near room/body temperatures have activation energies on the order of 50 kJ/mol, and thus the guideline that a 10 K increase in temperature can double the reaction rate is a useful rule of thumb. Of course, for reactions with activation energies that are far from 50 kJ/mol, this rule of thumb will not be accurate.

3.4   HETEROGENEOUS CHEMICAL REACTIONS

As discussed in the first part of this chapter, most reactions involving the solid state (and hence most reaction processes of interest in *materials* kinetics) are heterogeneous. In the following sections, we will introduce several examples of heterogeneous reaction processes involving gas/solid interfaces. These examples serve as a bridge connecting the homogeneous gas-phase reaction processes we learned about in the previous sections of this chapter with issues that are important to materials scientists and engineers. Later, in chapters 5 and 6 of this textbook, we will delve more deeply into heterogeneous gas–solid, liquid–solid, and solid–solid kinetic processes.

3.4.1   Effect of Catalyst

As is apparent from Equation 3.70, decreasing the size of the activation barrier ΔG_{act} will increase k and therefore increase the reaction rate. The most common way that this can be accomplished is by using a *catalyst*. A catalyst is a material which participates in the reaction process and thereby facilitates a faster reaction rate but is itself not consumed during the reaction. A catalyst increases the reaction rate by lowering the activation barrier for the reaction. This effect is illustrated schematically in Figure 3.6. Because ΔG_{act} appears as an exponent in the equation for k, even small decreases in the activation barrier can cause large effects. Using catalysts therefore provides a way to dramatically increase reaction rates.

There are several ways in which a catalyst can decrease the activation barrier for a reaction. In general, the basic idea is that the catalyst temporarily hosts (i.e., by absorbing on its surface) one or more of the reactant species participating in the reaction and, in so doing, provides a better, lower energy pathway for the reaction process to occur. For example, the catalyst can lower the activation energy by immobilizing reactant species in an optimal orientation to react. It can also help weaken (or even completely cleave) reactant bonds, thereby lowering the amount of additional energy that must be supplied to activate the reaction.

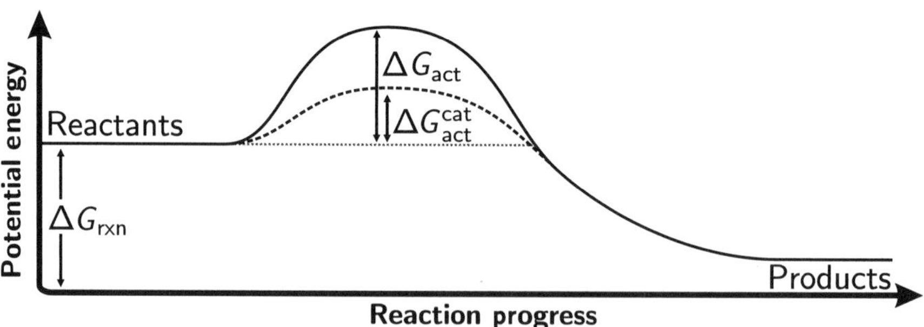

FIGURE 3.6   Schematic illustration of the effect of a catalyst. The use of a catalyst can significantly lower the activation barrier for a reaction ($\Delta G_{\text{act}}^{\text{cat}} < \Delta G_{\text{act}}$). Because k depends exponentially on ΔG_{act}, a catalyst can therefore cause a dramatic increase in reaction rate.

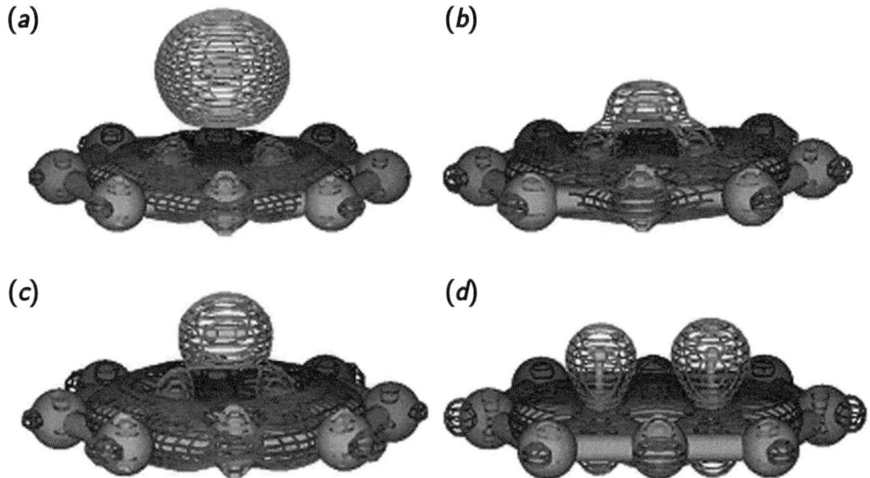

FIGURE 3.7 Evolution of electron orbitals as a hydrogen molecule approaches a cluster of platinum atoms. (*a*) Platinum and hydrogen molecule are not yet interacting. (*b*), (*c*) Atomic orbitals begin overlapping and forming bonds. (*d*) Complete separation of hydrogen atoms occurs almost simultaneously with reaching the lowest energy configuration. This figure was calculated using a quantum mechanical simulation technique known as density functional theory (DFT).

Reactions involving catalysts are *heterogeneous* because they proceed on the surface of the catalyst. In order to understand the effect of the catalyst, it is therefore important to understand the heterogeneous interaction between the reacting species and the catalyst phase. As an example of this interaction, consider the catalytic reaction depicted at the atomic scale in Figure 3.7, which shows the first step in the oxidation of hydrogen taking place on a platinum catalyst surface. As shown in Figure 3.7*a*, the hydrogen molecule consists of two hydrogen atoms strongly held together by an electron bond. The three-dimensional (3D) surface drawn around the hydrogen molecule in Figure 3.7*a* is a physical representation of the electron density in the molecule. In effect, the electron density distribution defines the spatial "extent" and "shape" of the molecule.

In Figure 3.7*b*, we watch as the hydrogen molecule begins to interact with a platinum catalyst cluster. As the hydrogen molecule gets closer and closer (Figures 3.7*b–d*) bonds between the hydrogen molecule and the platinum atoms are formed. The new emerging bonds between platinum and hydrogen lead to weakening of the hydrogen–hydrogen bond and ultimately to complete separation. Thus, the platinum catalyst facilitates the separation of the hydrogen molecule into hydrogen atoms, which thereby *activates* these atoms so that they can more easily react with oxygen (or other oxidant species) in subsequent reaction steps with very little additional activation energy required. By completely weakening the bonds between the hydrogen molecule reactants, the platinum catalyst therefore dramatically lowers the activation energy barrier for the hydrogen oxidation reaction.

In the absence of the platinum cluster, this reaction would not occur as easily; instead, significant energy input would be required to induce separation.

Example 3.7

Question: A reaction has $\Delta G_{act} = 50$ kJ/mol. Upon employing a catalyst, the activation energy is decreased to $\Delta G_{act}^{cat} = 20$ kJ/mol. Assuming all else is equal, how much faster is the catalyzed reaction at $T = 300$ K compared to the original, uncatalyzed reaction?

Solution: Once again, setting up a ratio enables an easy solution to this problem. Because reaction rates are directly proportional to the rate constant (this is true for all reactions, be they zero order, first order, second order, etc.), we can compare the reaction rates for the catalyzed and uncatalyzed reaction processes by comparing their rate constants:

$$\frac{\text{Catalyzed Rxn rate}}{\text{Uncatalyzed Rxn rate}}$$

$$= \frac{k_{cat}}{k_{uncat}} = \frac{k_0 e^{-\Delta G_{act}^{cat}/RT}}{k_0 e^{-\Delta G_{act}/RT}}$$

$$= \exp\left(-\frac{1}{RT}(\Delta G_{act}^{cat} - \Delta G_{act})\right)$$

$$= \exp\left(-\frac{1}{8.314 \text{ J/(mol} \cdot \text{K}) \cdot 300 \text{ K}}(20{,}000 \text{ J/mol} - 50{,}000 \text{ J/mol})\right)$$

$$= 1.7 \times 10^5$$

In other words, a 60% decrease in the activation energy has translated to a more than 100,000× increase in the reaction rate! This is why catalysts are so important—even small decreases in the activation energy for a reaction can lead to huge increases in reaction rate.

There are several important caveats to the simple example we have worked here. First, we have implicitly assumed that the reaction order does not change for the catalyzed reaction process compared to the uncatalyzed reaction process. However, because catalysts can provide completely new reaction pathways, the reaction order (and hence the rate law) can sometimes change significantly between the homogeneous uncatalyzed reaction and the heterogeneous catalyzed reaction process. In addition, compared to an uncatalyzed homogeneous gas-phase reaction that can occur anywhere within the volume of the gas phase, a heterogeneous solid surface catalyzed reaction can occur only on the surface of the catalyst. This fact would be reflected in different units for the rate constant for the homogeneous reaction process versus the heterogeneous reaction process (k_{hom} vs. k_{het}). Thus, the overall reaction rate *per unit volume* for the catalyzed process depends very strongly on the amount of catalyst surface area per unit volume. This is why catalysts are designed to have extremely high active surface areas for maximum effectiveness.

3.4.2 Gas–Solid Surface Reaction Processes

Gas–solid reactions are among the most common type of heterogeneous reaction processes. The platinum surface catalyzed oxidation of hydrogen, discussed in the previous example, is an excellent example of a heterogeneous gas–solid surface reaction process. In Chapter 5, we will study a number of different gas–solid kinetic processes in great detail. To prepare for those studies, in this section we will discuss a few more simple gas–solid surface reaction processes.

Perhaps appropriately, we will continue this discussion through the context of platinum catalysts, which have enormous commercial application in a variety of heterogeneous chemical reaction processes. Although Pt catalysts are extraordinarily effective in accelerating many reactions, they are expensive and are also susceptible to poisoning. For example, carbon monoxide (CO) permanently absorbs onto platinum, clogging up reaction sites. The CO-passivated Pt surface is thus "poisoned," and other desired reaction processes can no longer occur. Because CO is often present as a reactant or intermediate species in many of the reaction processes where Pt is used as a catalyst, this CO poisoning process can cause significant problems.

Using concepts from reaction kinetics, it is possible to understand this poisoning process and gain some insight into how it can be mitigated. The key question to address is how fast will a Pt surface be poisoned by CO and what are the main factors that impact this poisoning process? As we have done when treating the kinetics of other reaction processes in this chapter, a first step is to describe the CO poisoning process using a chemical equation:

$$CO(g) + \text{available Pt surface site} \rightleftharpoons Pt - CO_{(ads)} \tag{3.73}$$

This poisoning process is schematically illustrated at the atomic scale in Figure 3.8.

The next step is to write a rate equation (based on experimental evidence) that correctly captures the reaction order and hence the rate law for this reaction process.

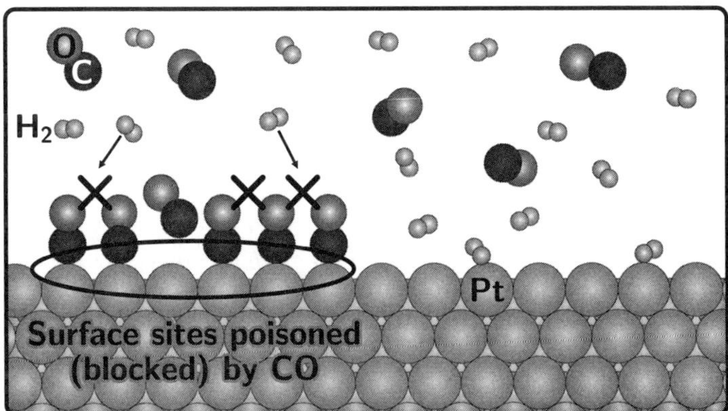

FIGURE 3.8 Schematic illustration of CO poisoning a Pt catalyst surface. This is an example of a heterogeneous gas–solid surface reaction process and can be described using reaction kinetic principles very similar to those we developed for homogeneous gas-phase reactions.

In this case, the rate law has been experimentally determined to be first order with respect to CO and also first order with respect to the Pt surface sites available for reaction (second order overall). Since we would like to know how fast the Pt surface is poisoned, we write the rate law in terms of the CO surface coverage, Φ:

$$\text{Surface coverage} = \Phi = \frac{\text{CO poisoned surface sites}}{\text{total surface sites}} \tag{3.74}$$

The concentration of available (unpoisoned) surface sites is therefore given by

$$\text{Available surface sites} = 1 - \Phi \tag{3.75}$$

The rate law can then be written as

$$\frac{d\Phi}{dt} = kc_{CO(g)}(1 - \Phi) \tag{3.76}$$

This gas–solid reaction is a *self-limiting* process that leads, at most, to a monolayer of CO gas coverage on the catalyst. Once all of the available Pt surface sites have reacted, no further CO will adsorb on the surface of the Pt. This is expressed by the fact that when the CO surface coverage (Φ) goes to 1, the reaction rate ($d\Phi/dt$) goes to zero.

Just as in previous cases, an integrated rate law can be obtained by integrating this equation in order to determine the CO poisoning as a function of time. To simplify this integration, we will assume that the CO concentration in the gas phase is significantly higher that the concentration of Pt surface sites that can be poisoned. Thus, even when CO fully covers the Pt surface, the depletion of CO from the gas phase is negligible. (Alternatively, we can assume that there is a constantly replenished flow of gas above the surface of the Pt catalyst which would also effectively fix the CO gas concentration at a constant value.) This simplification enables this second-order reaction to be reduced to a pseudo-first-order reaction and makes integration easy:

$$\frac{d\Phi}{dt} = -kc_{CO(g)}(1 - \Phi)$$

$$\int \frac{d\Phi}{1 - \Phi} = -kc_{CO(g)} \int dt$$

$$\ln \frac{1 - \Phi}{1 - \Phi_0} = -kc_{CO(g)}t$$

$$\Phi = 1 - (1 - \Phi_0)e^{-kc_{CO(g)}t} \tag{3.77}$$

where Φ is the fraction of the Pt surface that is poisoned by CO at time t and Φ_0 is the initial CO surface coverage at time $t = 0$. For an initially clean (unpoisoned) Pt surface, Φ_0 would be 0. For this CO poisoning process, the surface coverage increases with time, as shown in Figure 3.9, approaching a value of 1 (completely poisoned surface) at long times.

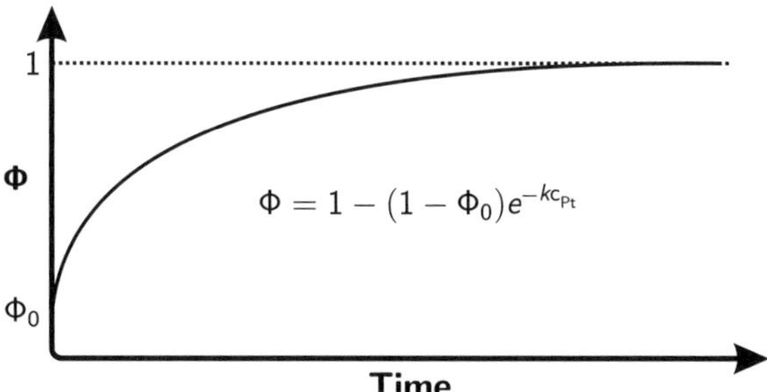

FIGURE 3.9   Change in surface coverage versus time for the irreversible adsorption (poisoning) of Pt by CO gas. Starting from an initial CO surface coverage (Φ_0) at time $t = 0$, Φ approaches a value of 1 (completely poisoned surface) at long times.

Example 3.8

Question: If the rate constant for the CO poisoning reaction on Pt is $k = 1.0 \times 10^5$ (mol/cm$^3 \cdot$ s)$^{-1}$ at $T = 300$ K, calculate how long it would take for an initially clean Pt surface to be 80% poisoned by CO at $T = 300$ K assuming it is exposed to a flowing gas stream containing 100 ppm CO at 1 atm total pressure.

Solution: We must first calculate the concentration of CO in the flowing gas stream using the ideal gas law and the fact that it contains 100 ppm (i.e., $100/10^6$) CO:

$$c_{CO} = \frac{100}{10^6} \frac{P_{tot}}{RT} = \frac{100}{10^6} \frac{1 \text{ atm} \cdot 101{,}300 \text{ Pa/atm}}{8.314 \text{ J/(mol} \cdot \text{K)} \cdot 300 \text{ K}} \frac{1 \text{ m}^3}{(100 \text{ cm})^3}$$

$$= 4.06 \times 10^{-9} \text{ mol/cm}^3$$

We can then use Equation 3.77 to solve for the time t when $\Phi = 0.8$ (80% CO coverage) given $\Phi_0 = 0$ (initially clean surface):

$$t = -\frac{1}{kc_{CO}} \ln \frac{1 - \Phi}{1 - \Phi_0}$$

$$= -\frac{1}{kc_{CO}} \ln(1 - \Phi) \qquad (\Phi_0 = 0)$$

$$= -\frac{1}{1.0 \times 10^5 \text{ (mol/cm}^3 \cdot \text{s)}^{-1} \cdot 4.06 \times 10^{-9} \text{ mol/cm}^3} \ln(1 - 0.80)$$

$$= 3964 \text{ s} = 66 \text{ min}$$

Thus, even for a relatively low level (100 ppm) of CO impurity in a flowing gas stream, this Pt surface would be 80% poisoned within about 1 h! Its effectiveness as a catalyst would then be greatly diminished as the adsorbed CO would block other gases from accessing catalytic Pt surface sites.

The analysis above detailing the kinetics of the CO-poisoning reaction implicitly assumed an irreversible (i.e., complete) reaction. In other words, the gas adsorption process is assumed to continue until the Pt surface is completely covered (saturated) with CO. This assumption is reasonable for the CO poisoning reaction on Pt and for many other gas–solid surface reactions, especially at lower temperatures where thermodynamics favors gas adsorption. (Quick quiz question: Can you provide a thermodynamic argument for why lower temperatures favors gas adsorption?) However, at higher temperatures and for certain gas–solid reactions, complete coverage may not occur. Instead, a balance between gas adsorption and gas desorption can occur, resulting in an equilibrium surface coverage somewhere between $\Phi = 0$ and $\Phi = 1$.

As was discussed in the context of Equation 3.65, for reactions that do not go to completion both the forward and backward reaction processes can be characterized with their own distinct rate constants and the equilibrium condition can be identified where the rates of the forward and backward reaction processes are equal. For the CO adsorption process on Pt, rate laws for the forward and reverse reactions can be written as

$$\text{Forward reaction:} \quad \frac{d\Phi}{dt} = k_{\mathrm{f}} c_{\mathrm{CO}(g)}(1 - \Phi) \tag{3.78}$$

$$\text{Reverse reaction:} \quad \frac{d\Phi}{dt} = k_{\mathrm{r}}\Phi \tag{3.79}$$

Note that the reverse reaction process only depends on the CO surface coverage and not the gas-phase CO concentration. Essentially, this expresses the fact that the probability of the desorption process only involves the concentration of CO atoms on the surface of the solid—it does not depend on the concentration of CO in the gas phase above the surface.

At equilibrium, the forward and reverse reaction rates must be equal, and so we have

$$k_{\mathrm{f}} c_{\mathrm{CO}(g)}(1 - \Phi_{\mathrm{eq}}) = k_{\mathrm{r}}\Phi_{\mathrm{eq}}$$

$$\Phi_{\mathrm{eq}} = \frac{K c_{\mathrm{CO}(g)}}{K c_{\mathrm{CO}(g)} + 1} \tag{3.80}$$

$$\text{where } K = \frac{k_{\mathrm{f}}}{k_{\mathrm{r}}}$$

Recall from our previous section on incomplete reaction processes that the ratio of the rate constants for the forward and reverse reaction processes yields the equilibrium constant K for the reaction. Thus, in the case of an incomplete reaction, this equilibrium constant, in concert with the gas-phase CO concentration, would determine the equilibrium surface coverage of CO. This equation is known as the *Langmuir isotherm* and it is one of a number of physical models that is frequently encountered when describing gas–solid adsorption processes.

3.5 CHAPTER SUMMARY

The purpose of this chapter was to introduce the basic concepts and tools used to understand and model the rate (or speed) of chemical reaction processes. The main points introduced in this chapter include:

- At the most basic level, a chemical reaction involves the breaking, forming, and/or rearrangement of chemical bonds. This process depends on random thermal motions/vibrations to bring reactants together so that bond breaking/forming can occur. It also requires overcoming an energy barrier (quantified by an activation energy) which impedes the conversion of reactants into products. Because of these factors, reaction rates tend to depend on (1) reactant concentration, (2) reaction complexity, and (3) temperature.
- Reaction processes can occur homogeneously or heterogeneously. A homogeneous chemical reaction is a reaction where the reactants and products are all in one phase and the reaction can proceed anywhere throughout the volume of the considered system.
- A heterogeneous chemical reaction is a reaction where more than one phase is involved. Heterogeneous reactions can generally only proceed at the interface between the involved phases.
- Mathematical rate laws can be developed to describe the rate at which a reaction process occurs. This rate is typically expressed in terms of the change in the concentration of a particular species taking place in the reaction as a function of time. Often, both differential (dc_i/dt) and integrated $[c_i(t)]$ rate laws are useful for answering questions about a chemical reaction process.
- The mathematical form of the reaction rate law for a specific chemical reaction depends on the *order of the reaction*, which is itself dictated by the reaction mechanism. Reaction order cannot be determined from a simple inspection of a stoichiometric chemical reaction; it must be determined empirically from experiment or from detailed knowledge about the underlying reaction mechanism.
- Analytical rate law expressions are available for zero-, first-, and second-order reaction processes. First- and second-order reaction processes are the most common in everyday occurrence.
- The integrated rate law for a zero-order reaction is $c_A = c_{A_0} - kt$, where c_A is the concentration of reactant species A at time t and c_{A_0} is the initial concentration of reactant species A at time $t = 0$. For a zero-order reaction, the reactant concentration decreases linearly with time. The rate constant k for a zero-order reaction has units of concentration/time.
- The integrated rate law for a first-order reaction is $c_A = c_{A_0} e^{-kt}$, where c_A is the concentration of reactant species A at time t and c_{A_0} is the initial concentration of reactant species A at time $t = 0$. For a first-order reaction, the reactant concentration decreases exponentially with time. The rate constant k for a first-order reaction has units of time^{-1}.

- A second-order reaction can be second order with respect to a single reactant or first order with respect to two distinct reactants. If the reaction is second order with respect to a single reactant (e.g., A), the integrated rate law is $1/c_A = 1/c_{A_0} + kt$, where c_A is the concentration of reactant species A at time t and c_{A_0} is the initial concentration of reactant species A at time $t = 0$. If the reaction is first order with respect to two distinct reactants (e.g., A and B), the integrated rate law is

$$\frac{c_A}{c_B} = \frac{c_{A_0}}{c_{B_0}} \exp\left[kt(\frac{b}{a}c_{A_0} - c_{B_0})\right]$$

  where c_A and c_B are the concentrations of reactant species A and B at time t and c_{A_0} and c_{B_0} are the initial concentrations of reactant species A and B at time $t = 0$. From the concentration decay profile, it is often difficult to tell the difference between many first- and second-order reactions. However, for second-order reactions, the rate constant k will have units of (concentration $\cdot$ time)$^{-1}$.

- The relative speed of a reaction is sometimes characterized in terms of its *half-life*, which quantifies the time required for a reaction to consume half of its initial reactant concentration. This is particularly true for first-order reactions, where the half-life is easy to obtain mathematically as $t_{1/2} = 0.693/k$.

- Under certain circumstances, second-order reactions can sometimes be approximated as first-order reactions. This is particularly true for a second-order reaction that is first order with respect to two distinct reactants if one of the reactants is present in *extreme excess*. If one of the rate-controlling reactant concentrations is much larger than the other rate-controlling reactant concentration, then it will remain essentially constant during the reaction process while the other reactant is fully consumed. In this situation, the second-order rate law can be rewritten as a pseudo-first-order rate law. For example, for the case where $c_B \gg c_A$,

$$\frac{dc_A}{dt} = -kc_Ac_B = k'c_A \qquad \text{(where } k' = kc_B)$$

  where k' is the pseudo-first-order rate constant, which has units of time^{-1}.

- The rate laws developed in this chapter all assume that the reactions go to completion. While this a reasonable assumption for many reactions, there are many others where the reaction proceeds only partway and an equilibrium state is reached where considerable concentrations of both the reactant and product species remain. The rate expressions in this chapter can be modified to approximately account for this by incorporating a mathematical offset into the rate equation: for example,

$$\text{First order:} \quad c_A - c_{A_{eq}} = (c_{A_0} - c_{A_{eq}})e^{-kt}$$

$$\text{Second order:} \quad \frac{1}{c_A - c_{A_{eq}}} = \frac{1}{c_{A_0} - c_{A_{eq}}} + kt$$

where $c_{A_{eq}}$ is the final (equilibrium) concentration of A when the reaction reaches its equilibrium state. In these equations, it is assumed that $c_{A_0} > c_{A_{eq}}$ so that the reaction proceeds in the forward direction and A is consumed until the equilibrium concentration of A is attained. If $c_{A_0} < c_{A_{eq}}$, the reaction would instead proceed in the reverse direction and A would be produced until the equilibrium concentration was attained. The equilibrium concentration of the reactants (e.g., $c_{A_{eq}}$) can be determined from thermodynamic equilibrium calculations (recall Chapter 2). It is important to note that this treatment of incomplete reaction kinetics is only a crude approximation. A more accurate treatment requires the solution of a simultaneous system of differential equations involving both the forward and backward reaction rates.

- The temperature dependence of the reaction rate is embedded inside of the reaction rate constant k. The rate constant depends on temperature because it captures the probability that reactants will have sufficient energy to undergo reaction. This probability increases exponentially with increasing temperature as $k = k_0 e^{-\Delta G_{act}/RT}$, where ΔG_{act} is the *activation energy* required to convert reactants into products.

- Catalysts can decrease the activation energy for a reaction and hence dramatically increase the reaction rate. A catalyst participates in a reaction process and thereby facilitates a faster reaction but is itself not consumed during the reaction. A catalyst can decrease the activation energy for a reaction, for example, by providing a better, lower energy pathway for the reaction process to occur or by helping to weaken (or even completely cleave) reactant bonds, thereby lowering the amount of additional energy that must be supplied to activate the reaction.

- Heterogeneous gas–solid surface adsorption reaction processes can frequently be treated using the same reaction rate law approach used for homogeneous chemical reactions. In such cases, surface sites are often a key reactant, and their concentration is often represented in terms of a fractional occupancy or availability [e.g., Φ or $(1 - \Phi)$]. Using these principles, as an example, the rate at which a Pt surface is poisoned by CO gas adsorption can be modeled as $\Phi = 1 - (1 - \Phi_0)e^{-kc_{CO_{(g)}}t}$, where Φ is the fraction of the Pt surface that is poisoned by CO at time t and Φ_0 is the initial CO surface coverage at time $t = 0$. For the case of incomplete or partial surface reactions, various kinetic expressions for the equilibrium surface coverage, such as the Langmuir Isotherm, can be derived.

3.6 CHAPTER EXERCISES

Review Questions

Problem 3.1. True (T) or False (F).

(a) $Fe(\gamma) \rightarrow Fe(\alpha)$ is a homogeneous reaction.

(b) $dc_A/dt = -kc_A c_B$ is a first-order reaction rate law.

(c) The "half-life" of a first-order reaction is the time it takes for the reaction to go to half of its final value.

(d) For the reaction $A + 2B \rightarrow C$, the rate equation must be: $dc_A/dt = -kc_A c_B^2$.

(e) Decreasing the activation energy of a reaction by half will increase the reaction rate by a factor of 2.

Problem 3.2. Define the following:

(a) Catalyst

(b) Homogeneous chemical reaction

(c) Half-life

(d) Phase

Problem 3.3. Consider the integrated second-order rate law for a reaction which is first order with respect to two distinct reactants (A and B). In the text, two different expressions were given for this rate law: Equations 3.52 and 3.53.

(a) Prove that these two expressions are equivalent.

(b) Using Equation 3.52, determine the limits for this expression at $t = 0$ and for $t \rightarrow \infty$.

Problem 3.4. An exact treatment of equilibrium reaction kinetics for reactions that do not go to completion was discussed in a dialog box in the text. Expressions 3.68 and 3.69 were provided as integrated rate laws for a simple equilibrium first-order reaction between A and B where the forward rate constant is given by k_f and the backward rate constant is given by k_r. Prove that as $t \rightarrow \infty$, these expressions yield the equilibrium concentrations of species A and B ($c_{A_{eq}}$ and $c_{B_{eq}}$).

Calculation Questions

Problem 3.5. If the rate constant k for a first-order reaction doubles with a change in temperature from $T = 300$ K to $T = 320$ K, what is ΔG_{act}?

Problem 3.6. In 2010, a stunning collection of Ice Age fossils, including several Wooly Mammoths, was found in an ancient high-altitude lake-bed near Snowmass, Colorado. Many of the fossils were "radiocarbon dead," meaning that they were so old nearly all traces of carbon-14 had decayed away. However, radiocarbon dating was successfully conducted on some of the "younger" mammoth fossils. If the half-life of $^{14}_{6}C$ is 5730 yr:

(a) Calculate the reaction rate constant (yr^{-1}) for this radioactive decay.

(b) Calculate the age of the youngest mammoth fossils given that the $^{14}_{6}C$ concentration in these remains had decreased to 0.14 of the initial value at the time of sampling in 2010.

(c) Calculate the minimum age of the "radiocarbon dead" fossils given that $^{14}_{6}C$ concentration can be measured reliably down to 1% of the initial level.

Problem 3.7. You have synthesized a new radioactive material which you have named OHarium (OHa) in honor of the professor that wrote your favorite undergraduate textbook on materials kinetics.

(a) You measure the rate of radioactive emission from 1 mol of OHa immediately after synthesis to be 6000 disintegrations/s. Exactly five days later, the rate has decreased to 5800 disintegrations/s. Based on this information, calculate the first-order reaction rate constant (s^{-1}) for the nuclear disintegration of OHa.

(b) What is the half-life of OHa (in years)?

CHAPTER 4

TRANSPORT KINETICS (DIFFUSION)

Kinetic processes almost always involve flows of energy or matter. For example, solidification involves the transport of atoms (matter) from the liquid phase to the solid–liquid interface, while the heat (energy) generated by solidification must be transported away from this interface. The transport of energy or matter to/from locations undergoing change is therefore a crucial and often rate-limiting step in many kinetic processes. This is especially true in the solid state, where transport rates are typically much slower than in the liquid or gaseous states of matter.

While there are many different types of transport processes, the transport of matter via diffusion will be the central focus of this chapter due to its particular importance in solid-state kinetic processes. Nevertheless, we will begin with a general overview of transport theory and discuss a number of transport processes, including electrical conduction, heat conduction, convection, and diffusion, as well as several interesting coupled transport processes such as electromigration, thermal diffusion, and thermal–electrical phenomena. We will then delve into the detailed phenomenological treatment of diffusion, where we will learn to apply mathematical equations (Fick's first and second laws) to describe both steady-state and transient diffusion processes. While we will focus on the phenomenological treatment of mass diffusion, the approaches that we develop can also be applied to describe heat and charge transport with very little modification. The chapter homework exercises will provide several opportunities to do so. After dealing with the phenomenological treatment of diffusion, we will examine the atomistic mechanisms underlying diffusion, with special emphasis on solid-state diffusion mechanisms. Finally, we will briefly discuss several important high-diffusivity pathways in solids, such as dislocations and grain boundaries, where diffusion can sometimes proceed at a much faster pace than the typical "bulk" rate. These high-diffusivity paths can therefore play a very important role in determining the overall mass transport behavior of microstructurally inhomogeneous solids, a theme that we will revisit in Chapter 7.

4.1 FLUX

The rate at which mass, charge, or energy moves through a material is generally quantified in terms of *flux* (denoted with the symbol J). Flux measures how much of a given quantity flows through a material *per unit area per unit time*. Figure 4.1 illustrates the concept of flux: Imagine water flowing down this tube at a volumetric flow rate of 10 L/s. If we divide this flow rate by the cross-sectional area of the tube (A), we get the volumetric flux J_A of water moving down the tube. In other words, J_A gives the per-unit-area flow rate of water through the tube. Be careful! Recognize that flux and flow rate are not the same thing. A flux represents a flow rate that has been "normalized" by a cross-sectional area.

The most common type of flux is a molar flux [typical units are mol/(cm$^2 \cdot$ s)]. Molar fluxes are convenient when discussing the transport of atoms or molecules by diffusion in the solid, liquid, or gas phase. In addition to molar flux, however, there are a variety of other commonly encountered types of flux:

Mass flux is another way to measure the flux of a species moving through a material in terms of mass rather than molar units. Mass flux and molar flux are related by

$$J_{\text{mass},i} = J_{\text{mol},i}M_i \tag{4.1}$$

where $J_{\text{mass},i}$ is the mass flux of species i [typical units are g/(cm$^2 \cdot$ s)], $J_{\text{mol},i}$ is the molar flux of species i, and M_i is the molecular weight of species i (e.g., g/mol).

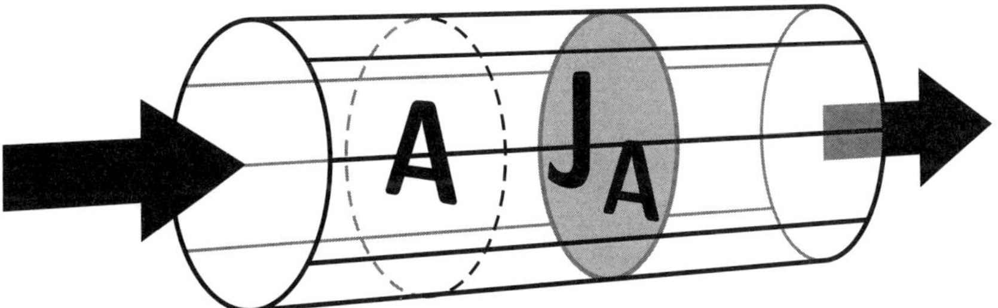

FIGURE 4.1 Schematic of flux. Imagine water flowing down this tube at a volumetric flow rate of 10 L/s. Dividing this flow rate by the cross-sectional area of the tube (A) gives the flux J_A of water moving down the tube. Fluxes of matter, especially for solid species, are more commonly measured in molar rather than volumetric terms. In this example, then, the liters of water could be converted to moles. Do you know how?

Number flux quantifies flux in terms of the raw number of atoms or molecules of the species that is moving. It is related to the molar flux by Avogadro's number:

$$J_{N,i} = J_{mol,i} N_A \tag{4.2}$$

where $J_{N,i}$ is the number flux of species i [typical units are no./cm$^2 \cdot$ s] and N_A is Avogadro's number (6.022×10^{23}/mol).

Volume flux is yet another way to quantify the flux of a species, using the volume rather than mass or moles of the species to describe the rate of transport. Volume flux is related to mass, molar, or number flux by the species density, molar volume, or atomic volume:

$$J_{vol,i} = J_{mass,i} \frac{1}{\rho_i} \tag{4.3}$$

$$= J_{mol,i} \frac{M_i}{\rho_i} = J_{mol,i} V_{m,i} \tag{4.4}$$

$$= J_{N,i} \frac{V_{m,i}}{N_A} = J_{mol,i} \Omega_i \tag{4.5}$$

where $J_{vol,i}$ is the volume flux of species i [typical units are cm^3/(cm$^2 \cdot$ s) $=$ cm/s], ρ_i is the mass density of species i (e.g., g/cm^3), $V_{m,i}$ is the molar volume of species i (typical units are cm^3/mol), and Ω_i is the atomic volume of species i ($\Omega_i = V_{m,i}/N_A$). The units for volume flux reduce to the units associated with velocity (e.g., cm/s), and, indeed, volume flux is sometimes described as a measure of species transport velocity. This is particularly common in gas or liquid media transport, where the volume flux is also known as the "superficial velocity." *Warning:* Please note that special care must be given when calculating densities/concentrations/volumes of species if they are present as part of a compound or mixture with other species. Revisit Section 2.10 and the accompanying Chapter 2 homework problems for further details about this issue.

Charge flux is a special type of flux that measures the *amount of charge* that flows through a material per unit area per unit time. Only the movement of charged species—e.g., electrons, holes, or ions—can give rise to a charge flux. Neutral atoms or molecules do not contribute. Typical units for charge flux are C/(cm$^2 \cdot$ s) $=$ A/cm^2. From these units, you may recognize that charge flux is the same thing as current density. The quantity $z_i F$ is required to convert from molar flux J_i to charge flux J_q, where z_i is the charge number for the carrier (e.g., z_i is $+1$ for Na$^+$, -2 for O^{2-}, etc.) and F is Faraday's constant:

$$J_q = z_i F J_i \tag{4.6}$$

Heat flux J_Q is a flux of thermal energy (heat). The typical units are J/(cm$^2 \cdot$ s). Because 1 J/s $=$ 1 W, the units for heat flux are often given as W/cm^2.

4.2 FLUXES AND FORCES

While there are many possible types of flux, a single common principle underpins all transport phenomena. This is the idea that a driving force must be present to cause the transport process to occur. If there is no driving force, there is no reason to move! This applies equally well to the transport of matter as it does to the transport of heat or charge. The governing equation for transport can be generalized (in one dimension) as

$$J_i = \sum_k M_{ik} F_k \tag{4.7}$$

where J_i represents a flux of species i, F_k represent the k different forces acting on i, and the M_{ik}'s are the coupling coefficients between force and flux. The coupling coefficients quantify the relative ability of a species to respond to a given force with movement as well as the effective strength of the driving force itself. The coupling coefficients are therefore a property both of the species that is moving and the material through which it is moving. This general equation is valid for any type of transport (charge, heat, mass, etc.) and can be used to capture the coupled effect of multiple driving forces acting simultaneously.

As an example of the application of Equation 4.7, consider the coupling between electrical conduction and heat conduction, which is at the heart of the well-known thermoelectric effect. In a system subjected to both a temperature gradient (dT/dx) and a voltage gradient (dV/dx), the resulting fluxes of heat (J_Q) and charge (J_q) are given as[1]

$$J_Q = -M_{QQ}\frac{1}{T}\frac{\partial T}{\partial x} - M_{Qq}\frac{\partial V}{\partial x} \tag{4.8}$$

$$J_q = -M_{qQ}\frac{1}{T}\frac{\partial T}{\partial x} - M_{qq}\frac{\partial V}{\partial x} \tag{4.9}$$

These equations express the fact that the charge flux and heat flux each depend on both the temperature gradient and the voltage gradient. This coupling means that a temperature gradient can induce current flow even in the absence of an applied voltage or a potential gradient can be used to pump heat even in the absence of (or possibly against) a temperature gradient! These two phenomena are known as the Seebeck effect and the Peltier effect, respectively. They enable a number of fascinating thermoelectric devices, including Peltier coolers (which harness the Peltier effect for solid-state refrigeration) and thermoelectric generators (which convert temperature gradients into electrical power). In addition to thermoelectric effects, coupled force/flux effects underly other fascinating transport phenomena that are sometimes important in materials kinetics, including thermodiffusion (mass diffusion due to a temperature gradient), electromigration (mass diffusion due to an electrical current), and stress-driven diffusion. These coupled diffusion processes will be detailed in Section 4.4.5.

---

[1]Fourier's law of heat conduction is generally written as $J_Q = -\kappa(\partial T/\partial x)$. However, the thermal conductivity κ contains a $1/T$ temperature dependence which is explicitly manifested when generalized mobilities are used, as in Equations 4.8 and 4.9.

4.3 COMMON TRANSPORT MODES (FORCE/FLUX PAIRS)

While the coupling of multiple fluxes and driving forces is important in certain situations, transport can frequently be described more simply in terms of sets of uncoupled fluxes that are each driven by a single direct (conjugate) dominant driving force. This results in a simplification of Equation 4.7 to a single term for each flux, where the flux of a species i can be related directly (and exclusively) to its *conjugate* driving force. This approximation is often valid because:

1. In many situations only one driving force is present or significant (e.g., when considering heat transport in many systems, there is often only a temperature gradient present, while pressure, voltage, and chemical potential gradients are not present).

2. Even when several gradients are present, often only the conjugate force/flux coupling coefficients are appreciable, while the cross-coefficients are too small to be meaningful. In the coupled charge and heat transport example expressed in Equations 4.8 and 4.9, this means that M_{QQ} and M_{qq} are large but M_{Qq} and M_{qQ} are small, often so small as to not be meaningful. Of course, for thermoelectric applications, materials are purposely selected/designed so that they have as large cross-coefficients as possible in order to magnify thermoelectric coupling effects.

You are likely already familiar with many of the simple direct force/flux pair relationships that are used to describe mass, charge, and heat transport—they include Fick's first law (diffusion), Ohm's law (electrical conduction), Fourier's law (heat conduction), and Poiseuille's law (convection). These transport processes are summarized in Table 4.1 using molar flux quantities. As this table demonstrates, Fick's first law of diffusion is really nothing more than a simplification of Equation 4.7 for

TABLE 4.1 Summary of Selected Transport Processes

Transport Process	Driving Force	Coupling Coefficient	Equation
Diffusion *Fick's law*	Concentration $\left(\dfrac{dc}{dx}\right)$	Diffusivity (D)	$J = -D\dfrac{dc}{dx}$
Electrical conduction *Ohm's Law*	Voltage $\left(\dfrac{dV}{dx}\right)$	Electrical Conductivity (σ)	$J_{e} = \dfrac{\sigma}{|z_{i}|F}\dfrac{dV}{dx}$
Heat conduction *Fourier's Law*	Temperature $\left(\dfrac{dT}{dx}\right)$	Thermal Conductivity (κ)	$J_{Q} = -\kappa\dfrac{dT}{dx}$
Convection *Poiseuille's law*	Pressure $\left(\dfrac{dp}{dx}\right)$	Viscosity (μ)	$J_{\text{conv}} = \dfrac{Gc}{\mu}\dfrac{dp}{dx}$

Note: The transport equation for convection in this table is based on Poiseuille's law, where G is a geometric constant and c is the concentration of the transported species. Convection flux is often calculated simply as $J = vc_{i}$, where v is the transport velocity.

the case where the mass diffusion of a species i can be directly and exclusively related to its chemical potential gradient; the direct coefficient relating the diffusion flux and its driving force in this case is known as the *diffusivity*. Similarly, electrical conductivity σ is nothing more than the name of the coefficient that describes how charge flux and electrical driving forces are related. For heat transport due to a temperature gradient, the relevant coefficient is called *thermal conductivity* while for transport due to a pressure gradient, the relevant coefficient is called *viscosity*.

Convection versus Diffusion

As both diffusion and convection involve the transport of mass, it is important to understand the differences between them:

- *Convection* refers to the transport of a species by bulk motion of a fluid under the action of a mechanical force, typically a pressure gradient.
- *Diffusion* refers to the transport of a species due to a gradient in chemical potential (concentration).

Figure 4.2 illustrates the difference between the two transport modes. In the solid state, mass is typically transported by diffusive mechanisms and convective transport is usually unimportant. In the gas and liquid state, however, both diffusion and convection can be important. Interestingly convection turns out to be far more "effective" at transporting species compared to diffusion in the gas and liquid states. Very small pressure gradients are typically sufficient to transport large fluxes of liquid or gas species down pipes or through channels. The field of fluid dynamics deals with the detailed understanding and treatment of convective transport processes. Because it is generally far less important for solid-state kinetic processes (except, for example, certain processes like molten metal casting), we will not tackle convective transport and fluid dynamics in this textbook.

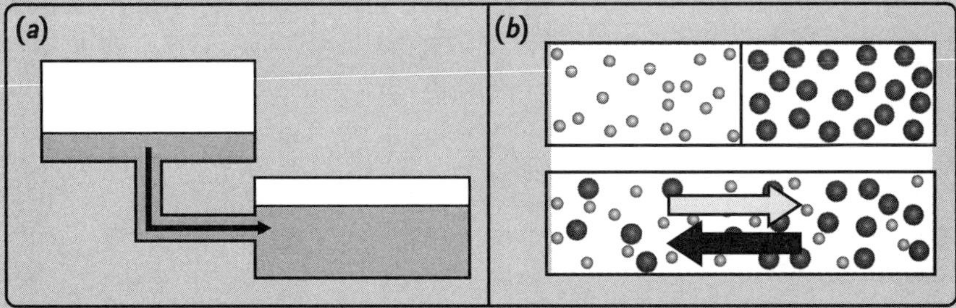

FIGURE 4.2 Convection versus diffusion. (*a*) Convective fluid transport in this system moves material from the upper tank to the lower tank. (*b*) A concentration gradient between white and gray particles results in net diffusive transport of gray particles to the left and white particles to the right.

4.4 PHENOMENOLOGICAL TREATMENT OF DIFFUSION

Under most circumstances, the transport of matter in materials can be treated using two mathematical equations known as Fick's first and second laws of diffusion. These two laws are quite general and can be applied to many diffusion problems in solids, liquids, and gases. Fick's first law deals with steady-state diffusion, while Fick's second law deals with transient diffusion. Figure 4.3 shows the differences between the two. This figure illustrates the solid-state diffusion of hydrogen through a Pd metal membrane, a process that can be modeled using Fick's first and second laws.

Palladium is an interesting metal because it permits the rapid transport of hydrogen through its lattice structure via an atomic diffusion mechanism. Thus, a thin Pd membrane can be used as a selective filter for the separation or purification of hydrogen gas. This technology has potentially important implications for a number of industrial chemical conversion applications.

Starting at time $t = 0$, the left side of the Pd membrane in Figure 4.3 is exposed to a gas stream containing H_2 mixed with undesired impurities. The right side of the membrane is exposed to a vacuum. Because of the difference in hydrogen chemical potential between the two sides of the membrane, there is a driving force for hydrogen to transport across the membrane. Because only hydrogen can diffuse through the membrane, this effect can be used to purify the hydrogen gas, hence eliminating the undesired impurities. Initially, before exposure, the concentration of hydrogen everywhere inside the membrane is zero. However, once hydrogen gas is introduced on the left-hand side of the membrane, some of this hydrogen will begin to diffuse

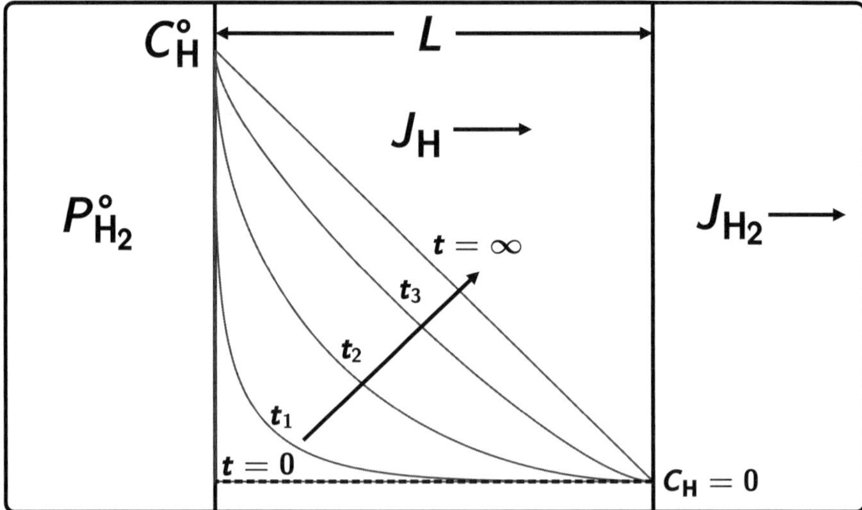

FIGURE 4.3 Diffusion of hydrogen across a Pd membrane illustrating both transient and steady-state diffusion processes. Both the *transient* evolution of the hydrogen concentration profile across the membrane during the initial stages of this process and the *steady-state* diffusion of hydrogen through the membrane once the final linear concentration gradient has been established can be mathematically modeled using Fick's second and first laws, respectively.

into the Pd. The hydrogen concentration on the left edge of the membrane rapidly reaches a limit given by the solubility of the hydrogen in the Pd metal (c_H°). Then, hydrogen atoms slowly begin to diffuse across the membrane. As more and more hydrogen diffuses into the membrane from the left-hand side, the hydrogen concentration profile inside the membrane gradually evolves as a function of time until finally a steady-state situation is reached where the hydrogen concentration varies linearly across the membrane. Once this steady-state concentration profile is reached, the hydrogen concentration no longer varies as a function of time across the membrane. At steady state, hydrogen continues to diffuse through the membrane from the left-hand side to the right-hand side. However, the rate at which fresh hydrogen enters the membrane on the left is exactly balanced by the rate at which hydrogen exits the membrane on the right. This steady state is reached because the hydrogen concentration on the left-hand side of the membrane is fixed at $c = c_H^\circ$ by the solubility of hydrogen in Pd while the hydrogen concentration on the right-hand side is fixed at $c = 0$ because the vacuum ensures that any hydrogen transporting all the way across the membrane is immediately pulled away as hydrogen gas.

A mathematical model can be constructed to predict the *transient* evolution of the hydrogen concentration profile across the membrane during the initial stages of this process. A model can also be constructed to predict the *steady-state* diffusion of hydrogen through the membrane once the final linear concentration gradient has been established. The steady-state process is modeled using Fick's first law of diffusion, while the transient process is modeled using Fick's second law of diffusion. In the sections below, you will learn how to apply both of these laws to model this and other diffusion problems. Because Fick's first law is mathematically simpler, we will begin there.

4.4.1 Steady-State Diffusion: Fick's First Law

Fick's first law of diffusion deals with the diffusional transport of matter under steady-state conditions. Steady state means that the concentration profile of the diffusing species does not vary as a function of time:

$$\text{Steady state:} \qquad \frac{\partial c_i}{\partial t} = 0 \qquad (4.10)$$

In one dimension (1D), Fick's first law is commonly given as[2]

$$J_i = -D_i \frac{\partial c_i}{\partial x} \qquad (4.11)$$

where J_i is the flux of species i, D_i is the diffusivity of species i (which depends both on the nature of species i and the medium through which it is diffusing), and $\partial c_i/\partial x$ is the concentration gradient of species i.

Fick's first law indicates that the flux of a diffusing species i is proportional to its concentration gradient. Fick's first law expresses the fundamental concept that matter

---

[2]The more precise version of Fick's first law relates the flux of a species i to its chemical potential gradient (e.g., $\partial \mu_i/\partial x$ in 1D). Under many circumstances, however, this can be simplified to the more common expression involving the concentration gradient.

tends to flow "down" a concentration gradient from regions of higher concentration to regions of lower concentration. Furthermore, Fick's first law indicates that the rate of diffusion (i.e., the size of the flux) depends on the magnitude of the diffusivity and the steepness of the concentration gradient. Higher diffusivities and steeper concentration gradients lead to larger fluxes.

Fick's first law (as well as Fick's second law) can be widely applied to model many diffusion processes in gases, liquids, or solids. The equations do not change between these phases—what changes is the diffusivity. Diffusivities tend to be high in gases, lower in liquids, and very low in solids, thereby capturing the differences in relative speed of diffusional transport between these three phases of matter. In solids, diffusivities tend to be on the order of 10^{-8}–10^{-20} cm^2/s, which means that even diffusion over relatively small distances (e.g., micrometers) can take hours, days, or even longer. A detailed derivation of Fick's first law based on an atomistic picture of diffusion is provided in Section 4.5.3.

A simple example showing how Fick's first law can be applied to model the steady-state diffusion of hydrogen in Pd is provided below.

Example 4.1

Question: We wish to investigate the steady-state rate at which a thin Pd permeation membrane can filter hydrogen. In this example, consider a Pd membrane $L = 10$ µm thick operated at 200 °C. The diffusivity of H in Pd at 200 °C is 10^{-5} cm^2/s. Assume that one side of the membrane is exposed to a gas stream containing a hydrogen gas partial pressure of $P^{\mathrm{I}}_{\mathrm{H}_2} = 1$ atm while the other side of the membrane is exposed to vacuum ($P^{\mathrm{II}}_{\mathrm{H}_2} = 0$ atm). At 200 °C and $P_{\mathrm{H}_2} = 1$ atm, the solubility of H in Pd is $c^{\circ}_{\mathrm{H}} = 10^{-2}$ mol H/cm^3. (a) What is the steady-state flux of hydrogen (J_{H_2}) across this membrane (in units of mol H/cm^{2}s)? (b) How many standard liters of hydrogen gas per minute (SLPM) could be produced with 1 m^2 of membrane area under these conditions? (c) If the membrane thickness is reduced by a factor of 2, will the flux of hydrogen through the membrane increase or decrease and by how much?

Solution: (a) We can apply Fick's first law to calculate the steady-state flux of hydrogen atoms (H) through the Pd membrane based on the steady-state concentration gradient in H that is established across the membrane and the diffusivity of H in Pd. At steady state, the hydrogen concentration profile across the membrane is linear, as shown in Figure 4.3, and thus the gradient is easily calculated from the concentration difference between the two sides of the membrane and the membrane thickness:

$$J_{\mathrm{H}} = -D_{\mathrm{H/Pd}}|_{200\,^{\circ}\mathrm{C}}\frac{\partial c_{\mathrm{H}}}{\partial x} = -D_{\mathrm{H/Pd}}|_{200\,^{\circ}\mathrm{C}}\frac{0 - c^{\circ}_{\mathrm{H}}}{L} \tag{4.12}$$

$$= -10^{-5}\ \mathrm{cm}^2/\mathrm{s}\cdot\frac{0 - 10^{-2}\ \mathrm{mol\ H/cm}^3}{10\ \mu\mathrm{m}(1\ \mathrm{cm}/10^{-4}\ \mu\mathrm{m})} = 10^{-4}\ \mathrm{mol\ H/cm}^2\ \mathrm{s}$$

This calculated flux is in terms of H atoms since the diffusion process across the Pd membrane involves individual H atoms (rather than H_2 molecules). However, we are asked to calculate the flux of H_2 across the membrane. The H_2 molecules dissociate and dissolve into the Pd on the high-pressure side of the membrane as H atoms; the reverse process occurs on the low-pressure side of the membrane:

$$\text{High } H_2 \text{ pressure side:} \quad H_2(g) \rightarrow 2H_{Pd}$$

$$\text{Low } H_2 \text{ pressure side:} \quad 2H_{Pd} \rightarrow H_2(g)$$

Thus, the H atom flux through the membrane and the resulting effective H_2 flux can be directly related as

$$J_{H_2} = (1/2)J_H \tag{4.13}$$

And so in this case we have

$$J_{H_2} = \tfrac{1}{2}10^{-4} \text{ mol H/cm}^2\text{s} = 5 \times 10^{-5} \text{ mol H/cm}^2\text{s}$$

(b) The total amount of hydrogen that can be filtered through 1 m^2 of Pd membrane per minute can be calculated from the flux as

$$\dot{n}_{H_2} = J_{H_2} \cdot A_{Pd} = 5 \times 10^{-5} \text{ mol H/cm}^2\text{s} \cdot 1 \text{ m}^2 \cdot \frac{(100 \text{ cm})^2}{1 \text{ m}^2} \cdot \frac{60 \text{ s}}{1 \text{ min}}$$

$$= 30 \text{ mol } H_2/\text{min}$$

We can then apply the ideal gas law to convert this into standard liters of hydrogen per minute (note that, by definition, a standard liter is calculated at $T = 300$ K and $P = 1$ atm even though the Pd membrane in this example is operated at 200 °C):

$$\dot{V}_{H_2} = \frac{\dot{n}_{H_2}RT}{P} = \frac{30 \text{ mol } H_2/\text{min} \cdot 8.134 \text{ J/(mol} \cdot \text{K)} \cdot 300 \text{ K}}{101300 \text{ Pa}} = 0.74 \text{ SLPM}$$

Thus it would take a little more than 2.7 min to fill a 2-L Coke bottle with pure H_2 filtered from 1 m^2 of this Pd membrane under these conditions.

(c) If the thickness of the membrane is decreased by half, the flux of hydrogen through the membrane will increase commensurately by a factor of 2. This can be seen by the inverse dependence between flux and thickness shown in Equation 4.12. One of the primary ways to increase the performance of permeation membranes is, therefore, to make them thinner.

4.4.2 Transient Diffusion: Fick's Second Law

Fick's second law of diffusion deals with the diffusional transport of matter under transient (time-dependent) conditions. Transient means that the concentration profile of the diffusing species varies as a function of time:

$$\text{Transient:} \quad \frac{\partial c_i}{\partial t} \neq 0 \tag{4.14}$$

In 1D, Fick's second law is commonly given as[3]

$$\frac{\partial c_i}{\partial t} = D_i \frac{\partial^2 c_i}{\partial x^2} \tag{4.15}$$

where $\partial c_i/\partial t$ is the time-dependent concentration profile of species i. We can use Fick's second law to "watch" how a non-steady-state diffusion profile evolves as a function of time. Under the condition that $D \neq f(c)$, Fick's second law indicates that the *rate of change* of a concentration gradient is proportional to its *curvature*. Regions of high curvature (i.e., "sharp" features) evolve quickly, while regions of low curvature evolve more slowly. Furthermore, Fick's second law indicates that a species will accumulate in regions where its concentration profile manifests positive curvature (concave up), while a species will dissipate from regions where its concentration profile is negative (concave down). Thus, Fick's second law predicts that abrupt concentration profile features tend to be smoothed out over time (see Figure 4.4).

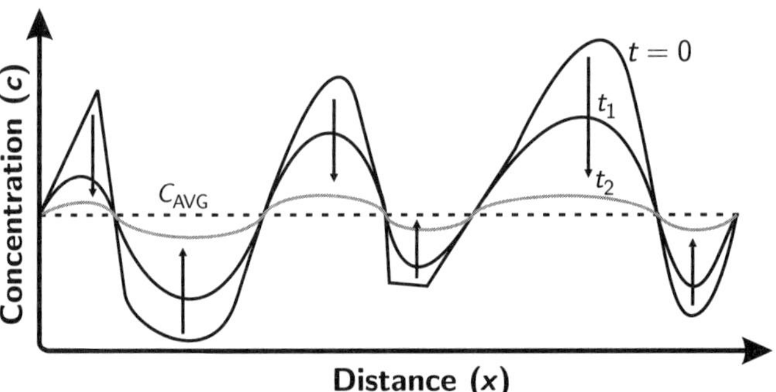

FIGURE 4.4 Schematic illustration of the evolution of a complex diffusion profile as a function of time. Regions of sharp curvature are rapidly "smoothed out" while regions of lower curvature change more slowly. The result is a gradual relaxation of the concentration profile until a steady-state (in this case uniform) concentration profile is achieved. Note that concentration *decreases* with time in regions of negative curvature and *increases* with time in regions of positive curvature. We can mathematically model this transient process as a function of time using Fick's second law of diffusion.

---

[3]A more precise version of Fick's second law includes the diffusivity inside the first spatial differential as $\partial c_i/\partial t = \partial/\partial x[D_i(\partial c_i/\partial x)]$. However, under the (common) assumption that the diffusion coefficient is independent of position, this expression reduces to Equation 4.15.

Derivation of Fick's Second Law

Fick's second law is essentially an expression of the principle of mass conservation. It can be derived directly from the conservation of mass and Fick's first law. Consider the 1D diffusion of a species i from left to right down a tube as shown in Figure 4.5. Mass conservation states that the rate at which species i accumulates (or depletes) in an infinitesimal slice of this tube is given by rate at which i enters into this slice from the left minus the rate at which i leaves this slice to the right. In other words,

$$\text{Rate of accumulation} = \text{rate of entry} - \text{rate of exit}$$

$$(A \ \Delta x) \cdot \frac{\partial c_i}{\partial t} = A \cdot J_{i,\text{in}} - A \cdot J_{i,\text{out}} \tag{4.16}$$

where $J_{i,\text{in}}$ and $J_{i,\text{out}}$ are the fluxes of species i into and out of the tube slice and A and Δx are respectively the cross-sectional area and thickness of the tube slice. In the limit of an infinitesimal tube slice thickness $(\Delta x \to 0)$, this expression reduces to

$$\frac{\partial c_i}{\partial t} = -\frac{J_{i,\text{out}} - J_{i,\text{in}}}{\Delta x} = -\frac{\partial J_i}{\partial x} \tag{4.17}$$

applying Fick's first law to this expression $[J_i = -D_i(\partial c_i/\partial x)]$ and assuming that D_i is independent of position lead to the familiar equation for Fick's second law:

$$\frac{\partial c_i}{\partial t} = -\frac{\partial}{\partial x}\left(-D_i\frac{\partial c_i}{\partial x}\right) = D_i\frac{\partial^2 c_i}{\partial x^2} \tag{4.18}$$

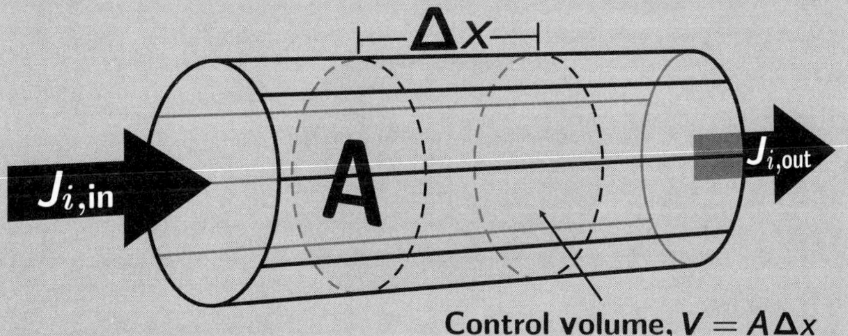

FIGURE 4.5   Derivation of Fick's second law. The rate at which species i accumulates within the control volume described in this figure is given by the difference between the rates at which species i enter and depart from the control volume. Based on the control volume geometry, the accumulation rate can be related to the flux of species i into and out of the control volume.

Equation 4.15 is a second-order partial differential equation. When treating diffusion phenomena with Fick's second law, the typical aim is to solve this equation to yield solutions for the concentration profile of species i as a function of time and space $[c_i(x, t)]$. By plotting these solutions at a series of times, one can then watch how a diffusion process progresses with time. Solution of Fick's second law requires the specification of a number of boundary and initial conditions. The complexity of the solutions depends on these boundary and initial conditions. For very complex transient diffusion problems, numerical solution methods based on finite difference/finite element methods and/or Fourier transform methods are commonly implemented. The subsections that follow provide a number of examples of solutions to Fick's second law starting with an extremely simple example and progressing to increasingly more complex situations. The homework exercises provide further opportunities to apply Fick's Second Law to several interesting "real world" examples.

Boundary Conditions and Initial Conditions

The solution of Fick's second law for any specific situation requires additional input information on the initial configuration and geometry of the diffusion problem. This required input information takes the form of *boundary conditions* and *initial conditions*. It is important to understand what these terms mean.

- *Boundary conditions* provide information about the behavior of the diffusion system at the physical edges (i.e., "boundaries") of the problem domain. Boundary conditions typically come in two basic forms:
  1. Specified concentration at a boundary: e.g., $c_i(x = 0, t) = c_i^\circ$
  2. Specified flux at a boundary: e.g., $J_i(x = 0, t) = J_i^\circ$

  Boundary conditions specify behavior at a *specific location* (i.e., for 1D problems, at a specific value of x).

- *Initial conditions* provide information about the initial concentration distribution within the system at some initial time. In order to predict the evolution of a concentration profile within a system as a function of time, it is necessary to know the starting profile at $t = 0$. This starting concentration profile information is supplied as the initial condition. In the simplest case, if the concentration of species i is initially zero everywhere in the problem domain, the initial condition would be $c_i(x, t = 0) = 0$. The initial concentration profile can be as simple or as complex as desired (consider, for example, the initial concentration profile shown in Figure 4.4), although obtaining solutions to Fick's second law for anything other than relatively simple initial concentration profiles generally requires numerical methods.

  Initial conditions specify behavior at a *specific time* (i.e., at a specific value of t, usually $t = 0$).

Transient Semi-Infinite Diffusion   The simplest transient diffusions problems are generally those that involve semi-infinite or infinite boundary conditions. Consider, for example, the situation illustrated in Figure 4.6, which represents diffusion of a substance from a surface into a semi-infinite medium.

In the real world, of course, no medium actually extends to infinity. However, infinite or semi-infinite boundary conditions are fully appropriate for many finite situations in which the length scale of the diffusion is much smaller than the thickness of the material. In such cases, the material appears infinitely thick relative to the scale of the diffusion—or, in other words, the diffusion process never reaches the far boundaries of the material over the relevant time scale of interest. Since typical length scales for solid-state diffusion processes are often on the micrometer scale, even diffusion into relatively thin films can often be treated using semi-infinite or infinite boundary condition approaches. Semi-infinite and infinite transient diffusion has therefore been widely applied to understand many real-world kinetic processes—everything from transport of chemicals in biological systems to the doping of semiconductor films to make integrated circuits.

When solving Fick's second law for any specific problem, the first step is always to specify the boundary and initial conditions. For the semi-infinite diffusion process illustrated in Figure 4.6 as an example, the concentration of species i is initially constant everywhere inside the medium at a uniform value of c_i°. At time $t = 0$, the surface is then exposed to a higher concentration of species i (c_i^*), which causes i to begin to diffuse into the medium (since $c_i^* > c_i^\circ$). It is assumed that the surface concentration of species i is held constant at this new higher value c_i^* during the entire transient diffusion process. Based on this discussion, we can mathematically specify the boundary and initial conditions as follows:

- Boundary condition: $c_i(x = 0, t) = c_i^*$
- Initial condition: $c_i(x \geq 0, t = 0) = c_i^\circ$

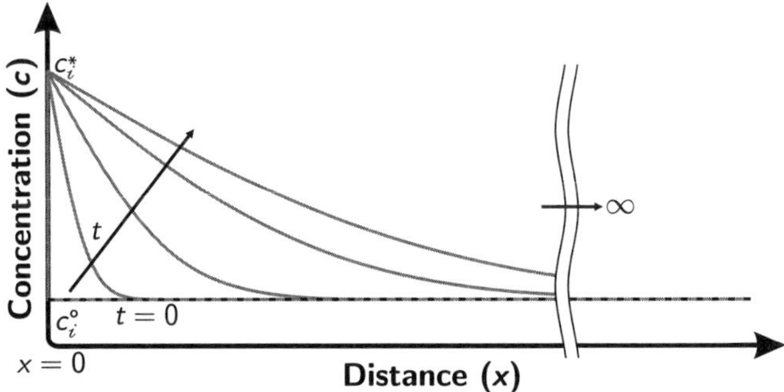

FIGURE 4.6   Schematic illustration of the transient semi-infinite diffusion of a species i from the surface into the bulk of a medium. The concentration of species i at the surface of the medium is assumed to be held fixed at c_i^* while the initial concentration of species i within the bulk of the medium is assumed to be c_i°. As time elapses, species i diffuses deeper and deeper into the medium from the surface. Since the medium is semi-infinitely thick, this process can proceed indefinitely and the concentration of species i never reaches c_i^* anywhere inside the medium except at the surface. This figure assumes that $c_i^* > c_i^\circ$; however, the reverse situation (which would involve out-diffusion of i from the bulk) could be similarly modeled.

Thus, the transient diffusion problem in this case is fully specified by the following:

$$c_i(x = 0, t) = c_i^*$$

$$c_i(x \geq 0, t = 0) = c_i^\circ$$

$$\frac{\partial c_i}{\partial t} = D_i \frac{\partial^2 c_i}{\partial x^2} \qquad x \geq 0 \qquad\qquad (4.19)$$

The mathematics to obtain the analytical solution to Fick's second law under these conditions [5] are actually fairly involved. However, generalized analytical solutions for this and many other diffusion problems have been obtained and compiled in extensive reference books. In particular, Crank's handy reference text, *The Mathematics of Diffusion* [5], provides solutions to a wide range of transient diffusion problems. For many diffusion problems, it is often sufficient to consult such a reference text in order to obtain a generalized solution and then apply a particular problem's specific boundary and initial conditions to obtain a full solution.

For the semi-infinite transient diffusion problem specified by Equation 4.19, the general solution is given by

$$c_i(x, t) = A + B \operatorname{erfc}\left[\frac{x}{2\sqrt{D_i t}}\right] \qquad x \geq 0 \qquad\qquad (4.20)$$

where A and B are constants and erfc is a mathematical function known as the *complementary error function*. The complementary error function and its relative, the error function (erf), are frequently encountered in the solutions to transient diffusion problems. We will be using these two functions extensively, so it is important to become intimately familiar with them. The dialog box below provides more information about erf and erfc. Example problems 4.2 and 4.3 then provide practice working with the error function and complementary error function in solutions to semi-infinite and infinite transient diffusion problems.

Error Function (erf[ω]) and Complementary Error Function (erfc[ω])

The error function is closely related to the integral of the standard normal (i.e., Gaussian) distribution and is defined by the equation

$$\operatorname{erf}[\omega] = \frac{2}{\sqrt{\pi}} \int_0^\omega e^{-s^2} \, ds \qquad\qquad (4.21)$$

The error function owes its name to its application in statistics, where it is used in the description and characterization of measurement errors. However, the error function and its cousin, the complementary error function, are frequently encountered in a number of other fields, including transient diffusion. The complementary error function is related to the error function as

$$\text{erfc}[\omega] = 1 - \text{erf}[\omega] \tag{4.22}$$

Figure 4.7 provides a graph of the error function and the complementary error function while Table 4.2 provides a look-up table of numerical values for the error function—you will find these to be handy when working diffusion problems that have error function solutions.

Table 4.3 summarizes several values of the error function and the complementary error function that are quite useful to remember when evaluating limits associated with diffusion problems involving the error function.

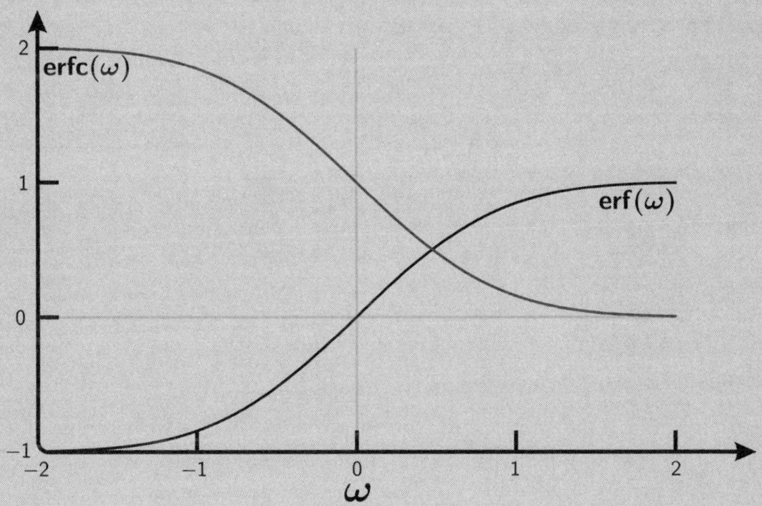

FIGURE 4.7 Graph of error function erf[ω] and complimentary error function erfc[ω].

TABLE 4.2 Important Values of Error Function and Complementary Error Function

erf[ω]	Value	erfc[ω]	Value
erf[$-\infty$]	-1	erfc[$-\infty$]	2
erf[0]	0	erfc[0]	1
erf[∞]	1	erfc[∞]	0

TABLE 4.3 Look-up Table of Error Function Values

ω	erf[ω]	ω	erf[ω]	ω	erf[ω]	ω	erf[ω]
0.00	0.000000	0.40	0.428392	0.80	0.742101	1.40	0.952285
0.02	0.022565	0.42	0.447468	0.82	0.753811	1.44	0.958297
0.04	0.045111	0.44	0.466225	0.84	0.765143	1.48	0.963654
0.06	0.067622	0.46	0.484655	0.86	0.776100	1.52	0.968413
0.08	0.090078	0.48	0.502750	0.88	0.786687	1.56	0.972628
0.10	0.112463	0.50	0.520500	0.90	0.796908	1.60	0.976348
0.12	0.134758	0.52	0.537899	0.92	0.806768	1.70	0.983790
0.14	0.156947	0.54	0.554939	0.94	0.816271	1.80	0.989091
0.16	0.179012	0.56	0.571616	0.96	0.825424	1.90	0.992790
0.18	0.200936	0.58	0.587923	0.98	0.834232	2.00	0.995322
0.20	0.222703	0.60	0.603856	1.00	0.842701	2.10	0.997021
0.22	0.244296	0.62	0.619411	1.04	0.858650	2.20	0.998137
0.24	0.265700	0.64	0.634586	1.08	0.873326	2.30	0.998857
0.26	0.286900	0.66	0.649377	1.12	0.886788	2.40	0.999311
0.28	0.307880	0.68	0.663782	1.16	0.899096	2.50	0.999593
0.30	0.328627	0.70	0.677801	1.20	0.910314	2.60	0.999764
0.32	0.349126	0.72	0.691433	1.24	0.920505	2.70	0.999866
0.34	0.369365	0.74	0.704678	1.28	0.929734	2.80	0.999925
0.36	0.389330	0.76	0.717537	1.32	0.938065	2.90	0.999959
0.38	0.409009	0.78	0.730010	1.36	0.945561	3.00	0.999978

Note: To determine corresponding values for the complementary error function, recall that erfc[ω] = 1 − erf[ω]

Exact solution to the transient diffusion problem illustrated in Figure 4.6 may be obtained by applying the boundary and initial conditions to the general solution provided in Equation 4.20. First, applying the initial condition $[c_i(x, t = 0) = c_i^{\circ}]$ yields

$$c_i(x, t = 0) = c_i^{\circ} = A + B \text{ erfc}[\infty]$$
$$c_i^{\circ} = A \tag{4.23}$$

Next, applying the boundary condition $[c_i(x = 0, t) = c_i^{*}]$ yields

$$c_i(x = 0, t) = c_i^{*} = A + B \text{ erfc}[0]$$
$$c_i^{*} = A + B$$
$$B = c_i^{*} - A = c_i^{*} - c_i^{\circ} \tag{4.24}$$

Thus, the final exact solution for this specific semi-infinite transient diffusion problem is given by

$$c_i(x, t) = c_i^\circ + (c_i^* - c_i^\circ)\,\mathrm{erfc}\left[\frac{x}{2\sqrt{Dt}}\right] \tag{4.25}$$

For the special case where $c_i^\circ = 0$ (i.e., initial concentration of i inside the medium is zero), the exact solution further simplifies to

$$c_i(x, t) = c_i^*\,\mathrm{erfc}\left[\frac{x}{2\sqrt{Dt}}\right] \tag{4.26}$$

In analogy to the "reaction half-life" that was discussed in Chapter 3, we can specify a "diffusion half-depth" in transient diffusion problems ($\delta_{1/2}$) which is the spatial position at which the concentration of the diffusing species reaches half of its surface value. As an example, the diffusion half-depth for the semi-infinite diffusion process in Equation 4.26 can be obtained as

$$c_i(\delta_{1/2}, t) := \frac{1}{2}c_i^* = c_i^*\,\mathrm{erfc}\left[\frac{\delta_{1/2}}{2\sqrt{Dt}}\right]$$

$$\mathrm{erfc}\left[\frac{\delta_{1/2}}{2\sqrt{Dt}}\right] = \frac{1}{2}$$

Since $\mathrm{erf}[\omega] = 1 - \mathrm{erfc}[\omega]$, we have

$$\mathrm{erf}\left[\frac{\delta_{1/2}}{2\sqrt{Dt}}\right] = 1 - \mathrm{erfc}\left[\frac{\delta_{1/2}}{2\sqrt{Dt}}\right]$$

$$= 1 - 1/2$$

$$= 1/2$$

Then, using Table 4.3 we can determine that if $\mathrm{erf}[\omega] = 1/2$, ω must be ≈ 0.475, and so

$$\frac{\delta_{1/2}}{2\sqrt{Dt}} \approx 0.475 \qquad \delta_{1/2} \approx \sqrt{Dt} \tag{4.27}$$

This expression reveals a characteristic *square-root dependence* between the spatial extent of a transient diffusion process and the time elapsed. This square-root dependence is very commonly observed in transient diffusion processes. Thus Equation 4.27 can be a helpful way to roughly approximate the extent of progress of a transient diffusion process in a material as a function of time.

Example 4.2

Question: Steel can be hardened and made more corrosion resistant by heat treating in a nitrogen atmosphere to diffuse nitrogen into the surface layer of the steel. This process can be roughly modeled using semi-infinite transient diffusion. Assume that a plate of steel is heat treated in a nitrogen environment at $500\,°C$, where the solubility of nitrogen in steel is $c_N^* \approx 0.5$ mol % and the initial background concentration of nitrogen in the steel is zero ($c_N^\circ = 0$).

(a) Provide a sketch for this problem with as much detail as possible about the problem geometry, the boundary and initial conditions, and your expectation for how the diffusion of nitrogen into the steel will proceed as a function of time (i.e., sketch curves for the nitrogen concentration profile inside the steel sheet for several times).

(b) Provide the boundary condition and initial condition for this problem.

(c) Provide the general solution $c_N(x, t)$ to this problem.

(d) Based on your boundary and initial conditions, provide the exact solution to this semi-infinite transient diffusion problem.

(e) Assuming the diffusivity of nitrogen in steel is $D_{N/steel} = 10^{-9}$ cm^2/s at the heat treatment temperature, determine how long it will take for the nitrogen concentration 10 μm deep into the steel to reach 70% of its value on the surface (i.e., 70% of c_N^*)?

(f) How long would it take the nitrogen concentration 20 μm deep into the steel to reach 70% of its value on the surface?

Solution:

(a)

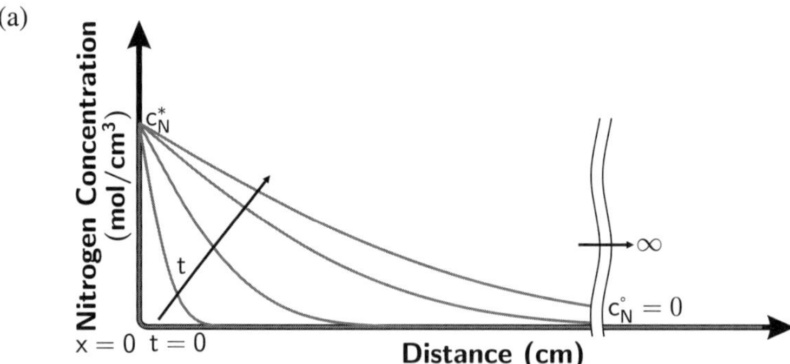

(b) Based on the information provided in the problem statement, we can stipulate the boundary and initial conditions for this problem as follows:

- Boundary condition: $c_N(x = 0, t) = c_N^*$
- Initial condition: $c_N(x \geq 0, t = 0) = 0$

(c) As the transient diffusion of nitrogen into steel can be modeled as a 1D semi-infinite diffusion problem, the general solution is given by Equation 4.20:

$$c_N(x, t) = A + B \operatorname{erfc}\left[\frac{x}{2\sqrt{D_i t}}\right] \qquad x \geq 0 \qquad (4.28)$$

(d) First, applying the initial condition $c_N(x, t = 0) = 0$ yields

$$c_N(x, t = 0) = 0 = A + B \operatorname{erfc}[\infty]$$

$$0 = A$$

Next, applying the boundary condition $c_N(x = 0, t) = c_N^*$ yields

$$c_N(x = 0, t) = c_N^* = 0 + B \operatorname{erfc}[0]$$

$$c_N^* = B$$

Thus, the final exact solution for this specific semi-infinite transient diffusion problem is given by

$$c_N(x, t) = c_N^* \operatorname{erfc}\left[\frac{x}{2\sqrt{Dt}}\right] \qquad (4.29)$$

(e) We wish to know the time t when the concentration of N reaches 70% of c_N^* at $x = 10\ \mu\text{m}$. Applying this criterion to Equation 4.29 yields

$$c_N(x = 10\ \mu\text{m}, t) = 0.7 c_N^* = c_N^* \operatorname{erfc}\left[\frac{10^{-3}\ \text{cm}}{2\sqrt{Dt}}\right]$$

$$0.7 = \operatorname{erfc}\left[\frac{10^{-3}\ \text{cm}}{2\sqrt{Dt}}\right]$$

$$0.7 = 1 - \operatorname{erf}\left[\frac{10^{-3}\ \text{cm}}{2\sqrt{Dt}}\right]$$

$$\operatorname{erf}\left[\frac{10^{-3}\ \text{cm}}{2\sqrt{Dt}}\right] = 0.3$$

Using Table 4.3 we can determine that if $\mathrm{erf}[\omega] = 0.3$, ω must be ≈ 0.28, and so

$$\frac{10^{-3}\ \mathrm{cm}}{2\sqrt{Dt}} \approx 0.28$$

$$t \approx \frac{1}{D}\left(\frac{10^{-3}\ \mathrm{cm}}{2\cdot 0.28}\right)^2$$

$$\approx \frac{1}{10^{-9}\ \mathrm{cm^2/s}}\left(\frac{10^{-3}\ \mathrm{cm}}{2\cdot 0.28}\right)^2$$

$$\approx 3200\ \mathrm{s} \approx 53\ \mathrm{min}$$

(f) In order to determine the time t when the concentration of N reaches 70% of c_N^* at $x = 20\ \mu\mathrm{m}$, we could repeat the calculations above using this new value for x. Alternatively, we can note the square-root relationship between x and t in the erfc argument to infer that if the diffusion distance increases by 2×, the time required will increase by 4×:

$$\frac{x_1}{2\sqrt{Dt_1}} = \frac{x_2}{2\sqrt{Dt_2}}$$

Thus:

$$\frac{x_1}{x_2} = \sqrt{\frac{t_1}{t_2}}$$

So, in this example, because we have increased the desired diffusion distance by a factor of 2 from 10 to 20 $\mu\mathrm{m}$, the time required increases by a factor of 4 from 53 to 212 min:

$$\frac{10\ \mu\mathrm{m}}{20\ \mu\mathrm{m}} = \sqrt{\frac{53\ \mathrm{min}}{t_2}}$$

$$t_2 = 2^2 \cdot 53\ \mathrm{min} = 212\ \mathrm{min}$$

Transient Interdiffusion in Two Semi-Infinite Bodies The transient diffusion problem illustrated in Figure 4.8, which involves the interdiffusion of two semi-infinite bodies in contact with one another, is closely related to the previous semi-infinite transient diffusion problem. In fact, if you consider just one-half of the problem domain (e.g., consider the evolution of the diffusion profiles for species A for $x > 0$), diffusion proceeds exactly like the previous semi-infinite diffusion problem. The only difference is that in this case the interfacial concentration of species A is assumed to be pinned at half of its bulk (i.e., pure material A) value,

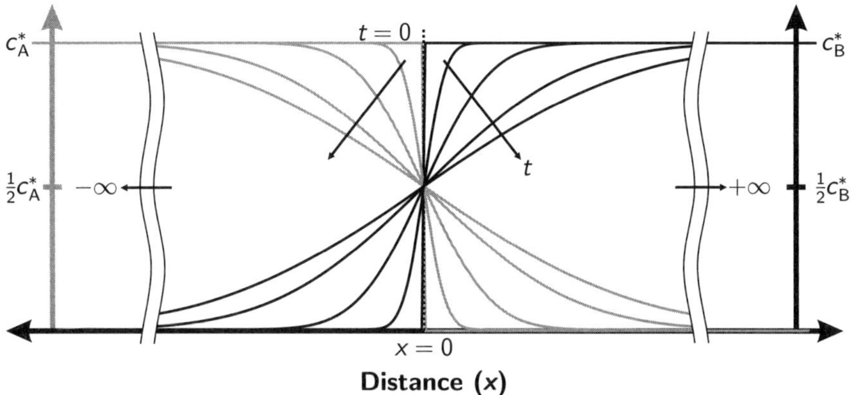

FIGURE 4.8 Transient interdiffusion of two semi-infinite bodies. Material A on the left ($x <$ 0), assumed to be initially composed purely of species A, is in contact with material B on the right ($x > 0$), which is assumed to be initially composed purely of species B. At time $t = 0$, species A and B begin to interdiffuse. If the initial bulk concentrations and diffusivities of A and B are equal, their concentrations at the interface will be pinned at $1/2\ c^*$. The materials will slowly diffuse into one another, but the position of the interface between them will not move.

which yields a slight change to the boundary condition compared to the previous problem:

$$c_A(x = 0, t) = \frac{1}{2}c_A^*$$

$$c_A(x \geq 0, t = 0) = 0$$

$$\frac{\partial c_A}{\partial t} = D_A \frac{\partial^2 c_A}{\partial x^2} \tag{4.30}$$

The general solution is therefore the same as before:

$$c_A(x, t) = A + B\ \text{erfc}\left[\frac{x}{2\sqrt{D_A t}}\right] \tag{4.31}$$

And application of the initial and boundary conditions yields

$$c_A(x, t = 0) = 0 = A + B\ \text{erfc}[\infty]$$

$$0 = A \tag{4.32}$$

$$c_A(x = 0, t) = 1/2c_A^* = A + B\ \text{erfc}[0]$$

$$1/2c_A^* = A + B$$

$$B = 1/2c_A^* - A = 1/2c_A^* \tag{4.33}$$

Thus, the final exact solution for this specific semi-infinite transient diffusion problem is given by

$$c_A(x, t) = \frac{1}{2}c_A^* \; \text{erfc} \left[\frac{x}{2\sqrt{Dt}} \right] \tag{4.34}$$

The analogous solution for the transient diffusion of species B is (by inspection)

$$c_B(x, t) = c_B^* - \frac{1}{2}c_B^* \; \text{erfc} \left[\frac{x}{2\sqrt{Dt}} \right]$$

$$= \frac{1}{2}c_B^* + \frac{1}{2}c_B^* \; \text{erf} \left[\frac{x}{2\sqrt{Dt}} \right] \tag{4.35}$$

This treatment implicitly assumes that species A and B are fully soluble in one another and the rate of diffusion of species A within material B is equal to the rate of diffusion of species B within material A, so that the spatial location of the "interface" between materials A and B remains fixed (although it clearly becomes less abrupt as time goes on). If these assumptions cannot be made, this interfacial diffusion problem must be modified and the solution becomes more complex. In particular, if the diffusion of A within B and that of B within A are not equal, the A/B interface itself can effectively move with time! See Section 4.4.3 for more details.

Example 4.3

Question: Two semi-infinite slabs, one made of pure species A and one of pure species B, are joined together and heated so that A and B begin to interdiffuse. Assuming that the diffusion of A into B and that of B into A are equal and independent of concentration, derive an expression for the thickness of the "interdiffusion region" as a function of time. For the purposes of this question, the thickness of the interdiffusion region is bounded by the locations where the concentrations of species A and B fall to 1/10 of their initial values as they diffuse into the opposing slabs.

Solution: To answer this question, it is helpful to first establish a sketch of the problem, as shown in Figure 4.9. Considering first the diffusion of A into B from left to right, we wish to determine the right-side boundary of the "interdiffusion region." As detailed in the problem statement, this corresponds to the location (x value) where the concentration of species A falls to 1/10 of its initial bulk value. As shown on the schematic illustration, we will call this location $x = \delta$. Applying this criteria to the solution for the transient interdiffusion of two semi-infinite bodies gives

$$c_A(x = \delta, t) = \frac{1}{10}c_A^* = \frac{1}{2}c_A^* \; \text{erfc} \left[\frac{\delta}{2\sqrt{Dt}} \right]$$

$$\frac{1}{5} = \text{erfc} \left[\frac{\delta}{2\sqrt{Dt}} \right] \tag{4.36}$$

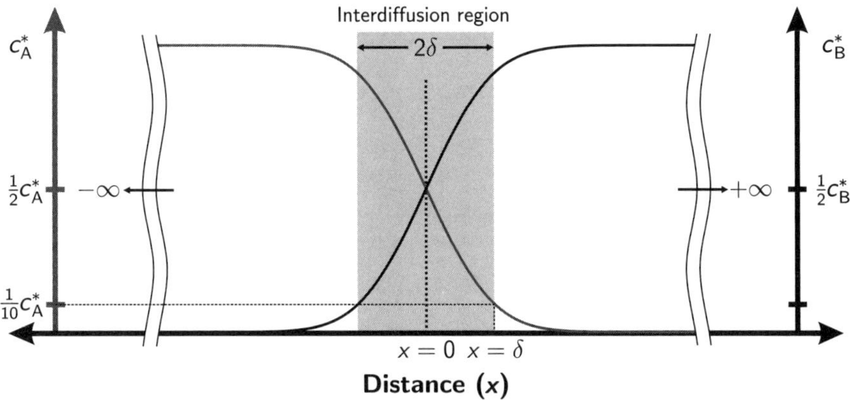

FIGURE 4.9   Transient semi-infinite interdiffusion of materials A and B. At time $t = 0$, species A and B begin to interdiffuse. The initial bulk concentrations and diffusivities of A and B are equal, so their concentrations at the interface will be pinned at $1/2c^*$. The materials will slowly diffuse into one another. The "interdiffusion region" (thickness = 2δ) is bounded by the locations where the concentrations of species A and B fall to 1/10 of their initial values as they diffuse into the opposing slabs. The interdiffusion region increases with time.

Consulting Table 4.3 we can determine that, if $\mathrm{erfc}[\omega] = 1/5 = 0.20$, then $\mathrm{erf}[\omega] = 0.80$, and so ω must be ≈ 0.92. Thus

$$\frac{\delta}{2\sqrt{Dt}} \approx 0.92 \qquad \delta \approx 1.84\sqrt{Dt} \tag{4.37}$$

Similarly, the left-side boundary of the interdiffusion region corresponds to the location where the concentration of species B falls to 1/10 of its initial bulk value. Owing to the symmetry of the problem, this occurs at the location $x = -\delta$. Thus, the total thickness of the interdiffusion region is 2δ, which increases with time according to

$$\text{Interdiffusion region thickness} = 2\delta = 3.68\sqrt{Dt} \tag{4.38}$$

Transient Infinite Diffusion of a Rectangular Source   As illustrated in Figure 4.10, the transient infinite diffusion of a rectangular concentration profile of thickness $2l$ can be determined by *superimposing* (in this case, subtracting) the solutions for two semi-infinite step functions located at $x = -l$ and $x = +l$,

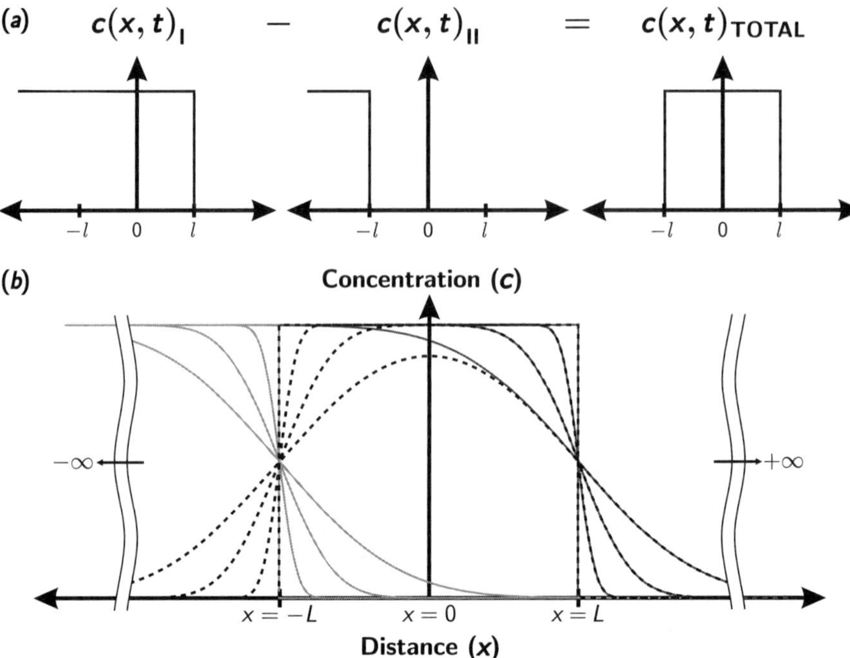

FIGURE 4.10   (*a*) The transient infinite diffusion of a rectangular concentration profile of thickness $2l$ can be obtained by subtracting the solutions for two semi-infinite step functions located at $x = -l$ and $x = +l$, respectively. (*b*) Illustration of the subtraction of the two semi-infinite step functions as they evolve with time, yielding the correct evolution of the transient diffusion profile for the rectangular source.

respectively. Mathematically

$$c_{i,1}(x,t) = \frac{1}{2}c_i^* \, \mathrm{erfc}\left[\frac{x-l}{2\sqrt{Dt}}\right]$$

$$c_{i,2}(x,t) = \frac{1}{2}c_i^* \, \mathrm{erfc}\left[\frac{x+l}{2\sqrt{Dt}}\right]$$

$$c_{i,\mathrm{tot}}(x,t) = c_{i,1}(x,t) - c_{i,2}(x,t) = \frac{1}{2}c_i^*\left(\mathrm{erfc}\left[\frac{x-l}{2\sqrt{Dt}}\right] - \mathrm{erfc}\left[\frac{x+l}{2\sqrt{Dt}}\right]\right)$$

$$= \frac{1}{2}c_i^*\left(\mathrm{erf}\left[\frac{x+l}{2\sqrt{Dt}}\right] - \mathrm{erf}\left[\frac{x-l}{2\sqrt{Dt}}\right]\right) \tag{4.39}$$

Transient Infinite Diffusion of a Thin Layer   In this example, we consider the transient diffusion of an infinitely thin layer of a diffusing species i placed in

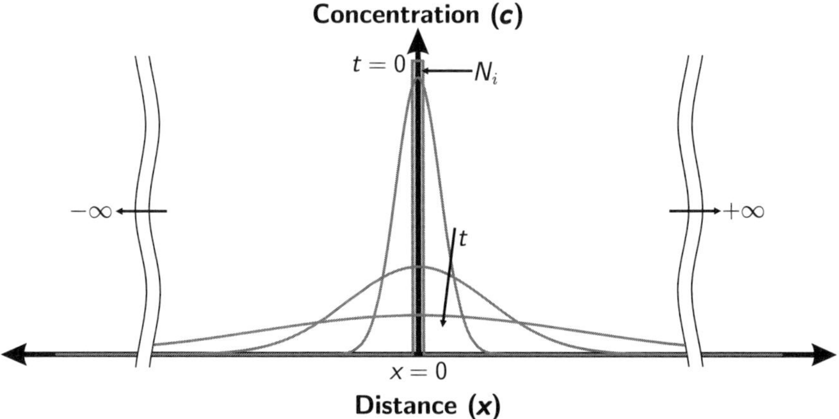

FIGURE 4.11   Transient diffusion of a thin layer between two semi-infinite bodies.

between two semi-infinite media, as shown in Figure 4.11. This situation is often a useful approximation for a number of real-world diffusion problems, and even more importantly it provides the basis for the construction of solutions to a variety of more complex diffusion situations through the use of linear superposition principles (to be discussed subsequently). The boundary and initial conditions for this "thin-film" solution are:

1. $c_i(x = -\infty, t) = 0$
2. $c_i(x = +\infty, t) = 0$
3. Initial "amount" of species i at $x = 0$, $t = 0$ is N_i (units of moles/area).

The solution to Fick's second law under these conditions is given by

$$c_i(x, t) = \frac{N_i}{\sqrt{4\pi Dt}} e^{-x^2/(4Dt)} \tag{4.40}$$

As shown in Figure 4.11 the thin-film solution yields a Gaussian concentration profile that gradually broadens as a function of time. The total amount of species i (the integral of the spatial concentration profile) stays constant, so as the profile broadens, the peak concentration decreases.

Transient Infinite Diffusion of an Arbitrary Concentration Profile   The thin-film solution developed in the last section can be used to build solutions to the transient diffusion of arbitrarily more complex starting diffusion profiles using the concept of *linear superposition*, as schematically illustrated in Figure 4.12. The basic idea is that the initial concentration profile can be built up by the linear superposition of a series of thin-film "point sources." The overall transient diffusion response of the system can then be calculated from the integrated transient diffusion response

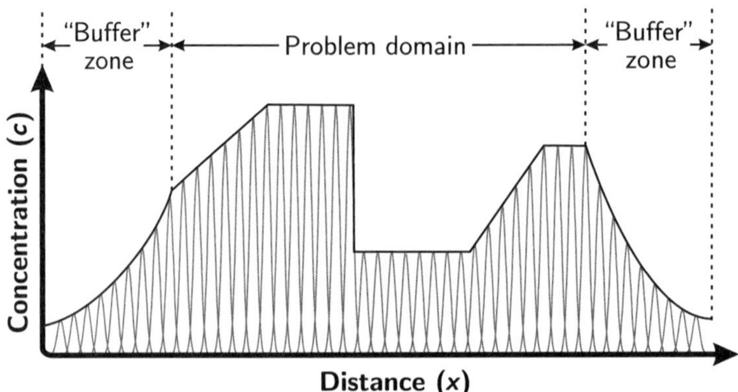

FIGURE 4.12   Schematic concept of the linear superposition of a series of thin-film point source solutions to model the transient diffusion of arbitrary concentration profiles.

of all of the point sources, each of which obeys the thin-film solution we previously obtained. This is an example of a general mathematical class of solutions known as *Green's functions.*

Applying this approach to a system with an arbitrary initial (starting) concentration profile given by $c_i^\circ(x)$ [i.e., $c_i^\circ(x) = c_i(x, t = 0)$], the solution to Fick's second law is given by

$$c_i(x, t) = \frac{1}{\sqrt{4\pi Dt}} \int_{-\infty}^{+\infty} c_i^\circ(x')e^{-(x-x')^2/(4Dt)}dx' \tag{4.41}$$

This solution is only valid under the following important conditions:

1. $D \neq f(c)$
2. $c_i(x = -\infty, t) = 0$
3. $c_i(x = +\infty, t) = 0$

The last two conditions mean that the problem must be framed within an infinite system. When this approach is applied to arbitrary (but finite) real concentration profiles in a numerical setting, the infinite system condition is typically met by including artificial concentration profile "buffer zones" at the edges of the finite problem domain that taper to zero concentration (e.g., see Figure 4.12). Since we are only interested in the evolution of the concentration profile well inside of these buffer zones, the response at the edges can be ignored.

Transient Finite (Symmetric) Planar Diffusion   In this section, we progress from transient *infinite* diffusion problems to transient *finite* diffusion problems. In many cases, the approaches and solutions to finite problems are quite similar to those just discussed in the context of infinite diffusion problems. Transient finite diffusion problems can often be solved using the separation-of-variables technique, which leads to trigonometric-series solutions (in Cartesian coordinate systems). Under certain boundary conditions (e.g., symmetric boundary conditions) and when $D \neq f(c)$, the spatial and time-dependent components of the solution can be separated:

$$c_i(x, t) = f(x)f(t) \tag{4.42}$$

As an example, consider the transient finite diffusion of material into or out of a thin plate of thickness L, as shown in Figure 4.13. The initial concentration profile of species i inside the plate is constant (given by c_i°), while the concentration of species i on both surfaces of the plate is held constant at c_i^* starting at time $t = 0$. If $c_i^* > c_i^\circ$, i will diffuse *into* the plate (Figure 4.13a, *infusion*); if $c_i^* < c_i^\circ$, i will diffuse *out* of the plate (Figure 4.13b, *effusion*). Such situations are frequently encountered in many real-world problems, including carbonization/nitridation (case hardening) of materials (infusion) and degassing or outgassing processes (effusion). The initial and boundary conditions for this problem are:

1. $c_i(0 < x < L, t = 0) = c_i^\circ$
2. $c_i(x = 0, t) = c_i^*$
3. $c_i(x = L, t) = c_i^*$

The solution to Fick's second law under these conditions is given by

$$c_i(x, t) = c_i^* + (c_i^\circ - c_i^*)\frac{4}{\pi} \sum_{j=0}^{\infty} \left\{ \underbrace{\frac{1}{2j+1} \sin\left[(2j+1)\pi\frac{x}{L}\right]}_{\substack{\text{Spatially} \\ \text{dependent term}}} \underbrace{\exp\left(-\frac{(2j+1)^2\pi^2 Dt}{L^2}\right)}_{\substack{\text{Time-} \\ \text{dependent term}}} \right\}$$

$$\tag{4.43}$$

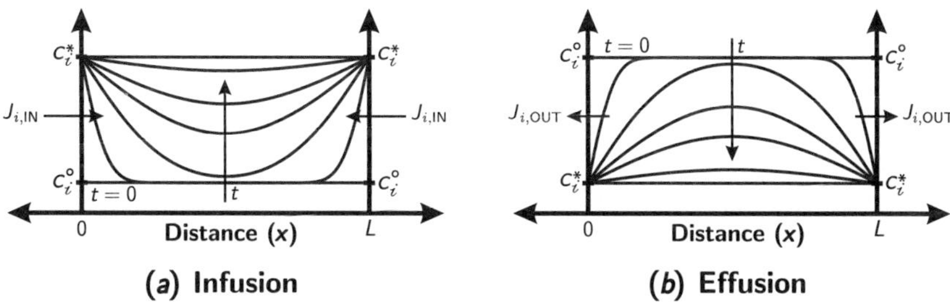

FIGURE 4.13   Transient finite diffusion of species i into (*a*) or out of (*b*) a plate of thickness L. Inside the plate, the initial concentration of species i is constant at c_i°. At time $t = 0$, the surface concentration of species (on both sides of the plate) i is set to c_i^* and this surface concentration is maintained constant during the subsequent diffusion process. If $c_i^* > c_i^\circ$, i will diffuse *into* the plate (*a*). This is termed "infusion." If $c_i^* < c_i^\circ$, i will diffuse *out* of the plate (*b*). This is termed "effusion."

Example 4.4

Question: NASA has asked us to model a transient out-gassing problem for a thin plastic plate containing a volatile substance that will gradually effuse from the plate over time when exposed to the vacuum of space. Obtain an approximate expression for the time it takes for the concentration of this substance to decrease to 1/10 of its initial value at the mid point of the plate upon exposure to the space vacuum. Assume that at time $t = 0$ exposure to the space vacuum fixes $c_i^* = 0$ on the two faces of the plate, while the initial concentration of the species inside the plate starts at a uniform value of c_i° and the plate has a thickness of L.

Solution: This is a transient 1D finite (symmetric) planar diffusion problem. The initial and boundary conditions for this problem are:

1. $c_i(0 < x < L, t = 0) = c_i^\circ$
2. $c_i(x = 0, t) = 0$
3. $c_i(x = L, t) = 0$

Compared to the general solution previously given in Equation 4.43, in this problem $c_i^* = 0$, and so Equation 4.43 simplifies to

$$c_i(x, t) = \frac{4c_i^\circ}{\pi} \sum_{j=0}^{\infty} \left\{ \frac{1}{2j+1} \sin\left[(2j+1)\pi\frac{x}{L}\right] \exp\left(-\frac{(2j+1)^2\pi^2 Dt}{L^2}\right) \right\}$$

(4.44)

This solution is composed of an infinite series of terms. However, because of the exponential function, each term in the series rapidly becomes less and less important, especially at longer times. This can be seen if we write out just the first three terms in the series:

$$c_i(x, t) = \frac{4c_i^\circ}{\pi} \left[\left(\sin\left[\frac{\pi x}{L}\right] e^{-\pi^2 Dt/L^2}\right) + \left(\frac{1}{3}\sin\left[\frac{3\pi x}{L}\right] e^{-9\pi^2 Dt/L^2}\right) \right.$$
$$\left. + \left(\frac{1}{5}\sin\left[\frac{5\pi x}{L}\right] e^{-25\pi^2 Dt/L^2}\right) + \cdots \right]$$

(4.45)

For $t \geq L^2/(\pi^2 D)$, approximating the solution using only the first term in the series yields less than a 0.01% error! Thus, except during the initial stages of a finite transient diffusion process, it is often sufficient to approximate such series solutions using only the first term in the series. Such approximate solutions are often called "long-time" solutions, because they are most valid when the transient diffusion process has already been proceeding for a "long time."[a] Truncation of Equation 4.44 to a single term and evaluation of this solution at the midpoint of the plate ($x = L/2$) yields

$$c_i\left(x = \frac{L}{2}, t\right) = \frac{4c_i^\circ}{\pi} \left[\left(\sin\left[\frac{\pi L}{2L}\right] e^{-\pi^2 Dt/L^2}\right) \right]$$
$$= \frac{4c_i^\circ}{\pi} [e^{-\pi^2 Dt/L^2}]$$

(4.46)

We wish to obtain an expression for the time $t_{1/10}$ when the concentration at the midpoint of the plate falls to $1/10$ of its initial value, or in other words, when $c_i(x = L/2, t) = \frac{1}{10}c_i^\circ$. Applying this criterion to Equation 4.46 and solving for time yields

$$c_i\left(x = \frac{L}{2}, t = t_{1/10}\right) = \frac{1}{10}c_i^\circ = \frac{4c_i^\circ}{\pi}\left[e^{-\frac{\pi^2 Dt_{1/10}}{L^2}}\right]$$

$$\frac{1}{10} = \frac{4}{\pi}\left[e^{-\frac{\pi^2 Dt_{1/10}}{L^2}}\right]$$

$$t_{1/10} = -\frac{L^2}{\pi^2 D}\ln\frac{\pi}{40}$$

$$\approx 0.258\frac{L^2}{D}$$

Note that this time is "long" compared with our long-time approximation criteria which stated that t should be $\geq 0.1[L^2/(\pi^2 D)] \approx 0.01(L^2/D)$ in order to reasonably truncate the series solution to a single term. Thus, our decision to approximate the series with a single term in this example was reasonable.

[a]The appropriate time scale for "long time" depends on the relative size of the system vs. the magnitude of the diffusion coefficient. Often, series truncation to a single term may be reasonable for $t \geq 0.1[L^2/(\pi^2 D)]$.

Alternative Boundary Conditions: What If Surface Concentration Is Not Fixed?

So far, all of our analyses of Fick's second law have been based on cases involving constant-concentration boundary conditions [e.g., $c_i(x = 0, t) = c_i^*$]. However, this will not always be the case. For example, consider outgassing from a finite plate of thickness L under a situation where the rate of removal of species i from the surface of the plate is much slower than the rate of diffusion of species i to the surface of the plate. In such a situation, it is often necessary to employ a flux-based boundary condition at the surface of the plate instead of a fixed concentration boundary condition to express the rate at which i is removed from the surface of the plate. In the limit that diffusion of i within the plate is much much faster than the flux of i out of the plate at the surface, the concentration profile will evolve with time, as shown in Figure 4.14a. In this case, the concentration of species i slowly and uniformly depletes from the plate. As the outgassing process proceeds, the concentration of i remains uniform inside the plate because i can rapidly diffuse (and hence distribute evenly throughout the plate) much more quickly than it is removed from the surfaces. If the rate of diffusion inside the plate and the rate of removal of i from the surface of the plate are similar, a situation like that illustrated in Figure 4.14b will occur instead. In this case, there is some curvature to the concentration profile of species i inside the plate, but the surface concentration of i is also changing in time. Finally, if the rate of diffusion inside the plate is much slower than the rate of removal of i from the surface, we are back to the standard case where a fixed concentration boundary condition is appropriate and a situation like that illustrated in Figure 4.14c occurs. Numerical methods are generally required to achieve solutions to Fick's second law when time-dependent or flux-based boundary conditions are invoked.

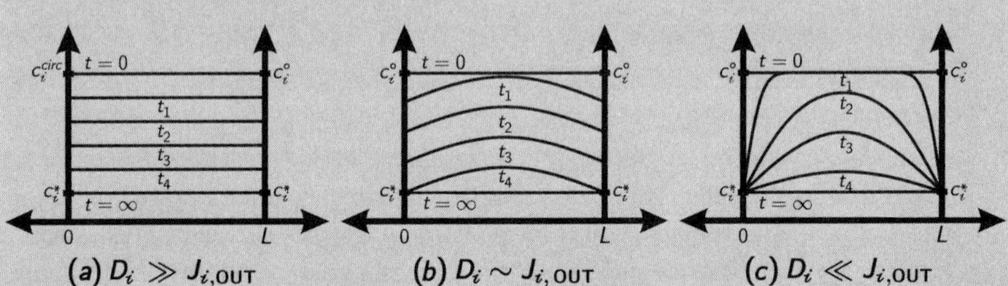

FIGURE 4.14   Transient finite diffusion from a plate when (a) diffusion of i within the plate is much much faster than the flux of i out of the plate at the surface, (b) rate of diffusion inside the plate and rate of removal of i from the surface of the plate are similar, and (c) rate of diffusion inside the plate is much slower than rate of removal of i from the surface.

Transient Finite (Symmetric) Spherical Diffusion   So far, we have only examined 1D (Cartesian) examples of Fick's second law. Solving Fick's second law in alternative coordinate systems (e.g., for radial, spherical, 2D, or 3D problems) is not really any different. As an example, we examine here the case of transient finite spherical diffusion, which is essentially analogous to the transient finite planar diffusion problem that we just finished discussing.

Fick's second law in spherical coordinates is given by

$$\frac{\partial c_i}{\partial t} = \frac{D_i}{r^2}\frac{\partial}{\partial r}\left[r^2\frac{\partial c_i}{\partial r}\right] \tag{4.47}$$

Consider the transient finite spherical diffusion problem illustrated in Figure 4.15, which describes the diffusion of H_2 into a spherical particle.

This process can be used, for example, to model how quickly hydrogen can be stored inside a metal hydride powder. Metal hydrides are intriguing metal alloy materials (often based on magnesium or aluminum) that can store and release large quantities of hydrogen gas. In fact, they can store much more hydrogen (volumetrically) than a compressed gas cylinder or even liquid hydrogen! As such, they are interesting for hydrogen storage applications. However, the *speed* at which they can store and release their hydrogen is crucial for commercial applications and thus must be understood and optimized. For the finite transient spherical hydrogen diffusion process illustrated in Figure 4.15, the boundary and initial conditions are:

1. $c_{H_2}(r < R, t = 0) = c_{H_2}^o$
2. $J_{H_2}(r = 0, t) = 0$ (this is a "no-flux" boundary condition imposed by the symmetry of the problem)
3. $c_{H_2}(r = \pm R, t) = c_{H_2}^*$

The full solution to Fick's second law under these conditions is similar to that of the finite plate:

$$c_{H_2}(r, t) = c_{H_2}^o + (c_{H_2}^* - c_{H_2}^o)\left[1 + \frac{2R}{\pi r}\sum_{j=0}^{\infty}\left(\frac{(-1)^j}{j}\sin\left[j\pi\frac{r}{R}\right]e^{-j^2\pi^2 Dt/R^2}\right)\right] \quad (4.48)$$

For the example illustrated in Figure 4.15, $c_{H_2}^o = 0$; thus we have

$$c_{H_2}(r, t) = c_{H_2}^*\left[1 + \frac{2R}{\pi r}\sum_{j=0}^{\infty}\left(\frac{(-1)^j}{j}\sin\left[j\pi\frac{r}{R}\right]e^{-j^2\pi^2 Dt/R^2}\right)\right] \quad (4.49)$$

An approximate solution can be obtained that is valid for all but the earliest time scale of this diffusion problem by truncating the solution to include only the first term in

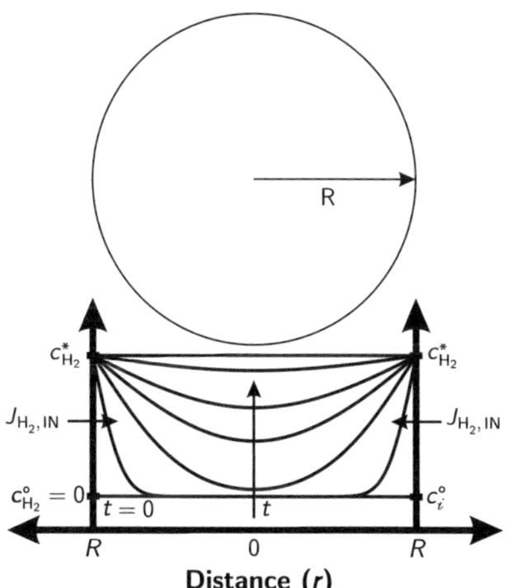

FIGURE 4.15   Transient finite diffusion of H_2 into a spherical particle. The initial concentration of H_2 in the particle is zero $[c_{H_2}(r, t = 0) = 0]$. At time $t = 0$, the particle is exposed to a sufficiently high hydrogen gas pressure to fix the H_2 concentration on the exterior surface of the particle at the solid-solubility limit, given by $c_{H_2}^*$. As time evolves, hydrogen gradually diffuses into the particle as shown.

the series[4]:

$$c_{H_2}(r, t) = c_{H_2}^* \left[1 - \frac{2R}{\pi r} \sin\left[\pi \frac{r}{R} \right] e^{-Dt(\pi/R)^2} \right] \tag{4.50}$$

This solution consists of two pieces, a *position-dependent* piece, given by the preexponential term, and a *time-dependent* piece, given by the exponential. The fact that the position dependence and the time dependence can be separated from one another embodies the concept of *self-similarity*. This concept came up previously in our discussion of transient finite diffusion in a thin plate (Equation 4.42). Self-similarity is a common and important property of many transient diffusion problems. Self-similarity means that the concentration at each point in space along the profile evolves with time in precisely the same way. For the example discussed here, this means that everywhere inside the sphere the concentration of hydrogen increases exponentially in time at a rate given by $e^{-Dt\left(\frac{\pi}{2R}\right)^2}$. Because of this self-similarity property, it is often possible to answer general questions about the progress of diffusion in such systems without ever having to explicitly determine $c(r)$. See example problem 4.5 for details.

Example 4.5

Question: For the spherical diffusion problem given in Figure 4.15, determine the time required for the concentration of H_2 in the center of the particle to reach $\frac{1}{2}c_{H_2}^*$ (i.e., the time when the particle is "half full" with H_2 at its center).

Solution: To tackle this problem, we will begin by assuming that the time required to fill the particles half full with H_2 can be evaluated using a long-time solution. Then, the change in concentration as a function of time at the center of the particle can be obtained by finding the limit of this long-time solution (Equation 4.50) as $r \to 0$. Since the limit of $\sin[\pi r/R]/r = \pi/R$ as $r \to 0$, we can obtain

$$c_{H_2}(r = 0, t) = c_{H_2}^*[1 - 2e^{-Dt(\pi/R)^2}] \tag{4.51}$$

The concentration of hydrogen at the center of the particle will reach $\frac{1}{2}c_{H_2}^*$ at a time $t_{1/2}$ given by

$$c_{H_2}(r = 0, t = t_{1/2}) = \frac{1}{2}c_{H_2}^* = c_{H_2}^*[1 - 2e^{-Dt_{1/2}(\pi/R)^2}]$$

$$\frac{1}{4} = e^{-Dt_{1/2}(\pi/R)^2}$$

$$t_{1/2} = -\frac{1}{D}\left(\frac{R}{\pi}\right)^2 \ln \frac{1}{4} \tag{4.52}$$

---

[4]The concentration profile is extremely sharp at the start of the diffusion process, with the hydrogen concentration jumping from zero inside the particle to $c_{H_2}^*$ at the surface. Modeling the initial evolution of this sharp concentration profile requires using the full Fourier series solution. However, as we have seen previously, the sharpest parts of the concentration profile decay most rapidly, and so the profile quickly evolves to a smooth decay that can be modeled using just the first term of the series.

Note that the time required to fill the particle half full with hydrogen depends on the square of the particle radius (R^2). Thus, a particle that is twice as large would require four times longer to fill. This R^2 dependence provides a powerful motivation to use nanoscale particles to improve the kinetics of hydrogen storage/release in metal hydrides and also motivates the use of nanoscale particles in applications such as Li ion batteries, where diffusion of Li ions into and out of the electrode particles often controls the rate capacity of the battery.

Rewriting Equation 4.52 in terms of R yields

$$R = \pi \sqrt{\frac{1}{\ln 4}} \sqrt{Dt_{1/2}}$$

$$\approx 2.67 \sqrt{Dt_{1/2}} \tag{4.53}$$

This result is remarkably similar to that determined for the transient diffusion process in Equation 4.27.

What If D Is a Function of Concentration?

All of the solutions presented in the previous sections made the assumption that the diffusion coefficient is not a function of concentration, that is, $D \neq f(c)$. This is a reasonable assumption for many cases, especially in the limit of dilute concentrations. However, in situations where there are large variations in concentration, this assumption may not be valid. Under such circumstances, numerical methods are generally required to achieve solutions to Fick's first and second laws. Concentration-dependent diffusion can result in strange behavior. For example, while we are used to thinking that a linear concentration gradient is established during steady-state diffusion across a finite membrane, if $D = f(c)$, the steady-state concentration profile can have positive or negative curvature, as shown in Figure 4.16. When $D = f(c)$, superposition methods cannot be used, since Fick's second law in such cases is no longer a linear partial differential equation.

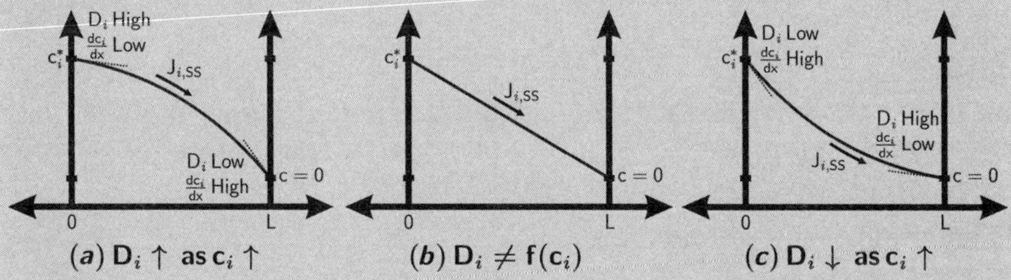

FIGURE 4.16 Steady-state diffusion across a finite membrane when (a) diffusivity increases with concentration, (b) diffusivity is independent of concentration, and (c) diffusivity decreases with concentration. At steady state, Fick's first law requires that the flux is constant at all spatial positions. Thus, if D increases as c_i increases, the concentration gradient (dc_i/dx) must decrease in order to keep the flux constant. Likewise, if D decreases as c_i increases, the concentration gradient must increase in order to keep the flux constant.

4.4.3 Kirkendal Effect and Moving Interface Problems

Kirkendal Effect When we previously discussed the transient interdiffusion of two semi-infinite bodies (material A and material B, respectively), we explicitly specified that the diffusion of A in B and that of B in A were identical and could therefore be described by a single diffusion coefficient. In many solids, however, this is not true. For example, the diffusivity of zinc in copper is much larger than the diffusivity of copper in zinc. If a block of brass (a copper–zinc alloy) and a block of pure copper are bonded together at high temperatures, the zinc atoms will diffuse out of the brass and into the copper at a much faster rate than the copper atoms diffuse into the brass block. *The net result is that the effective interface between the brass and copper blocks moves toward the brass, as illustrated schematically in Figure* 4.17. This phenomenon is known as the *Kirkendal effect* and it occurs in many solid-state systems.

For the interdiffusion of two species A and B with unequal diffusivities $D_A \neq D_B$, the speed at which the A/B interface moves (v) can be calculated by

$$v = \frac{1}{c_A + c_B}(D_A - D_B)\frac{dc_A}{dx} \tag{4.54}$$

Fick's first and second laws must be modified when describing a system that manifests the Kirkendal effect. Fick's first law becomes (written in terms of species A)

$$J_A = -D_A\frac{dc_A}{dx} + vc_A \tag{4.55}$$

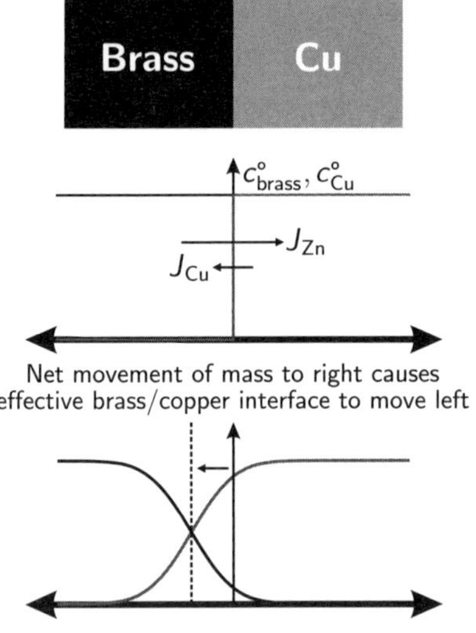

FIGURE 4.17   Kirkendal effect illustrated for a brass/copper diffusion couple. Since the diffusion of zinc into copper is much faster than the diffusion of copper into brass, a net flow of mass occurs from the brass side to the copper side. As a result, the location of the interface between the brass and the copper moves toward the brass side.

while Fick's second law becomes

$$\frac{dc_A}{dt} = \frac{d}{dx}\left[\left(\frac{c_B}{c_A + c_B}D_A + \frac{c_A}{c_A + c_B}D_B\right)\frac{dc_A}{dx}\right] \tag{4.56}$$

Diffusion at Phase Boundaries: Eutectic Interfaces   In all previous discussion of diffusion, we have assumed that diffusing components are completely soluble in one another. In fact, this is rarely true. When large concentration gradients are present, it is common that species will be segregated into multiple phases. A typical example is eutectic behavior, as illustrated by the phase diagram in Figure 4.18*a*. In a eutectic system, species A and B have limited solubility in one another and two phases can exist: an A-rich α phase and a B-rich β phase. Thus, when a diffusion couple is created, an abrupt change in composition will occur at the interface that marks the transition from the α phase to the β phase, as shown in Figure 4.18*b*.

As A and B diffuse across the eutectic interface into the β and α phases, respectively, the interface will move (just as we discussed previously in the context of the Kirkendal effect). This will occur even if the diffusivities of A and B in α and β are equal, because the local concentration gradients on the two sides of the interface will likely be different. (Remember, the flux depends on the diffusivity *and* the concentration gradient.)

As with the Kirkendal effect, an expression for the velocity of the interface can be derived, which yields (written in terms of species B)

$$v = \sqrt{\frac{\pi}{t}}\left[\frac{\sqrt{D_{B,\beta}}(c_B^\circ - c_{B,\beta}) - \sqrt{D_{B,\alpha}}(c_{B,\alpha})}{(\pi - 2)(c_{B,\beta} - c_{B,\alpha}) + 2c_B^\circ}\right] \tag{4.57}$$

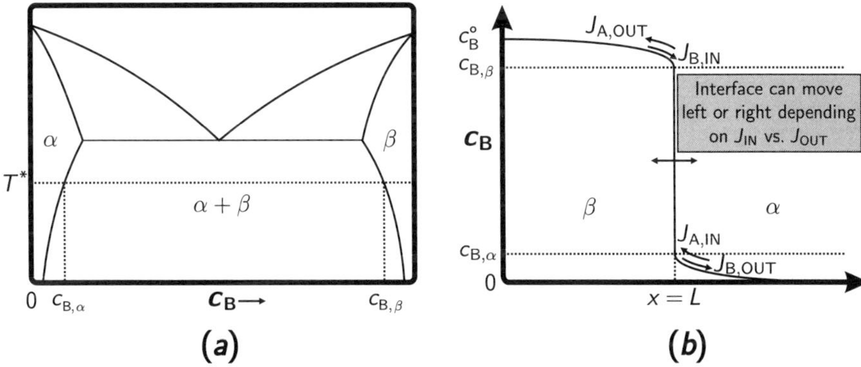

FIGURE 4.18   (*a*) Eutectic phase diagram involving an A-rich α phase and a B-rich β phase. At the indicated temperature (T^*), the maximum solubility of B in α is given by $c_{B,\alpha}$ while the minimum solubility of B in β is given by $c_{B,\beta}$. (*b*) Concentration profile that results when A and B are bonded together in a diffusion couple at temperature T^*. At the interface ($x = L$) there is a phase change from the B-rich β phase to the A-rich α phase. The concentrations of A and B therefore change abruptly at this interface. As in our previous discussions of the Kirkendal effect, the location of this eutectic interface will move as a function time since the rate of diffusion of B into α and the rate of diffusion of A into β will almost certainly be unequal.

where c_{B}° is the concentration of species B in pure B and the concentration of species B in pure A is zero. The terms $c_{\mathrm{B},\beta}$ and $c_{\mathrm{B},\alpha}$ represent the maximum/minimum solubility limits of species B in the β and α phases at the temperature at which the diffusion is occurring (as illustrated in Figure 4.18).

4.4.4 Summary of Transient Diffusion Problems

Table 4.4 summarizes the solutions to the various types of transient diffusion problems that we have discussed in this chapter while the bullet points below summarize the recommended step-by-step approach to solve typical transient diffusion problems:

1. Determine the geometry of the problem (e.g., 1D, 2D, 3D, radial, spherical) and identify any symmetries.

2. Make a sketch of the physical problem and show the initial concentration profile and your expectation for how the concentration profile will evolve as a function of time

3. Specify the boundary condition(s), initial condition, and governing equation (i.e., Fick's second law in the appropriate 1D, 2D, 3D, radial, spherical, etc., form as needed for your problem).

4. Obtain the general solution to the problem (by looking it up in a reference text, consulting your friend in the math department, or solving the partial differential equation using your amazing mathematical skills).

5. Apply the boundary and initial conditions to the general solution obtained above in order to obtain the exact solution.

6. Investigate limiting solutions (e.g., behavior at the boundaries or at short/long times) to ensure that the solution is reasonable and behaves as expected.

4.4.5 Coupled Diffusion Processes

While this chapter has focused almost exclusively on diffusion, it is important to remember that in any system that is out of equilibrium a variety of different transport processes can be simultaneously occurring (e.g., simultaneous flows of mass, heat, and charge). Furthermore, as we alluded to in Section 4.2, these different transport processes can couple, or interact, with one another. The interaction between transport processes can be collectively and comprehensively dealt with under the framework of *nonequilibrium thermodynamics* (NET). This theory, originally developed by Onsager [6, 7], deals with the thermodynamically enforced coupling between the driving forces experienced by a nonequilibrium system and its resulting response as it attempts to return toward equilibrium (assuming that the system is not too far from equilibrium). NET theory can capture the effects of multiple simultaneous driving forces acting on a nonequilibrium system and predict how these driving forces induce flows (fluxes) of multiple chemical species and/or heat and/or charge in response.

Because of the interwoven coupling between forces and fluxes stipulated by NET, the mathematical treatment of transport can quickly get complicated. For example,

TABLE 4.4 Summary of the Various Transient Solutions Discussed in This Chapter

1) Transient Semi-Infinite Diffusion from Constant Surface Source

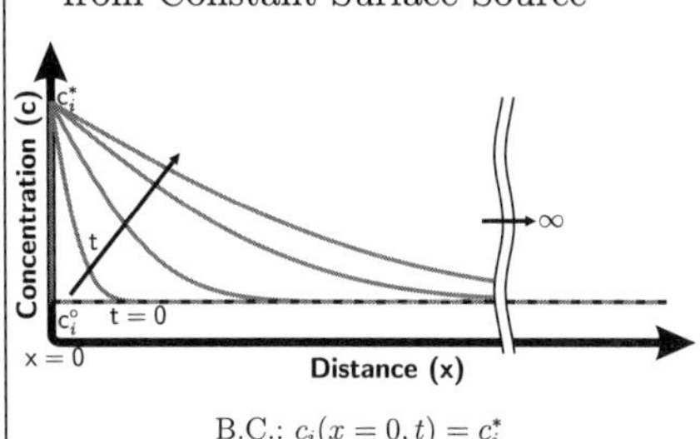

B.C.: $c_i(x = 0, t) = c_i^*$
I.C.: $c_i(x > 0, t = 0) = c_i^\circ$

$$c_i(x, t) = c_i^\circ + (c_i^* - c_i^\circ)\,\mathrm{erfc}\left[\frac{x}{2\sqrt{Dt}}\right]$$

2) Transient Interdiffusion in Two Semi-Infinite Bodies

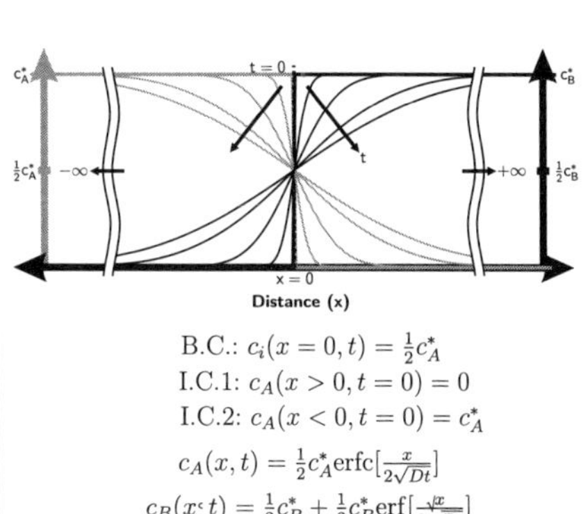

B.C.: $c_i(x = 0, t) = \frac{1}{2}c_A^*$
I.C.1: $c_A(x > 0, t = 0) = 0$
I.C.2: $c_A(x < 0, t = 0) = c_A^*$

$$c_A(x, t) = \frac{1}{2}c_A^*\,\mathrm{erfc}\left[\frac{x}{2\sqrt{Dt}}\right]$$

$$c_B(x, t) = \frac{1}{2}c_B^* + \frac{1}{2}c_B^*\,\mathrm{erf}\left[\frac{x}{2\sqrt{Dt}}\right]$$

3) Transient Infinite Diffusion of a Rectangular Concentration Profile

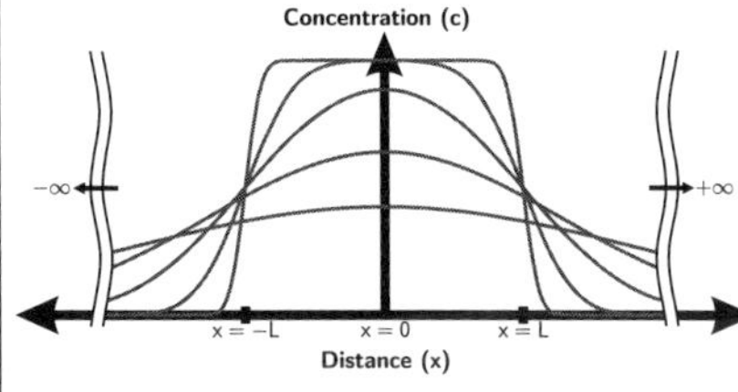

B.C.1: $c_i(x = -\infty, t) = 0$
B.C.2: $c_i(x = \infty, t) = 0$
I.C.: $c_i(-L < x < L, t = 0) = c_i^*$

$$c_i(x, t) = \frac{1}{2}c_i^*\left[\mathrm{erf}\left[\frac{x+L}{2\sqrt{Dt}}\right] - \mathrm{erf}\left[\frac{x-L}{2\sqrt{Dt}}\right]\right]$$

4) Transient Infinite Diffusion of a Thin Layer

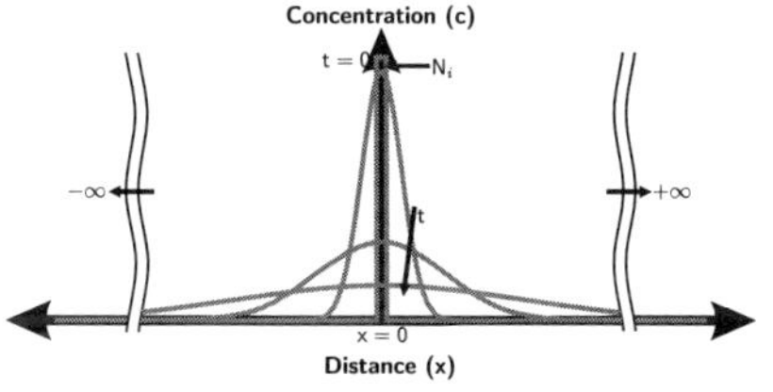

B.C.1: $c_i(x = -\infty, t) = 0$
B.C.2: $c_i(x = \infty, t) = 0$
I.C.: "Amount of Species" i at $x = 0$ and $t = 0$; N_i

$$c_i(x, t) = \frac{N_i}{\sqrt{4\pi Dt}}e^{-\frac{x^2}{4Dt}}$$

5) Transient Finite (Symmetric) Planar Diffusion

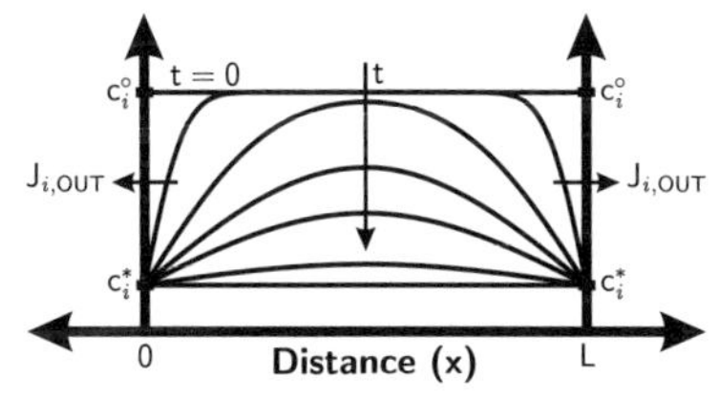

B.C.1: $c_i(x = 0, t) = c_i^*$
B.C.2: $c_i(x = L, t) = c_i^*$
I.C.: $c_i(0 < x < L, t = 0) = c_i^{circ}$

$$c_i(x, t) = c_i^* + (c_i^\circ - c_i^*)\frac{4}{\pi}\sum_{j=0}^{\infty}\left(\frac{1}{2j+1}\sin\left[(2j+1)\pi\frac{x}{L}\right]e^{-\frac{(2j+1)^2\pi^2 Dt}{L^2}}\right)$$

6) Transient Finite (Symmetric) Spherical Diffusion

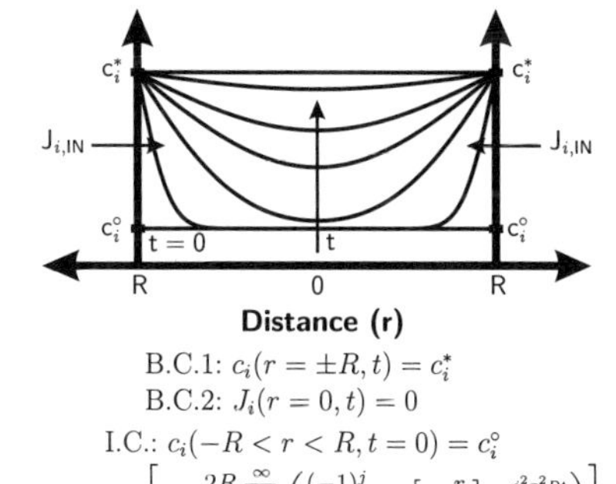

B.C.1: $c_i(r = \pm R, t) = c_i^*$
B.C.2: $J_i(r = 0, t) = 0$
I.C.: $c_i(-R < r < R, t = 0) = c_i^\circ$

$$c_i(r, t) = c_i^*\left[1 + \frac{2R}{\pi r}\sum_{j=0}^{\infty}\left(\frac{(-1)^j}{j}\sin\left[j\pi\frac{r}{R}\right]e^{-\frac{j^2\pi^2 Dt}{R^2}}\right)\right]$$

in a system with temperature, voltage, and concentration gradients, a 3×3 matrix (or larger) of coupled flux–force terms can easily be required. Fortunately, however, it is often possible to simplify the dominant transport process in a given system to a single-term force–flux relationship. Thus, as we have seen throughout most of this chapter, mass transport in many solid-state kinetic processes can often be adequately described by simple Fickian diffusion. While simple Fickian diffusion is often adequate, it is useful to examine some important instances where more complex coupled diffusion phenomena are encountered. Here are a few of the most common coupled diffusion processes:

- Electrodiffusion
- Thermodiffusion
- Stress driven diffusion

Figure 4.19 summarizes these important coupled diffusion processes, each of which is briefly treated below.

Electrodiffusion   Electrodiffusion, or electromigration, occurs when an applied electrical field provides an additional driving force for diffusion. There are two major categories of electric-field-assisted diffusion:

1. In a material containing mobile charged ions (e.g., solid-state ion conductors or electrolytes), an applied electric field will act as an additional (often dominating) driving force for the transport of these charged species. Therefore, when both concentration and electric potential gradients are present, the diffusion equation must be modified to account for both driving forces:

$$J_i = -D_i \left(\frac{dc_i}{dx} + \frac{c_i q_i}{RT} \frac{d\phi}{dx} \right) \tag{4.58}$$

where q_i is the charge associated with the diffusing ion and $d\phi/dx$ is the electric potential gradient (electric field). Positively charged ions will be driven in the direction of the field (_down_ the electric potential gradient), while negatively charged ions will be driven against the field. Thus, depending on its direction, an electric field can either assist or counteract concentration-driven diffusion. A sufficiently strong electric field acting in the opposite direction to diffusion can even force ions to migrate up their concentration gradient.

2. In metals, an applied electric field can induce a large electronic current to flow through the material, which can in turn induce movement of the metal atoms. This effect is known as electromigration and is sometimes referred to as an "electron wind." Electromigration occurs because current-carrying electrons scatter off of metal atoms or defects in the lattice as they move. This scattering can distort the electron density of the metal atoms, thereby inducing temporary local dipoles that can lead to migration under the force of the applied field. Electromigration can be treated essentially the same way that the electrodiffusion

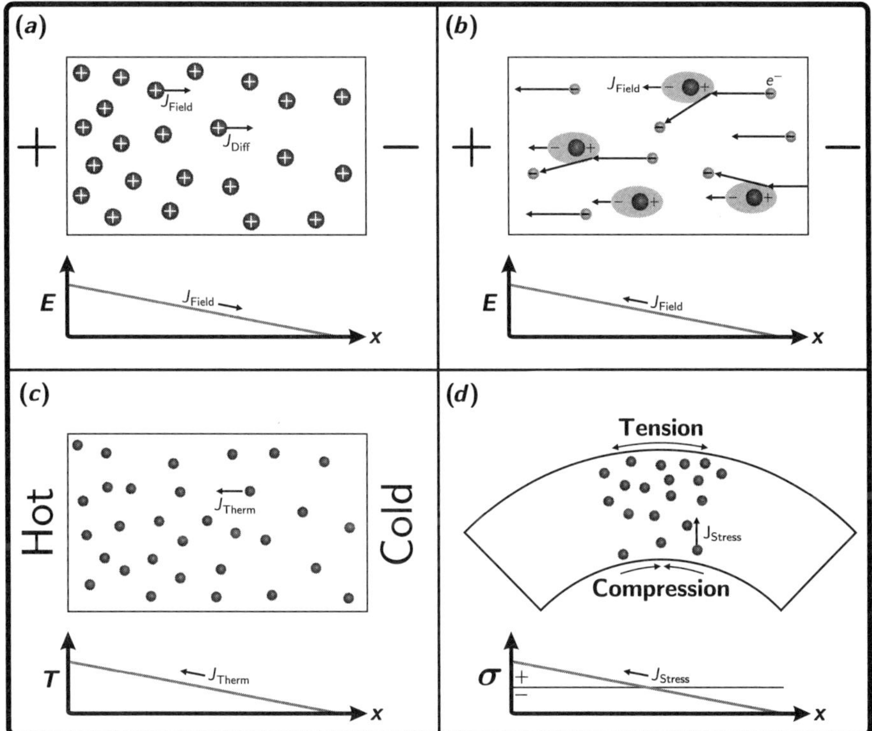

FIGURE 4.19 Overview comparison of various coupled diffusion processes. (*a*) Electrodiffusion in an ion conductor. Application of an electric field creates an additional driving force for the transport of charged ions in a material. Positive ions will be driven down the electric potential gradient. (*b*) Electromigration in a metal. Current-carrying electrons occasionally scatter off the metal atoms, creating temporary charge dipoles that enable the electric field to exert a force on the charged nuclei of the atoms, causing them to move. (*c*) Thermodiffusion. A temperature gradient can also induce atomic transport. In the schematic, atoms are driven up the temperature gradient due to a negative heat of transport. (*d*) Stress-driven diffusion. A stress gradient, here induced by bending of a bar, can also induce atomic transport. In this case, interstitial atoms are driven away from the side of the bar that is in compression and toward the side of the bar that is in tension.

of charged ions is treated, except q_i is replaced with an *effective charge* term (β_i) that reflects the average dipole strength manifested by the metal atoms:

$$J_i = -D_i \left(\frac{dc_i}{dx} + \frac{c_i \beta_i}{RT} \frac{d\phi}{dx} \right) \tag{4.59}$$

where β_i is typically small and thus the effect of electromigration is typically negligible except under certain circumstances such as extremely high electric fields or high electric currents. Electromigration can be important in integrated circuits, where the extremely tiny current-carrying metal vias must sustain extremely high current densities. Local constrictions in the vias can lead to localized areas of even higher current density, leading to mass loss from these regions. This mass loss causes even greater electrical current constriction, thereby further increasing the current density and thus the mass loss and ultimately leading to a runaway effect that can cause complete failure of the via.

Thermodiffusion   Thermodiffusion reflects the fact that a temperature gradient can act as a driving force for diffusion. Incorporating both concentration gradient and temperature gradient driving forces into the 1D diffusion equation yields

$$J_i = -D_i \left(\frac{dc_i}{dx} + \frac{c_i Q_i^*}{RT^2} \frac{dT}{dx} \right) \tag{4.60}$$

where Q_i^* is known as the *heat of transport*. The heat of transport for a diffusing species can be either negative or positive, and thus thermal effects can cause diffusion *up* or *down* a temperature gradient. In steel, for example, interstitial carbon atoms will diffuse up a temperature gradient (i.e., toward the "hot" side of a sample). Multicomponent alloys, oxides, or other materials that are held for long periods of time under large temperature gradients can experience thermal "unmixing," whereby components that possess positive and negative heats of transport will diffuse away from one another, causing demixing or even phase separation. Such instabilities are a particular issue for nuclear reactor cladding materials and nuclear fuel pellets.

Stress-Driven Diffusion   Stress and diffusion can be coupled in a number of ways. In a uniform stress field, the *diffusivity* of the diffusing species can become directionally dependent. This is because the stress field can affect the amount work required for the species to move in different directions (e.g., parallel vs. perpendicular to the stress field). Movements in directions that cause the greatest distortions to the stress field will be penalized, while movements in directions that minimize the distortion to the stress field will be favored.

When a *gradient* in the stress field is present, the *driving force* for diffusion is also modified. In this case, both the concentration gradient and the stress gradient must be included when modeling diffusion. Assuming a uniaxial hydrostatic stress gradient (dP/dx), the resulting modified 1D diffusion equation may be written as

$$J_i = -D_i \left(\frac{dc_i}{dx} + \frac{c_i \Delta\Omega_i}{RT} \frac{dP}{dx} \right) \tag{4.61}$$

where $\Delta\Omega$ is the molar dilatation associated with the defect (compared to the "perfect" lattice). Depending on the direction of the stress gradient relative to the concentration gradient, the diffusion of a species can either be enhanced or impeded. As with the other coupled transport effects, a sufficiently strong stress gradient acting in the opposite direction to diffusion can even force ions to transport up their concentration gradient. Stress-driven diffusion effects are of critical importance for a number of real-world devices, including Li ion batteries, where the large stresses that arise upon lithiation/delithiation of the electrode materials can significantly affect the Li diffusion and hence the speed at which the battery can be charged/discharged.

4.5 ATOMISTIC TREATMENT OF DIFFUSION

4.5.1 Overview of Diffusion in Gases Versus Liquids Versus Solids

As we have seen, the macroscopic treatment of diffusion using Fick's first and second laws makes no distinction about whether the diffusion process occurs in a solid, liquid, or gaseous medium. In general, these macroscopic laws apply fairly well to all three phases. The differences between solid-, liquid-, and gas-phase diffusion mostly show up in the magnitude of the diffusion coefficient D_i. This parameter quantifies the relative ease with which atoms or molecules can be transported via diffusion in a material. Because diffusion occurs by a series of discrete random movements, for example, as a species jumps from lattice site to lattice site in a solid, or veers from one collision event to another collision event in a liquid or gas, both the speed (v_i) at which a species moves and the average distance traveled during each movement (λ) are embedded in the diffusion coefficient. In a gas or liquid, this dependence is often expressed as

$$D \approx \tfrac{1}{3} \lambda v_{\text{RMS}} \tag{4.62}$$

where λ is the mean free path (average distance traveled) between collisions and v_{RMS} is the root-mean-squared speed of the diffusing species. The factor of $\tfrac{1}{3}$ accounts for the fact that only one-third of the average random 3D motion occurs along a given orthogonal direction.

For a solid, the analogous expression is

$$D \approx \tfrac{1}{6} a(a\Gamma) = \tfrac{1}{6} a^2 \Gamma \tag{4.63}$$

where a is the atomic jump distance and Γ is the hopping rate. Here, $a\Gamma$ can be thought of as a "speed," enabling a more direct comparison of the solid-phase diffusivity expression to the expression for gas- and liquid-phase diffusivity.[5]

Typically, gas-phase diffusivities are the highest, while solid-phase diffusivities are the lowest, leading to huge differences in the typical rates of diffusion in these phases. In addition, the temperature dependence of the diffusivity in gases versus liquids and solids tends to be different, with diffusion in gases showing only a weak temperature dependence while diffusion in liquids and solids tends to show an exponentially activated temperature dependence. Figure 4.20 compares the key features of solid-, liquid-, and gas-phase diffusion. In the sections that follow, we will explore in more detail the atomic underpinnings of diffusion in the gas and solid phases.

---

[5]The $\tfrac{1}{6}$ term appearing in this expression compared to the $\tfrac{1}{3}$ term appearing in the case of gases and liquids reflects the additional constraint imposed by diffusion in a crystalline lattice, where only certain specific "jump" directions are allowed.

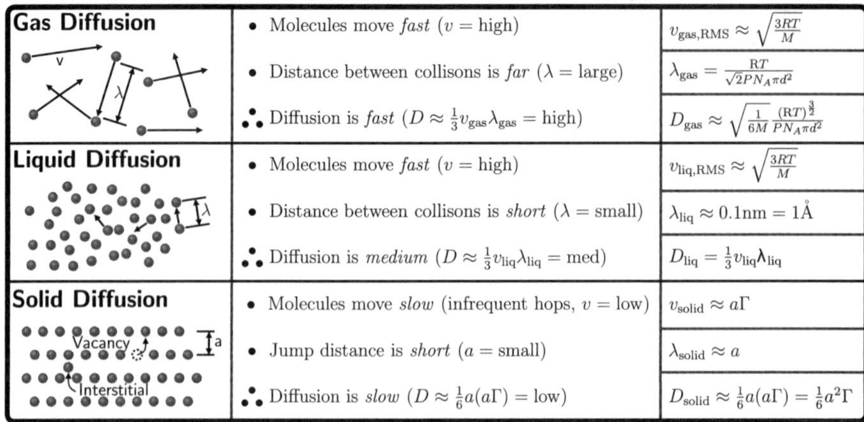

FIGURE 4.20   Overview comparison of gaseous, liquid, and solid-state diffusion.

4.5.2   Diffusion in Gases: Kinetic Theory of Gases

The atomic underpinnings of gas-phase diffusion are captured in the kinetic theory of gases, which uses statistical concepts to describe the distribution of energies and speeds of molecules in the gas phase, as well as their frequency of collision.

Calculating v_{RMS} The kinetic theory of gases begins with the concept that the temperature of a gas is a statistical measure of the kinetic energy, and hence average speed, of the atoms or molecules making up the gas. From the ideal gas law and consideration of the force (and hence pressure) exerted by molecules in a gas phase as they randomly collide with the walls of their container, the following relationship between kinetic energy and temperature may be obtained:

$$\text{Kinetic energy:}\quad \tfrac{1}{2}mv_{\text{RMS}}^2 = \tfrac{3}{2}kT \tag{4.64}$$

where k is the Boltzmann constant and m is the mass of the gas atom or molecule. From this relationship, an expression for the root-mean-squared speed of the gas can be obtained:

$$v_{\text{RMS}}^2 = \frac{3\,kT}{m}$$

$$= \frac{3\,RT}{M}$$

$$v_{\text{RMS}} = \sqrt{\frac{3\,RT}{M}} \tag{4.65}$$

where we have converted from atomic units (k, m) to molar units (R, M). Here, M is the molar mass (kg/mol) of the gas-phase atom or molecule, R is the gas constant, and SI units are used.

Equation 4.65 reveals that the root-mean-squared speed increases as the temperature increases and decreases as the molar mass of the gas-phase species increases. Although it is beyond the scope of this textbook, the kinetic theory of gases can also be used (through the application of statistical mechanics principles) to derive the complete distribution of velocities for a gas-phase species:

$$f(v) = 4\pi v^2 \left(\frac{M}{2\pi RT} \right)^{\frac{3}{2}} \exp\left[-\frac{Mv^2}{2\,RT} \right] \tag{4.66}$$

where $f(v)$ is the probability density function for the speed. Figure 4.21 provides an example of such a distribution (known as Maxwell–Boltzmann distribution) for two temperatures $(T_1 < T_2)$ and for two different molar masses $(M_1 < M_2)$. As temperature increases or molecular mass decreases, the distribution flattens and spreads to higher overall speeds. Note that due to the characteristic shape of the Maxwell–Boltzmann distribution (long tail at higher speeds) the root-mean-squared speed is greater than both the average speed (v_{AVG}) and the most probable speed. In fact, the average speed can be computed from $f(v)$, yielding

$$v_{AVG} = \sqrt{\frac{8\,RT}{\pi M}} \tag{4.67}$$

which is 92% of v_{RMS}.

Calculating λ Assuming that atoms or molecules in a gas can be modeled as spheres that interact only when they collide and that all collisions are perfectly elastic, the kinetic theory of gases can also be used to predict the frequency at which gas atoms will collide and hence the average distance (or mean free path, λ) between collisions.

Consider a gas molecule of diameter d moving through space at a velocity v as shown in Figure 4.22. As this molecule sweeps through space in a time period of Δt,

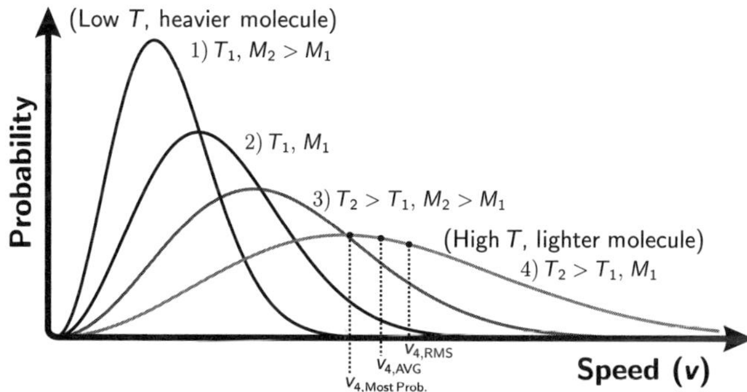

FIGURE 4.21   Distribution of speeds in a gas as predicted by the kinetic theory of gases (Maxwell–Boltzmann distribution). As T increases or M decreases, the distribution flattens and spreads to higher overall speeds.

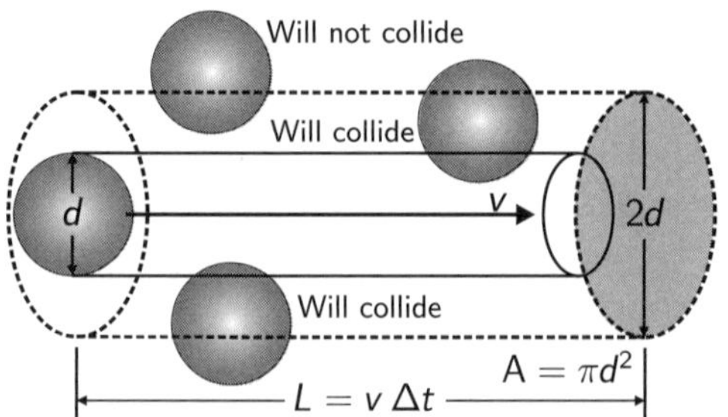

FIGURE 4.22 Derivation of the mean free path λ between collisions for a gas-phase molecule. A molecule of diameter d traveling at a speed v will collide with any molecules that it encounters within its "interaction volume" of $v\,\Delta t\,A$ in a time Δt.

it will collide with any molecules that it encounters within its "interaction volume," which can be visualized as a cylinder of radius d and length $v\,\Delta t$. The size of this interaction volume is therefore given by

$$V = L \times A = (v\,\Delta t)(\pi d^2) \tag{4.68}$$

If the pressure of the gas is given by P, then, using the ideal gas law, the number of atoms encountered by the molecule in its interaction volume can be calculated as

$$N = N_A \frac{PV}{RT} = N_A \frac{Pv\,\Delta t\,\pi d^2}{RT} \tag{4.69}$$

where N_A is used to convert from number of moles to number of atoms. The mean free path λ can then be calculated from the ratio of the total distance traveled by the molecule during the time interval Δt divided by the total number of atoms encountered during this time interval:

$$\lambda = \text{average distance traveled between collisions}$$

$$= \text{total distance traveled/atoms encountered}$$

$$= \frac{v\,\Delta t}{N_A[Pv\,\Delta t\pi d^2/(RT)]}$$

$$= \frac{RT}{PN_A\pi d^2} \tag{4.70}$$

It should be noted that this simplified derivation neglects the relative motions of the other molecules (which in reality are not fixed in space during the time when our test particle is sweeping out its interaction volume). A complete derivation that takes the relative motion of the other molecules into account results in a modest $\sqrt{2}$ adjustment to the result:

$$\lambda = \frac{RT}{\sqrt{2}PN_A\pi d^2} \tag{4.71}$$

Both equations show that the mean free path increases as temperature increases, decreases as pressure increases, and decreases as the size of the molecule increases.

Calculating D   Having obtained expressions for both v_{RMS} and λ, we can now combine these results to obtain an expression for gas-phase diffusivity:

$$D \approx \frac{1}{3}\lambda v_{RMS}$$

$$\approx \frac{1}{3}\left(\frac{RT}{\sqrt{2}PN_A\pi d^2}\right)\left(\sqrt{\frac{3\,RT}{M}}\right)$$

$$\approx \sqrt{\frac{1}{6M}\frac{(RT)^{3/2}}{PN_A\pi d^2}} \tag{4.72}$$

While this equation is algebraically straightforward, *the units often present students with significant difficulties.* Please keep in mind that SI units must be used when evaluating this expression. Thus, M is in kg/mol, d is in m, P is in Pa, T is in K, and so on. Example problem 4.6 provides practice dealing with this expression.

Example 4.6

Question: Use Equation 4.72 to estimate the gas-phase diffusivity of pure O_2 gas at $T = 25\,°C$ and 1 atm pressure. Assume that the diameter of the oxygen molecule, d_{O_2}, is 3 Å.

Solution: This is a very straightforward "plug-and-chug" problem. As such, it should offer no significant difficulties. However, it is essential to use SI units to avoid problems. Equation 4.72 is

$$D \approx \sqrt{\frac{1}{6M}\frac{(RT)^{\frac{3}{2}}}{PN_A\pi d^2}}$$

To evaluate this expression using SI units, we need to supply the molar mass of O_2 gas (in kg/mol), the gas constant (in J/(mol · K)), the temperature [in K], the pressure (in Pa), and the diameter of the O_2 molecule (in m):

$$D \approx \sqrt{\frac{1}{6 \cdot 32 \times 10^{-3}\ \text{kg/mol}}\frac{(8.314\ \text{J/(mol · K)} \cdot 298.15\ \text{K})^{\frac{3}{2}}}{101{,}300\ \text{Pa} \cdot 6.022 \times 10^{23}/\text{mol} \cdot \pi(3 \times 10^{-10}\ \text{m})^2}}$$

$$\approx 1.65 \times 10^{-5}\ \text{m}^2/\text{s} = 0.165\ \text{cm}^2/\text{s}$$

The initial answer is given in SI units (m^2/s) but can be easily converted into other units (e.g., cm^2/s) after the fact. The actual experimentally measured value for the O_2 gas self-diffusion coefficient at $T = 25\,°C$ and 1 atm pressure is 0.178 cm^2/s. *The very close agreement between theory and experiment is remarkable!*

Binary Gas Diffusivity   The above discussion of gas-phase diffusion implicitly assumes that the gas is composed of a *single* species. Diffusion of a pure species (e.g., diffusion of O_2 gas molecules in pure O_2 gas) is known as *self-diffusivity*. Very frequently, however, we would like to calculate the diffusivity of a gas species in a mixture—for example, the diffusivity of O_2 in air (which is essentially a mixture of N_2 and O_2). For mixtures of two gas-phase species, a binary gas-phase diffusivity can be estimated in analogy to the pure-species diffusivity:

$$D_{AB} = D_{BA} \approx \sqrt{\frac{1}{12}\left(\frac{M_A + M_B}{M_A M_B}\right)} \frac{(RT)^{3/2}}{PN_A \pi \left[\frac{1}{2}(d_A + d_B)\right]^2} \qquad (4.73)$$

For binary (two-species ONLY) gas-phase diffusivities, the diffusivity of species A in B must be equal to the diffusivity of species B in A. Note that this is not generally true for solid-state diffusion or multicomponent (more than two species) gas-phase diffusion, where more complicated expressions (e.g., the Stefan–Maxwell equation) must be used to describe diffusion. However, such treatments are beyond the scope of this textbook.

4.5.3   Diffusion in Solids: Atomistic Mechanisms of Solid-State Diffusion

In this section, we develop an atomistic picture to understand solid-state diffusion in more detail. At the most fundamental level, a solid-state diffusion coefficient D is a measure of the intrinsic rate of the hopping process by which atoms/molecules can move from one site to another in a solid medium. Even in the absence of any driving force, hopping of atoms from site to site within the lattice still occurs at a rate that is characterized by the diffusivity. Of course, without a driving force, the net movement of atoms is zero, but they are still exchanging lattice sites with one another. This is another example of a *dynamic equilibrium*; compare it to the dynamic reaction equilibrium processes that we discussed in Chapter 3.

Broadly viewed, there are two main mechanisms of solid-state diffusion in crystalline materials:

1. Vacancy diffusion
2. Interstitial diffusion

These two types of solid-state diffusion processes are illustrated schematically in Figure 4.23.

The interstitial mechanism is generally favored for small atoms (e.g., impurity cations such as Na in Si or C in Fe) that can fit into the interstitial sites in a crystal lattice. Interstitial diffusion is generally faster than vacancy diffusion because bonding of interstitials to the surrounding atoms is normally weaker and there are generally many more available interstitial sites than vacancy sites to jump to. Larger atoms, for example the oxygen anions in most oxide ceramics, must diffuse via a vacancy

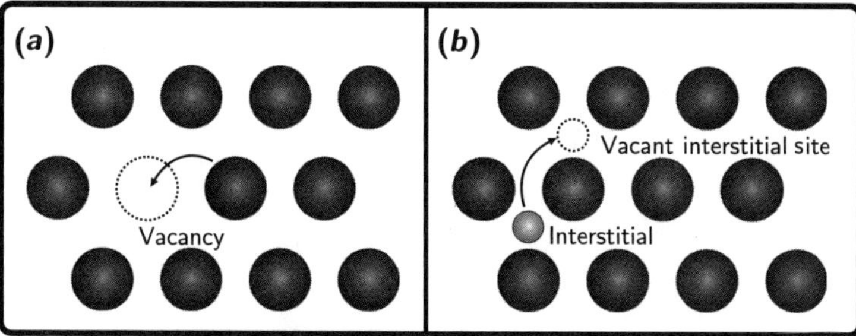

FIGURE 4.23 Two principal solid-state diffusion mechanisms: (*a*) vacancy diffusion; (*b*) interstitial diffusion.

mechanism instead. Because both vacancy and interstitial diffusion require the presence of point defects (vacancy and interstitial defects, respectively) to occur, they are highly sensitive to the degree of crystalline imperfection. High concentrations of point defects, as well as high concentrations of extended defects such as dislocations, grain boundaries, and surfaces, generally lead to higher rates of solid-state diffusion.

A number of other chemical and structural factors can also affect solid-state diffusion. In general, the following factors often lead to higher solid-state diffusivities:

- Interstitial diffusion
- Lower density materials
- Smaller diffusing atoms
- Cations
- Lower melting point materials
- Open crystal structures

In contrast, the following factors often lead to lower solid-state diffusivities:

- Vacancy diffusion
- Higher density materials
- Larger diffusing atoms
- Anions
- Higher melting point materials
- Close-packed crystal structures
- Strong covalent bonding

Theory of Solid-State Diffusion Using the schematic in Figure 4.24*b*, we can derive an atomistic model of diffusion in a crystalline solid. The atoms in this figure are arranged in a series of parallel atomic planes. We would like to calculate the net flux (net movement) of gray atoms from left to right across the imaginary plane labeled *A* in Figure 4.24 (which lies between two real atomic planes in the material).

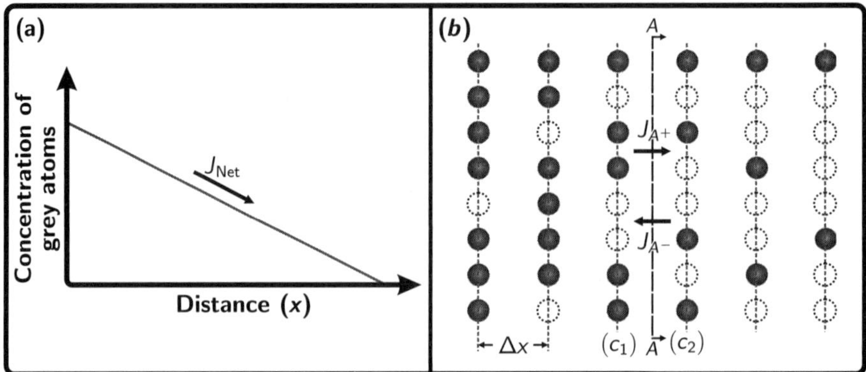

FIGURE 4.24   (*a*) Macroscopic picture of diffusion. (*b*) Atomistic view of diffusion. The net flux of gray atoms across an imaginary plane *A* in this crystalline lattice is given by the flux of gray atoms hopping from plane 1 to plane 2 minus the flux of gray atoms hopping from plane 2 to plane 1. Since there are more gray atoms on plane 1 than plane 2, there is a net flux of gray atoms from plane 1 to plane 2. This net flux will be proportional to the *concentration difference* of gray atoms between the two planes.

Examining atomic plane 1 in the figure, we assume that the flux of gray atoms hopping in the forward direction (and therefore through plane *A*) is simply determined by the number (concentration) of gray atoms available to hop times the hopping rate:

$$J_{A+} = \tfrac{1}{2}\Gamma c_1\,\Delta x \tag{4.74}$$

where J_{A+} is the forward flux through plane *A*, Γ is the hopping rate, c_1 is the volume concentration (mol/cm^3) of gray atoms in plane 1, Δx is the atomic spacing required to convert volume concentration to planar concentration (mol/cm^2), and the $\tfrac{1}{2}$ accounts for the fact that on average only half of the jumps will be "forward" jumps. (On average, half of the jumps will be to the left, half of the jumps will be to the right.)

Similarly, the flux of gray atoms hopping from plane 2 backward through plane *A* will be given by

$$J_{A-} = \tfrac{1}{2}\Gamma c_2\,\Delta x \tag{4.75}$$

where J_{A-} is the backward flux through plane *A* and c_2 is the volume concentration (mol/cm^3) of gray atoms in plane 2.

The net flux of gray atoms across plane *A* is therefore given by the difference between the forward and backward fluxes through plane *A*:

$$J_{\text{net}} = \tfrac{1}{2}\Gamma\,\Delta x(c_1 - c_2) \tag{4.76}$$

We would like to make this expression look like the familiar Fick's first law expression for diffusion: $J = -D(dc/dx)$. We can express Equation 4.76 in terms of a concentration gradient as

$$J_{\text{net}} = -\frac{1}{2}\Gamma(\Delta x)^2 \frac{c_2 - c_1}{\Delta x}$$

$$= -\frac{1}{2}\Gamma(\Delta x)^2 \frac{\Delta c}{\Delta x}$$

$$= -\frac{1}{2}\Gamma(\Delta x)^2 \frac{dc}{dx} \quad \text{(for small } x\text{)} \tag{4.77}$$

Comparison with the traditional diffusion equation $J = -D(dc/dx)$ allows us to identify what we call the diffusivity as

$$D = \frac{1}{2}\Gamma(\Delta x)^2 \tag{4.78}$$

We therefore recognize that the diffusivity embodies information about the intrinsic hopping rate for atoms in the material (Γ) and information about the atomic length scale (jump distance) associated with the material.

As mentioned previously, the hopping rate embodied by Γ is exponentially activated. Consider Figure 4.25b, which shows the free-energy curve encountered by an atom as it hops from one lattice site to a neighboring lattice site via a vacancy-based diffusion mechanism. Because the two lattice sites are essentially equivalent, in the absence of a driving force a hopping atom will possess the same free energy in its initial and final positions. However, an activation barrier impedes the motion of the atom as it hops between positions. We might associate this energy barrier with the displacements that the atom causes as it squeezes through the crystal lattice between lattice sites. (See Figure 4.25a, which shows a physical picture of the hopping process.)

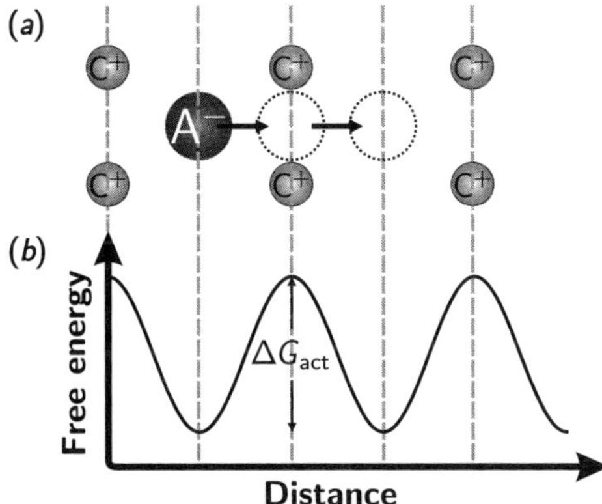

FIGURE 4.25   Atomistic view of hopping process. (a) Physical picture of the hopping process. As the anion (A^-) hops from its original lattice site to an adjacent, vacant lattice site, it must squeeze through a tight spot in the crystal lattice. (b) Free-energy picture of the hopping process. The tight spot in the crystal lattice represents an energy barrier for the hopping process.

In a treatment analogous to the reaction rate theory developed in the previous chapter, we can write the hopping rate as

$$\Gamma = \Gamma_0 e^{-\Delta G_{act}/RT} \tag{4.79}$$

where ΔG_{act} is the activation barrier for the hopping process and Γ_0 is the jump attempt frequency.

Based on this activated model for diffusion, we can then write a complete expression for the diffusivity as

$$D = \frac{1}{2}(\Delta x)^2 \Gamma_0 e^{-\Delta G_{act}/RT} \tag{4.80}$$

or, lumping all the preexponential constants into a D_0 term,

$$D = D_0 \, e^{-\Delta G_{act}/RT} \tag{4.81}$$

Example 4.7

Question: Silicon is an intriguing potential anode material for Li ion batteries because it can store and release more than four Li atoms per Si atom. Quantifying the diffusivity of Li in Si is an important aspect for determining the charge/discharge kinetics of such a battery material. If the diffusivity of Li in Si is 4.0×10^{-14} cm^2/s at 300 K and 2.0×10^{-11} cm^2/s at 400 K, determine ΔG_{act} and D_0 for the diffusion of Li in Si.

Solution: As we have done for a number of problems, the easiest way to solve this problem is to set up a ratio. We have been given two diffusivities, D_1 and D_2, corresponding to two temperatures, T_1 and T_2. Applying these to Equation 4.81, we can write

$$\frac{D_1}{D_2} = \frac{D_0 \, e^{-\Delta G_{act}/RT_1}}{D_0 e^{-\Delta G_{act}/RT_2}}$$

$$\frac{D_1}{D_2} = e^{-(\Delta G_{act}/R)(1/T_1 - 1/T_2)} \tag{4.82}$$

Solving this expression for ΔG_{act} yields

$$\Delta G_{act} = R \frac{T_1 T_2}{T_1 - T_2} \ln \left[\frac{D_1}{D_2} \right] \tag{4.83}$$

Finally, inserting values into this expression gives

$$\Delta G = 8.314 \text{ J/(mol} \cdot \text{K)} \frac{300 \text{ K} \cdot 400 \text{ K}}{300 \text{ K} - 400 \text{ K}} \ln \left[\frac{4.0 \times 10^{-14} \text{ cm}^2/\text{s}}{2.0 \times 10^{-11} \text{ cm}^2/\text{s}} \right]$$

$$= 62{,}000 \text{ J/mol} = 62 \text{ kJ/mol}$$

To determine D_0, we can again make use of Equation 4.81 and the value of D at *either* of the two temperatures. They should both give approximately the same result. We use T_1 and D_1:

$$D_1 = D_0 e^{-\Delta G_{act}/RT_1}$$

$$D_0 = D_1 e^{+\Delta G_{act}/RT_1}$$

$$= (4.0 \times 10^{-14}\ \text{cm}^2/\text{s}) \exp\left(\frac{62{,}000\ \text{J/mol}}{8.314\ \text{J}/(\text{mol} \cdot \text{K}) \cdot 300\ \text{K}} \right)$$

$$= 2.5 \times 10^{-3}\ \text{cm}^2/\text{s}$$

4.5.4 Diffusion in Solids: High-Diffusivity Paths

So far, our discussion of solid-state diffusion has focused exclusively on diffusion through the relatively "perfect" bulk volume of a solid. However, real solids are far from perfect. They typically possess a variety of extended defects, including dislocations, grain boundaries, pores, and surfaces, as shown schematically in Figure 4.26.

In general, the diffusivity scales with the relative degree of constraint facing atoms in the close-packed 3D lattice. Thus, the following relationship is often a good rule of thumb: $D_{\text{lattice}} < D_{\text{dislocation}} < D_{\text{grain boundary}} < D_{\text{surface}}$. On the surface of a solid, the diffusivity can sometimes even approach that of a liquid! The activation energy

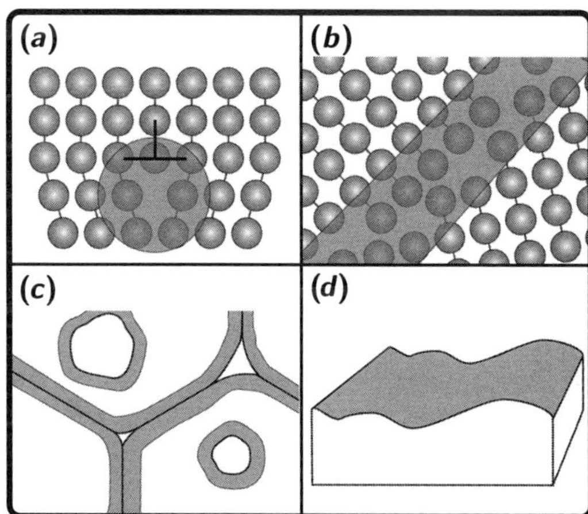

FIGURE 4.26 Schematic illustration of various extended defects in solids including (*a*) dislocations, (*b*) grain boundaries, (*c*) internal pores and (*d*) surfaces. Diffusion through these defective regions often proceeds more rapidly than through the "perfect" bulk volume of the solid, as the atomic disruption and relaxed constraint associated with defective regions lead to faster diffusivities. Shaded regions are areas of enhanced diffusion.

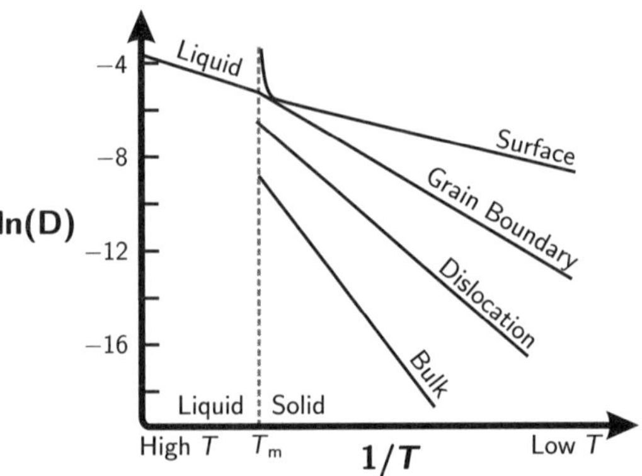

FIGURE 4.27   Schematic Arrhenius plot showing the typical relationships between diffusion through the bulk lattice, along dislocations, grain boundaries, and surfaces. The activation energy for diffusion tends to decrease with decreasing constraint. Thus, transport through defective regions becomes even more important at lower temperatures. At the melting point, the various diffusivities begin to (but do not entirely) converge with the liquid diffusivity.

for diffusion tends to decrease with decreasing constraint. Thus, transport through defective regions becomes even more important at lower temperatures. This is shown schematically in Figure 4.27.

While diffusion through extended defect regions can occur much more rapidly than diffusion through the bulk solid volume, the overall impact of this effect depends on the relative *amount* (density) of extended defects in the solid. A material with 100 nm average grain size will have 2000× more grain boundary area per unit volume of solid than a material with a grain size of 100 μm. Thus, the impact of grain boundary diffusion should be expected to be at least 2000× greater in the nanometer-grained solid. Nevertheless, it is important to keep in mind that the "enhanced region" associated with extended defects typically extends only a few atomic layers from the defect core itself, and thus even if diffusion through these regions is much faster, their overall impact on diffusion through the solid will only be felt if these defects are present at a significant density.

In order to illustrate this point, consider an idealized model of diffusion through a solid with square-shaped grains as shown in Figure 4.28. The total amount of species *i* passing through this slice per unit time is given by the sum of the fluxes passing through the grain bulk (J_{bulk}) and grain boundary (J_{bdry}) regions multiplied by their respective cross-sectional areas. Considering diffusion through a single "representative unit" of this geometry, which includes one grain plus half of the thickness of the surrounding grain boundaries and applying Fick's first law, we have

$$N_{tot} = A_{bulk}J_{bulk} + A_{bdry}J_{bdry}$$

$$= L^2 D_{bulk}\frac{dc}{dx} + 4L\left(\tfrac{1}{2}\delta\right)D_{bdry}\frac{dc}{dx} \tag{4.84}$$

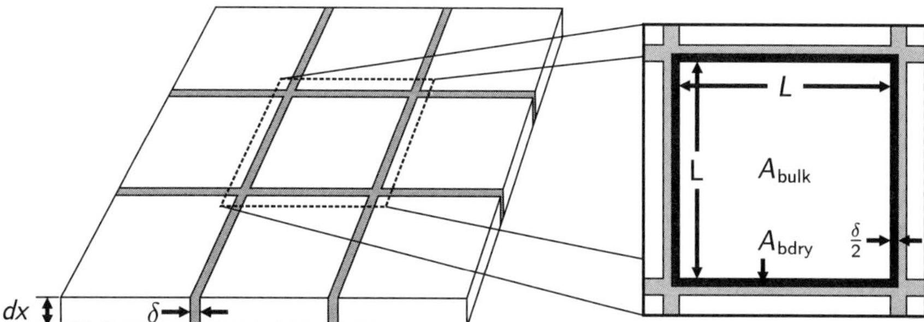

FIGURE 4.28 Diffusion occurring perpendicularly through a slice (of thickness dx) of an idealized solid with square grains of size L separated by grain boundaries of thickness δ. The total amount of species i passing through this slice per unit time is given by the sum of the fluxes passing through the grain bulk (J_{bulk}) and grain boundary (J_{bdry}) regions multiplied by their respective cross-sectional areas.

The relative contribution of grain boundary versus bulk diffusion can be determined by taking a ratio of the two terms in Equation 4.84:

$$\frac{N_{\text{bulk}}}{N_{\text{bdry}}} = \frac{L^2 D_{\text{bulk}}(dc/dx)}{4L\left(\frac{1}{2}\delta\right)D_{\text{bdry}}(dc/dx)} = \frac{LD_{\text{bulk}}}{2\delta D_{\text{bdry}}} \tag{4.85}$$

Grain boundary diffusivities can often be three orders of magnitude (or more) greater than bulk diffusivities, while grain boundary thicknesses are typically on the order of 1–2 nm. Thus, when the grain size approaches a few micrometers, grain boundary diffusion can often become dominant over bulk diffusion. The crossover also depends on the temperature, as the bulk and grain boundary diffusivities have different activation energies. Example 4.8 below extends our simple model to examine these issues in more detail. More complicated heterogeneous diffusion geometries can also be modeled, often using resistor network-type models. Such models can successfully capture the effects of heterogeneous diffusion through a variety of microstructural inhomogeneities, including grain/grain boundary effects, dislocation networks, internal pores, and second-phase inclusions.

There are several important instances where the generalities about high-diffusivity extended defect paths do not hold true. One of the most important is in certain solid-state ion conductors, where grain boundaries often lead to *decreased* ionic diffusion rather than enhanced ionic diffusion. This occurs when the grain boundary traps charge (either positive or negative charge) which can then repel (or trap) diffusing ions, thereby slowing them down compared to their motion in the bulk lattice. Eliminating these "resistive" grain boundaries is a grand challenge for improving the performance of many solid-state electrolyte and ion conductor applications.

Example 4.8

Question: The goal of this problem is to use the simple grain bulk/grain boundary diffusion model discussed above to estimate the temperature above which bulk diffusion becomes more important than grain boundary diffusion. (*a*) Assuming that the activation energy for grain boundary diffusion is exactly half that of bulk diffusion and the D_0 values for both diffusivities are equal, derive a generalized expression for the "crossover" temperature T^* at which bulk diffusion becomes more important than grain boundary diffusion. (*b*) Using this simple model, estimate T^* for a polycrystalline material assuming $\delta = 2.0$ nm, $L = 4.0$ μm, and $\Delta G_{\text{act,bulk}} = 60$ kJ/mol. (*c*) If the grain size is reduced, will T^* increase or decrease?

Solution: (*a*) Inserting the equation for diffusivity (Equation 4.81) into the equation comparing the relative contributions from bulk–grain boundary diffusion (Equation 4.85) gives

$$\frac{N_{\text{bulk}}}{N_{\text{bdry}}} = \frac{LD_0\, e^{-\Delta G_{\text{act,bulk}}/RT}}{2\delta D_0\, e^{-\Delta G_{\text{act,bdry}}/RT}} = \frac{L e^{-\Delta G_{\text{act,bulk}}/RT}}{2\delta e^{-\Delta G_{\text{act,bulk}}/2RT}}$$

$$= \frac{L}{2\delta} e^{-\Delta G_{\text{act,bulk}}/2RT} \tag{4.86}$$

When $T = T^*$, $N_{\text{bulk}} = N_{\text{bdry}}$; thus

$$\frac{N_{\text{bulk}}}{N_{\text{bdry}}} = 1 = \frac{L}{2\delta} e^{-\Delta G_{\text{act,bulk}}/2RT^*}$$

$$T^* = \frac{\Delta G_{\text{act,bulk}}}{2R} \frac{1}{\ln\,[L/2\delta]} \tag{4.87}$$

(*b*) Applying the values given in the problem statement to Equation (4.87) and taking care to use consistent units yields

$$T^* = \frac{60{,}000\ \text{J/mol}}{2 \cdot 8.314\ \text{J/(mol} \cdot \text{K)}} \frac{1}{\ln\,[4.0 \times 10^{-6}\ \text{m}/(2 \cdot 2.0 \times 10^{-9}\ \text{m})]} = 520\ \text{K}$$

(*c*) Intuitively, we can reason that if the grain size is reduced, the amount of grain boundary diffusion increases, and thus grain boundary diffusion will remain dominant up to a higher overall temperature before bulk diffusion overtakes it. Thus, we would expect T^* to increase. The same conclusion can be drawn by examining Equation 4.87, which shows that as L decreases, the natural log term decreases, and thus 1/ln increases as does T^*.

4.6   CHAPTER SUMMARY

The purpose of this chapter was to introduce the basic concepts and tools used to understand and model the rate (or speed) of diffusion processes. The main points introduced in this chapter include the following:

- *Diffusion* is a process by which mass is transported down a gradient in chemical potential (often simplified to a gradient in concentration) by random thermal motion. It is distinguished from alternative mass transport processes such as convection in that it does not require bulk motion or mechanical action to move particles from one place to another.
- *Flux* (J) quantifies the rate at which mass, charge, or energy moves through a material. Flux measures how much of a given quantity flows through a material per unit area per unit time. While the most common units for flux are $mol/(cm^2 \cdot s)$, flux can also be expressed in terms of atomic, mass, volumetric, or coulometric units, among others. It is important to be able to convert easily between the various potential flux units depending on the circumstances.
- A *driving force* is required in order for any type of transport process to occur. The governing equation for transport can be generalized (in one dimension) as

$$J_i = \sum_k M_{ik} F_k \qquad (4.88)$$

where J_i represents a flux of species i, F_k represent the k different forces acting on i, and the M_{ik}'s are the coupling coefficients between force and flux. The coupling coefficients quantify the relative ability of a species to respond to a given force with movement as well as the effective strength of the driving force itself. The coupling coefficients are therefore a property both of the species that is moving and the material through which it is moving. This general equation is valid for any type of transport (charge, heat, mass, etc.) and can be used to capture the coupled effect of multiple driving forces acting simultaneously.
- While the coupling of multiple fluxes and driving forces is important in certain situations, transport can frequently be described more simply in terms of a number of direct one-to-one force–flux relationships. These include Fick's first law (diffusion), Ohm's law (electrical conduction), Fourier's law (heat conduction), and Poiseuille's law (convection), with the corresponding direct coupling coefficients known as the diffusivity (D), conductivity (σ), thermal conductivity (κ), and viscosity (ν), respectively.
- *Fick's first law* of diffusion deals with the diffusional transport of matter under steady-state conditions. Steady state means that the concentration profile of the diffusing species does not vary as a function of time ($\partial c_i / \partial t = 0$). In 1D, Fick's first law is commonly given as $J_i = -D_i(\partial c_i / \partial x)$. Fick's first law indicates that

the flux of a diffusing species i is proportional to its concentration gradient. Fick's first law expresses the fundamental concept that matter tends to flow "down" a concentration gradient, from regions of higher concentration to regions of lower concentration. Furthermore, Fick's first law indicates that the rate of diffusion (i.e., the size of the flux) depends on the magnitude of the diffusivity and the steepness of the concentration gradient. Higher diffusivities and steeper concentration gradients lead to larger fluxes.

- *Fick's second law* of diffusion deals with the diffusional transport of matter under transient (time-dependent) conditions. Transient means that the concentration profile of the diffusing species varies as a function of time ($\partial c_i/\partial t \neq 0$). In 1D, Fick's second law is commonly given as: $\partial c_i/\partial t = D_i(\partial^2 c_i/\partial x^2)$. We can use Fick's second law to "watch" how a non-steady-state diffusion profile evolves as a function of time. Fick's second law indicates that the *rate of change* of a concentration gradient is proportional to its *curvature*. Regions of high curvature (i.e., "sharp" features) evolve quickly, while regions of low curvature evolve more slowly. Furthermore, Fick's second law indicates that a species will accumulate in regions where its concentration profile manifests positive curvature (concave up), while a species will dissipate from regions where its concentration profile curvature is negative (concave down). Thus, Fick's second law predicts that abrupt concentration profile features tend to be smoothed out over time (see Figure 4.4).

- Fick's second law is a second-order partial differential equation. Solving it in order to predict transient diffusion processes can be fairly straightforward or quite complex, depending on the specific situation. In this chapter, analytical solutions were discussed for a number of cases, including 1D transient infinite and semi-infinite diffusion, 1D transient finite planar diffusion, and transient spherical finite diffusion as summarized in Table 4.4. In all cases, solution of Fick's second law requires the specification of a number of boundary conditions and initial conditions.

- *Boundary conditions* provide information about the behavior of a system at the physical edges (i.e., "boundaries") of the problem domain. Thus, boundary conditions specify behavior at a *specific location* (i.e., a specific value of x).

- *Initial conditions* provide information about the initial concentration distribution (i.e., the "starting" concentration profile) within the system at some initial time. Thus, initial conditions specify behavior at a *specific time* (i.e., a specific value of t, usually $t = 0$).

- A characteristic square-root dependence is frequently observed between the spatial extent of a transient diffusion process and the time elapsed, for example, $\delta \approx \sqrt{Dt}$, where δ is a measure of the spatial extent of the diffusion process. This equation can be used as a helpful way to roughly estimate the extent of progress of many different types of transient diffusions processes in materials as a function of time.

- Fick's first and second laws can apply equally well to diffusion in solids, liquids, or gases. The differences between these phases show up in the magnitude of the diffusion coefficient D. The diffusion coefficient, or diffusivity, quantifies the relative ease at which atoms or molecules can be transported via diffusion in a material. Gas-phase diffusivities are typically on the order of 0.1–1.0 cm^2/s, while liquid-phase diffusivities are typically on the order of 10^{-4}–10^{-5}cm^2/s and solid-phase diffusivities range from 10^{-8} cm^2/s (for high-diffusivity solids near their melting point) to 10^{-30} cm^2/s (for refractory ceramics, glasses, or strong covalently bonded network solids far from their melting point). Gas-phase diffusivities show a weak temperature dependence, while solid-phase diffusitivites tend to be exponentially temperature activated.

- Electric field, temperature gradients, and/or stress gradients can act as additional driving forces that can either accelerate or impede diffusion. Fick's laws can be rewritten to accommodate these additional driving forces, resulting in more complex expressions for diffusion that are most often treated numerically.

- The kinetic theory of gases can be used to predict gas-phase diffusivities based on the root-mean-squared speed and mean free path of the gas molecules as they move and collide in the gas phase. This analysis results in the following expression, which relates the self-diffusivity of a gas-phase species to fundamental characteristics such as its size and mass:

$$D_i \approx \sqrt{\frac{1}{6M_i}} \frac{(RT)^{\frac{3}{2}}}{PN_\text{A}\pi d_i^2}$$

  While this equation is algebraically straightforward, *the units often present students with significant difficulties*. Please keep in mind that SI units must be used when evaluating this expression. Thus, M is in kg/mol, d is in m, P is in Pa, T is in K, and so on.

- In crystalline solids, diffusion occurs by the hopping of atoms from lattice site to lattice site. This hopping process can occur by the hopping of either atoms into unoccupied (i.e., "vacant") sites in the lattice or between unoccupied interstitial spaces in the lattice. Interstitial diffusion is more common for small impurity atoms that can fit into the interstitial spaces in the lattice while vacancy diffusion is more common for larger atoms that can only occupy the regular lattice sites.

- Because solid-state diffusion is a hopping process that requires the moving atoms to overcome an activation barrier as they squeeze through the crystal between lattice sites, the hopping rate (and hence the diffusivity) is exponentially temperature dependent:

$$D = D_0 e^{-\Delta G_\text{act}/RT}$$

- In general, solid-state diffusivities scale with the relative degree of constraint facing atoms in the close-packed 3D lattice. Thus, typically

$$D_\text{lattice} < D_\text{dislocation} < D_\text{grain boundary} < D_\text{surface}$$

Since the activation energy for diffusion also tends to decrease with decreasing constraint, transport through defective regions becomes even more important at lower temperatures. While diffusion at extended defects can be greatly enhanced relative to diffusion through the bulk lattice, it is important to keep in mind that the "enhanced region" associated with extended defects typically extends only a few atomic layers from the defect core itself. Thus, the overall impact of extended defects on diffusion will only be felt if these defects are present at a significant density.

4.7 CHAPTER EXERCISES

Review Questions

Problem 4.1. Define the following:

(a) Diffusion

(b) Steady state

(c) Transient

(d) Self-similarity

(e) Boundary condition

(f) Initial condition

(g) Kirkendal effect

Problem 4.2. Equation 4.40 in the text provides the solution for the transient diffusion of a "thin layer" of material between two semi-infinite bodies:

$$c_i(x, t) = \frac{N_i}{\sqrt{4\pi Dt}} e^{-x^2/(4Dt)} \tag{4.89}$$

This thin-film solution yields a Gaussian profile that gradually broadens as a function of time. Derive a mathematical expression quantifying how this Gaussian profile "broadens" as a function of time. Use the peak width at half maximum (i.e., the distance between the two points where the concentration is one-half of its peak value at any given time) as the definition of peak breadth.

Problem 4.3. Equation 4.43 provides the solution for transient finite (symmetric) planar diffusion in a plate of thickness L starting from a uniform initial concentration of c_i° when the concentrations at the edges of the plate are set to c_i^* at time $t = 0$:

$$c_i(x, t) = c_i^* + (c_i^\circ - c_i^*)\frac{4}{\pi} \sum_{j=0}^{\infty} \left(\frac{1}{2j+1} \sin\left[(2j+1)\pi\frac{x}{L}\right] e^{-(2j+1)^2\pi^2 Dt/L^2} \right)$$

At "long times," this solution can be reasonably approximated by using only the first term in series without significant error. In order to determine a reasonable value for "long time" and to visualize the effect that this approximation has on the solution compared to retaining additional terms in the series, use a software program of your choice (e.g., Excel, Matlab, Mathematica) to make a set of plots for this solution using $j = 1, j = 3$, and $j = 10$ terms in the series at each of the following three times:

(a) $t = 0.01[L^2/(\pi^2 D)]$

(b) $t = 0.1[L^2/(\pi^2 D)]$

(c) $t = 1[L^2/(\pi^2 D)]$

Assume $c_i^* = 0$. Comment on the results.

Problem 4.4. For the moving interface involving the eutectic system discussed in Section 4.4.3, Equation 4.57 was provided to calculate the velocity of the interface. For this system, derive the criteria under which the interface will move to the right (i.e., the interface will move toward the α phase).

Problem 4.5.

(a) Provide the general equation that relates gas-phase diffusivity to gas velocity and mean free path.

(b) Define mean free path and give the equation relating mean free path to temperature, pressure, and molecule size.

(c) Define the root-mean-square gas velocity and give an equation that shows how it depends on temperature and molecular weight.

(d) Sketch curves showing the distribution of velocities for gas molecules at two temperatures, T_1 and T_2 ($T_2 > T_1$). Indicate v_{RMS} on both curves.

(e) Combine the equations for mean free path and root-mean-squared velocity to arrive at the general equation describing how gas-phase diffusivity depends on temperature and pressure.

Problem 4.6. Based on your knowledge of the approximate order of magnitude for gas-versus liquid-versus solid-state diffusivity, identify the most likely (only one) value for diffusivity for each of the following situations:

(a) The self-diffusion coefficient of water molecules in water at room temperature is about:

(1) 10^{-16} cm^2/s; (2) 10^{-8} cm^2/s; (3) 10^{-4} cm^2/s; (4) 0.1 cm^2/s; (5) 100 cm^2/s

(b) The diffusion coefficient of water molecules in air at $28\,^{\circ}$C is about:

(1) 10^{-16} cm^2/s; (2) 10^{-8} cm^2/s; (3) 10^{-4} cm^2/s; (4) 0.1 cm^2/s; (5) 100 cm^2/s

(c) The diffusion coefficient of oxygen vacancies in solid ZrO_2 at 300 K is about:

(1) 10^{-16} cm^2/s; (2) 10^{-8} cm^2/s; (3) 10^{-4} cm^2/s; (4) 0.1 cm^2/s; (5) 100 cm^2/s

(d) The diffusion coefficient of oxygen vacancies in solid ZrO_2 at 1000 K is about:

(1) 10^{-16} cm^2/s; (2) 10^{-8} cm^2/s; (3) 10^{-4} cm^2/s; (4) 0.1 cm^2/s; (5) 100 cm^2/s

(e) The diffusion coefficient of Cu atoms in liquid copper at its melting point is about:

(1) 10^{-16} cm^2/s; (2) 10^{-8} cm^2/s; (3) 10^{-4} cm^2/s; (4) 0.1 cm^2/s; (5) 100 cm^2/s

Calculation Questions

Problem 4.7. A fuel cell is provided with a molar flux of H_2, $J_{H_2} = 10^{-5}$ mol H_2/cm^2s. Convert this molar H_2 flux into the following sets of units, given $T = 300$ K, $P = 1$ atm. Assume H_2 behaves as an ideal gas.

(a) Volumetric flux $[cm^3/(cm^2 \cdot s)]$

(b) Mass flux $[g/(cm^2 \cdot s)]$

(c) Molecular flux $[molecules/(cm^2 \cdot s)]$

(d) Current density (A/cm^2) (Assume each hydrogen molecule is oxidized in a fuel cell to produce two electrons.)

Problem 4.8. The aim of carburization is to increase the carbon concentration in the surface layers of steel in order to achieve a harder wear-resistant surface. This is usually done by holding the steel in a gas mixture containing CH_4 and/or CO at a temperature where austenite is present. A thick plate of 0.3 wt % C steel is carburized at 930 °C. The carburizing gas used in the treatment holds the surface concentration at 1.0 wt % C. Assume the plate is so thick that the diffusing carbon does not reach the opposite side of the plate over the course of the treatment. Given that the diffusivity of carbon in iron at this temperature is $D = 1.1 \times 10^{-9}$ m²/s, determine how long it will take for the carbon concentration to reach 0.6 wt % C at a depth of 0.3 mm?

Problem 4.9. Heat transfer obeys the same basic laws as diffusion. For example, a spherical transient heat transfer process can be described with the following equation:

$$\frac{\partial T}{\partial t} = \frac{\alpha}{r^2} \frac{\partial}{\partial r} \left[r^2 \frac{\partial T}{\partial r} \right] \tag{4.90}$$

where α is the thermal diffusivity (units of cm²/s). The military has asked you to properly size the insulation thickness around a delicate sensor that must survive submersion in freezing water ($T_f = 273$ K) for a length of time given by t_c. The sensor is surrounded by spherical insulation of thickness δ. The insulation has a thermal diffusivity α. The initial temperature of the sensor unit is T_0. The sensor (located at the center of the sphere) will fail if its temperature drops below a critical temperature T_c. Assume the sensor itself occupies negligible volume.

(a) Provide a sketch for this problem with as much detail as possible about the problem geometry, the boundary and initial conditions, and your expectation for how the temperature profile inside the spherical apparatus will evolve in time.

(b) How is steady state mathematically defined for this process?

(c) Does this problem require steady-state or transient analysis?

(d) Is this a finite, semi-infinite, or infinite boundary value problem?

(e) Provide two boundary conditions and one initial condition for this problem.

(f) Based on analogy to spherical mass diffusion, provide the approximate solution $[T(r, t)]$ to this problem at long times.

(g) Provide the approximate long-time solution for the temperature at the sensor $[T(r = 0, t)]$ as a function of t, α, δ_c, T_0, and T_f.

(h) Provided that the sensor temperature must not drop below T_c after a length of time t_c, solve this equation in terms of δ_c, the minimum thickness of insulation required.

(i) Given $t_c = 2.0$ h, $T_c = 17\,^\circ\mathrm{C}$ (290 K), $T_0 = 27\,^\circ\mathrm{C}$ (300 K), $T_f = 0\,^\circ\mathrm{C}$ (273 K), and $\alpha = 0.10$ cm^2/s, what is δ_c in cm?

Problem 4.10.   Steel often contains trace amounts of H_2, which can lead to embrittlement. To avoid embrittlement, steel is often degassed prior to use in order to remove these trace H_2 impurities. Degassing steel involves placing the steel in a vacuum, where the H_2 concentration in the vacuum can be considered to be zero at all times. Degassing proceeds by three steps: (1) solid-state diffusion of H_2 from the steel bulk to the steel surface; (2) desorption of H_2 from the surface of the steel; (3) gas-phase diffusion of the H_2 away from the steel surface.

(a) Based on what you know about the typical rates of the three steps involved in the steel degasification process, which of these three steps is likely to be the rate-determining step and why?

(b) Consider a steel plate that initially contains a uniform H_2 concentration $(c_{H_2}^\circ)$ of 10^{15} molecules/cm^3, as shown in Figure 4.29. Assuming that desorption of H_2 from the surface of the steel is the rate-determining step in steel degasification, draw a diagram (similar to Figure 4.29) illustrating the H_2 concentration profile across the steel plate at five times: $t = 0 < t_1 < t_2 < t_3 < t_\infty$.

(c) Assuming that solid-state diffusion of H_2 from the bulk to the surface of the steel is the rate-determining step in steel degasification, draw a diagram (similar to Figure 4.29) illustrating the H_2 concentration profile across the steel plate at five times: $t = 0 < t_1 < t_2 < t_3 < t_\infty$.

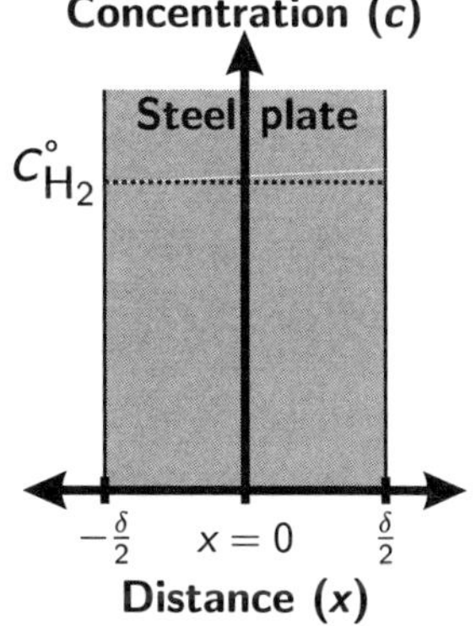

FIGURE 4.29   Schematic diagram of the cross section of a steel plate of thickness δ with a constant H_2 concentration of $c_{H_2}^\circ$.

(d) Assuming that solid-state H_2 diffusion is the rate-determining step, is this a finite, semi-infinite, or infinite boundary value diffusion problem?

(e) Assuming that solid-state H_2 diffusion is the rate-determining step, provide the two boundary conditions and one initial condition for this transient diffusion problem. Note that $x = 0$ corresponds to the center of the steel plate. Assume the steel plate is 1 mm thick.

(f) By employing your boundary conditions and initial conditions to evaluate this transient diffusion problem, you arrive at the following solution for the time-dependent concentration of H_2 at the center of the steel plate $[c(x = 0, t)]$:

$$c(0, t) = c_{H_2}^{\circ} e^{-Dt(\pi/\delta)^2}$$

where δ is the thickness of the steel plate and $c_{H_2}^{\circ}$ is the initial $(t = 0)$ concentration of H_2 in the steel plate. The H_2 embrittlement is avoided if the H_2 concentration in the steel is reduced below 10^{13} molecules/cm^3. If the diffusivity of H_2 in steel under degassing conditions is 10^{-6} cm^2/s, how long must the steel be degassed to ensure the H_2 concentration at the center of the plate falls below 10^{13} molecules/cm^3? Assume the steel plate is 1 mm thick.

(g) How long must a 4-mm-thick steel plate be degassed to avoid H_2 embrittlement?

Problem 4.11. The experimentally measured diffusion coefficients as a function of temperature for hydrogen diffusion in SiO_2 are given in Table 4.5.

TABLE 4.5 Hydrogen Diffusion in SiO_2

T (K)	D (cm^2/s)	T (K)	D (cm^2/s)	T (K)	D (cm^2/s)
373	8.10×10^{-10}	873	1.50×10^{-6}	1373	1.30×10^{-5}
473	9.99×10^{-9}	973	4.21×10^{-6}	1473	1.68×10^{-5}
573	3.22×10^{-8}	1073	4.54×10^{-6}	1573	3.32×10^{-5}
673	2.58×10^{-7}	1173	6.85×10^{-6}	1673	2.56×10^{-5}
773	3.20×10^{-7}	1273	7.19×10^{-6}	1773	2.85×10^{-5}

(a) Attach a plot of $\log D$ versus $1/T$ (K^{-1}).

(b) From these data, calculate ΔG_{act} (kJ/mol).

(c) From these data, calculate D_0 (cm^2/s).

(d) Calculate the value of D (cm^2/s) at $27\,^{\circ}$C.

Problem 4.12.

(a) Calculate the velocity (cm/s) of nitrogen molecules ($M_{N_2} = 28$ g/mol) at 1 atm at $1000\,^{\circ}$C.

(b) Calculate the number of nitrogen molecules per cubic centimeters at $1000\,^{\circ}$C and 1 atm.

(c) Calculate the mean free path (μm) of nitrogen molecules at $1000\,^{\circ}$C and 1 atm.

(d) Calculate the nitrogen diffusion coefficient at $1000\,^{\circ}$C and 1 atm.

Problem 4.13. For interstitial lithium ion diffusion in pure SiO_2 (silica) glass, the following important diffusion parameters are given: $D_0 = 0.24$ cm^2/s and $\Delta G_{act} = 34.2$ kJ/mol.

(a) Calculate the diffusion coefficient (cm^2/s) of lithium ion impurities in silica at 900 °C.

(b) Make a sketch of free energy versus distance for an interstitial lithium ion atom making a jump from one interstitial site in silica glass to another interstitial site.

Problem 4.14. Using Equation 4.73, calculate the binary O_2/N_2 gas diffusivity at $T = 300$ K and compare it to the pure O_2 self-diffusivity. The molecular diameter of N_2 is ≈ 3.2 Å (slightly larger than O_2).

PART II

APPLICATIONS OF MATERIALS KINETICS

CHAPTER 5

GAS–SOLID KINETIC PROCESSES

The first half of this textbook introduced the basic tools needed to understand most kinetic processes. Specifically, we learned how to calculate the main thermodynamic driving forces behind kinetic transformations (Chapter 2), we learned how to calculate the rates of reaction processes (Chapter 3), and we learned how to calculate the rates of transport processes (Chapter 4). In the second half of this textbook, we will use these tools to model and understand a number of real-world kinetic processes involving gas–solid, solid–liquid, and solid–solid transformations. In this chapter, we begin with gas–solid kinetic processes.

5.1 ADSORPTION/DESORPTION

Gas–solid kinetic processes are fundamentally *heterogeneous* as they involve both a gas phase and a solid phase. As a gas–solid kinetic process proceeds, atoms must pass from the gas phase to the solid phase or vice-versa. Thus, one question of fundamental importance in the study of gas–solid kinetic processes is: *How fast can atoms from the gas phase impinge upon a solid surface, or conversely, how fast can atoms from the solid surface evaporate into the gas phase?* In other words, we wish to know, in the absence of other limiting kinetic factors, what is the maximum rate at which atoms can move from the gas to the solid phase or vice-versa.

The answer to this question has its roots in the kinetic theory of gases, which we introduced in Chapter 4. The flux of atoms impinging on a solid surface per unit area of surface per unit time, J_s, is determined by the density (ρ) of the gas and the average velocity (v_{AVG}) of the gas:

$$J_s = \rho \frac{v_{AVG}}{4} \tag{5.1}$$

The factor of 4 in the denominator of this expression accounts for the fact that only a certain fraction of the average gas molecule's movement occurs in the direction perpendicular to the solid surface and thus contributes to the flux impinging upon the surface.

Applying the ideal gas law and the expression for v_{AVG} (Equation 4.67) yields

$$J_s = \frac{P}{4RT} \sqrt{\frac{8RT}{\pi M}}$$

$$= \frac{P}{\sqrt{2\pi MRT}} \tag{5.2}$$

Thus, the molar flux of gas impinging on the surface of a solid depends on the pressure and temperature of the gas as well as the molecular weight (M) of the gas. As with previous expressions involving the kinetic theory of gases, it is important to recall that SI units should be used with this expression to avoid error.

Adsorption Rate   In general, the actual rate at which atoms (or molecules) build up on the surface of a solid will differ from the impingement rate given by expression 5.2 above because not all gas molecules impinging on the surface will stick (or react). For certain relatively simple cases, this issue can be taken into account by incorporating a sticking coefficient (α) into Equation 5.2:

$$J_s' = \alpha J_s = \frac{\alpha P}{\sqrt{2\pi MRT}} \tag{5.3}$$

The sticking coefficient is a number between 0 and 1. It quantifies the fraction of impinging gas atoms that stick to the surface. If $\alpha = 0$, the surface is *perfectly reflective* and no impinging gas atoms stick to the surface. If $\alpha = 1$, the surface is *perfectly adsorbing* and all impinging gas atoms stick to the surface.

When applying Equation 5.3 to calculate gas adsorption on a solid surface, it is important to recognize that this expression assumes that the surface does not show a limiting adsorption behavior—in other words, the "sticking" probability is independent of surface coverage. Note that this is fundamentally different from the Pt/CO poisoning surface reaction example that we examined in Chapter 3 (Section 3.4.2). With a constant-sticking-coefficient model, gas molecules can continue to "stick" to the surface even after one or more monolayers of molecules have deposited. Thus, this treatment is not appropriate for modeling all types of gas–solid interactions. It is most appropriate for modeling the condensation/evaporation of gas molecules onto/from a solid surface of the same substance (e.g., high-temperature evaporation or condensation of W atoms from a W filament in a lightbulb). It is also highly useful for modeling gas adsorption/desorption processes occurring in a vacuum, where the concentration of atoms in the gas phase is very low, and hence the flux of those atoms to the surface controls the rate of adsorption/desorption.

Desorption Rate   The maximum rate of evaporation (desorption) from a surface is also described by Equation 5.3. This may not seem very intuitive at first. However, consider a surface that is in equilibrium with the gas above it. At equilibrium, the number of atoms evaporating from the surface per unit time must be equal to the number of atoms that strike the surface and stick to it per unit time, and both of these rates are described by Equation 5.3. The pressure of the gas above the surface in this case must be the equilibrium vapor pressure for the gas (P^{eq}) at the temperature at which the process is occurring (otherwise, the adsorption and desorption rates will not be in equilibrium). Now, consider that the gas is completely removed—in other words, replaced with a vacuum. Atoms will still be evaporating from the surface at the rate given by Equation 5.3 even though new atoms will no longer be impinging on the surface. Thus, the maximum rate of evaporation of a surface into a vacuum must be given by

$$J_s' = \frac{\alpha P^{eq}}{\sqrt{2\pi MRT}} \tag{5.4}$$

where P^{eq} is the equilibrium pressure (vapor pressure) of the evaporating species at the surface temperature.

Impingement versus Diffusion

The impingement rate discussed here characterizes the maximum intrinsic rate at which gas molecules immediately above a solid can strike the surface of a solid. The impingement rate does not take concentration gradients into effect, and thus it represents the maximum rate of transport for a gas-phase species to a surface when diffusion is unimportant. When calculating the rate of transport of a gas-phase species to a solid surface, the decision to use the impingement rate versus the diffusion equation depends on the length scale of the transport relative to the mean free path of the transporting gas molecules.

The impingement rate is associated with *ballistic* transport (i.e., direct line-of-sight transport) where the atoms can transport directly to the solid surface without undergoing multiple collisions in the gas phase and without a spatial variation in concentration. Under vacuum pressures and for small distances (where the mean free path is smaller than the distance of travel), direct line-of-sight impingement can determine the transport rate and hence the impingement rate can be used to quantify the flux of a species to a surface. However, at higher pressures, where molecules undergo many collisions along their journey and their concentration varies spatially, transport is instead determined by diffusion and Fick's laws should be used to quantify the flux. Figure 5.1 schematically illustrates the differences between these two transport modes.

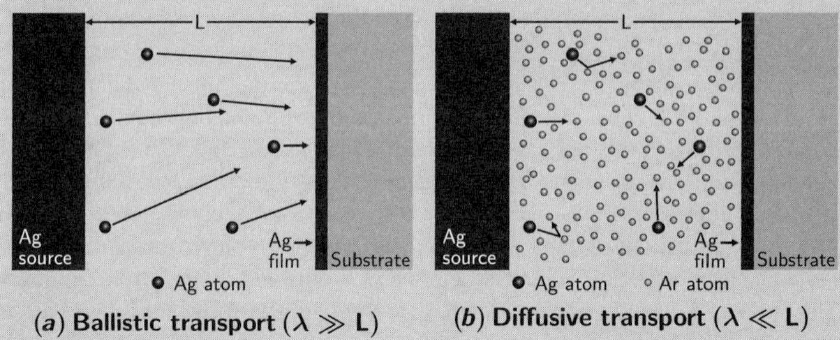

FIGURE 5.1   Schematic illustration of the evaporation of Ag vapor from a source and its transport (via the gas phase) to a substrate under two limiting conditions. (*a*) Ballistic transport under high-vacuum conditions where the direct line-of-sight impingement of the Ag atoms from the source to the substrate determines the transport rate. (*b*) Diffusive transport in a 1 atm inert gas environment where Fick's first law of diffusion determines the transport rate.

Example 5.1

Question: Silver (Ag) is being evaporated from a large-area source for deposition onto window glass as part of a low-*e* (low-emissivity) coating process that greatly improves the energy efficiency of the window. The Ag source is held at a temperature of 1300 °C, yielding an equilibrium Ag vapor pressure of 10^{-3} atm. Assuming deposition from the Ag source to the window glass (separated from one another by 10 cm), calculate the maximum expected rate of deposition [mol/(cm^2·s)] under the following two scenarios: (a) The Ag is deposited in a vacuum deposition process, where the deposition rate is controlled solely by the rate of evaporation from the Ag source. (b) The Ag is deposited under atmospheric conditions in an inert argon gas environment. In this case, because the chamber pressure is much greater than the Ag vapor pressure, the deposition rate is controlled by the diffusion of the Ag vapor through the Ar gas. In both scenarios, assume that any Ag atoms which reach the glass surface stick. For part (b), assume the diffusivity of Ag(g) in Ar(g) is 0.10 cm^2/s.

Solution: (a) In this first case, both the rate of Ag evaporation from the source and the rate of deposition onto the window glass are equal and given by the impingement rate formulation, since the flux of Ag atoms is controlled by ballistic line-of-sight transport. Assuming that both the source and the glass substrate are large in area relative to their distance of separation, there should be no angular dependence on the transport. Applying Equation 5.4, the flux of Ag atoms evaporating from the source is

$$
J'_{s,Ag} = \frac{\alpha P^{eq}_{Ag}}{\sqrt{2\pi M_{Ag} R T}}
$$

$$
= \frac{(1 \cdot 10^{-3}\ \mathrm{atm})(101{,}300\ \mathrm{Pa}/1\ \mathrm{atm})}{\sqrt{2\pi \cdot 0.1079\ \mathrm{kg/mol} \cdot 8.314\ \mathrm{J/(mol \cdot K)} \cdot 1573\ \mathrm{K}}}
$$

$$
= 1.08\ \mathrm{mol/(m^2 \cdot s)} = 1.08 \times 10^{-4}\ \mathrm{mol/(cm^2 \cdot s)}
$$

Thus, this also represents the deposition flux onto the window glass substrate.

(b) In this second case, the rate at which the Ag atoms arrive at the window glass is controlled by their rate of diffusion through the gas separating the source from the substrate. At the evaporation source, the partial pressure of Ag atoms in the gas phase is given as 10^{-3} atm. At the glass substrate, the partial pressure of Ag atoms in the gas phase is essentially zero, since all Ag atoms that arrive to the substrate stick. Thus, the Ag atoms diffuse through the Ar gas environment from the Ag source to the glass substrate down their concentration gradient. We can use Fick's first law to calculate the steady-state flux of Ag atoms to the substrate:

$$
J_{\mathrm{diff},Ag} = -D_{Ag/Ar} \frac{dc_{Ag}}{dx}
$$

$$
= -D_{Ag/Ar} \frac{c_{Ag,glass} - c_{Ag,source}}{\Delta x} \tag{5.5}
$$

We are told that $D_{Ag/Ar} = 0.10\ \mathrm{cm^2/s}$. If this information had not been provided, we could have estimated $D_{Ag/Ar}$ using Equation 4.73 from Chapter 4. We can use the ideal gas law to calculate $c_{Ag,source}$ based on P^{eq}_{Ag} at the source:

$$
c_{Ag,source} = \frac{n_{Ag}}{V} = \frac{P^{eq}_{Ag}}{RT}
$$

$$
= \frac{(10^{-3}\ \mathrm{atm})(101{,}300\ \mathrm{Pa}/1\ \mathrm{atm})}{8.314\ \mathrm{J/(mol \cdot K)} \cdot 1573\ \mathrm{K}}
$$

$$
= 7.75 \times 10^{-3}\ \mathrm{mol/m^3} = 7.75 \times 10^{-9}\ \mathrm{mol/cm^3}
$$

Finally, inserting the values for $D_{Ag/Ar}$, $c_{Ag,source}$, and $c_{Ag,glass} = 0$ into Equation 5.5 yields

$$J_{diff,Ag} = -0.10 \text{ cm}^2/\text{s} \cdot \frac{0 - 7.75 \times 10^{-9} \text{ mol/cm}^3}{10 \text{ cm}}$$

$$= 7.75 \times 10^{-11} \text{ mol}/(\text{cm}^2 \cdot \text{s})$$

Thus, the flux of Ag onto the window glass substrate when deposited in 1 atm Ar gas is more than *six orders of magnitude lower* than the flux in vacuum! This huge difference in deposition rates is a key reason why deposition processes such as evaporation, sputtering, and pulsed-laser deposition are carried out in vacuum.

It is important to understand that the crossover between the rate-limiting behavior described in scenario (a) versus that in (b) is determined by the mean free path of Ag atoms in the chamber relative to the distance separating the Ag source from the window substrate. If the mean free path of the Ag atoms is much greater than the distance between the source and the substrate (e.g., as may happen under vacuum deposition), then the Ag atoms can rapidly travel without collision in a direct line-of-sight deposition process from the source to the substrate. In this case, the deposition rate is controlled solely by the rate at which Ag is evaporated from the source and the geometry of the problem. On the other hand, if the mean free path of the Ag atoms is much smaller than the distance between the source and the substrate (e.g., if the chamber is filled with inert Ar gas), then the Ag atoms will undergo many collisions on their way from the source to the substrate and so the deposition rate will be limited by the rate of diffusion of the Ag atoms through the gas phase. Under this condition, deposition will be much slower. This type of crossover between two different rate-controlling kinetic processes is a common phenomenon that we will encounter throughout this chapter.

Surface Reactions at Atomic Scale

At the atomic scale, the surface of a solid is a diverse and fascinating place. As schematically illustrated in Figure 5.2, solid surfaces are generally not perfectly flat but contain a number of important features such as steps and ledges, kink sites, surface vacancies (or "holes"), and surface adatoms. When atoms are deposited onto or removed from the surface of a solid, the process will often proceed along such features, because these features represent higher energy sites that possess dangling bonds and thus are more susceptible to reaction. In fact, it is quite common for evaporation of a surface to proceed row by row along step edges, as the step edge atoms evaporate more easily. Thus, the surface gradually "unzips" as these step edges are etched away atom by atom. Another interesting phenomenon is the "spiral" growth or etching of atomic layers at sites corresponding to the surface termination of screw dislocations (not shown in the figure), which represent another type of high-energy surface site.

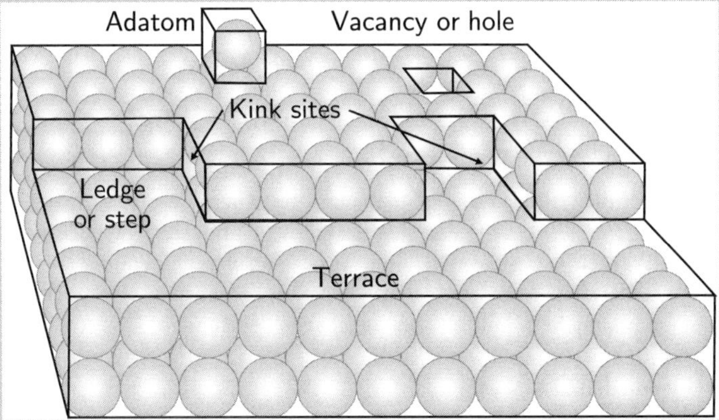

FIGURE 5.2 Schematic illustration of a surface at the atomic scale. Atomic-scale features of a surface include steps or ledges, kinks, adatoms, and vacancies. Surface reaction processes such as adsorption and desorption often proceed along step edges or at kink sites, as these locations represent higher energy sites on the surface with a larger number of dangling bonds.

5.2 ACTIVE GAS CORROSION

In the previous section, we considered one of the most basic gas–solid kinetic processes: the simple adsorption or desorption of atoms to/from a surface under the assumption that the rate is limited by the impingement of atoms from the gas phase to the surface. In this section, we consider a more complex situation in which a gas species actively *etches* or *corrodes* a solid surface via a chemical reaction process, thereby continuously removing material from the surface over time. Consider, for example, the corrosion of a Ti metal surface with HCl acid vapor:

$$\text{Ti(s)} + 4\text{HCl(g)} \leftrightarrow \text{TiCl}_4(\text{g}) + 2\text{H}_2(\text{g}) \tag{5.6}$$

As illustrated in Figure 5.3, this process involves three sequential steps:

1. Transport of HCl gas reactants to the Ti surface
2. Surface reaction between the HCl gas and Ti, creating TiCl_4 and H_2 gas
3. Transport of the $\text{TiCl}_4(\text{g})$ and $\text{H}_2(\text{g})$ products away from the surface

The slowest of the three steps will control the overall rate of the process. At low temperatures, the surface reaction (step 2) may be the slowest. However, since reaction

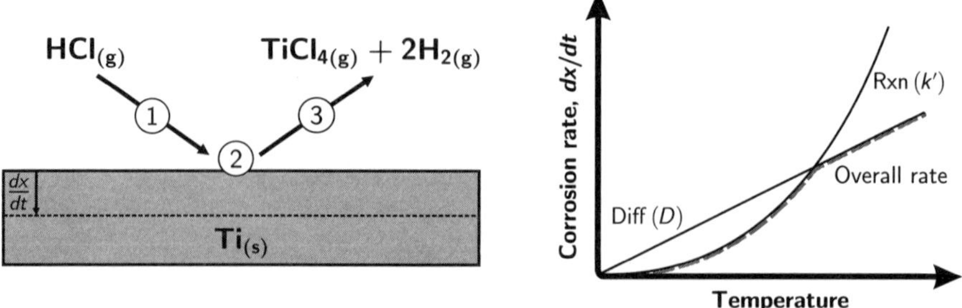

FIGURE 5.3 Schematic illustration of the active gas corrosion of a Ti metal surface with HCl gas. This process involves three sequential steps: (1) transport of HCl gas reactants to the Ti surface; (2) reaction between HCl and Ti on the surface, creating $TiCl_4$ and H_2 gas; and (3) transport of the $TiCl_4$ and H_2 gas products away from the surface. As shown in the accompanying graph, the rate-limiting step can change depending on the conditions.

rates generally increase exponentially with increasing temperature while gas-phase diffusion only increases weakly with temperature, diffusion of reactants/products to/from the surface may become rate limiting at high temperatures.

Surface Reaction Control We will first examine the behavior of this system under the assumption that the surface reaction step (step 2) is the slowest step and hence controls the overall corrosion rate. To start, we must write the rate law for this surface reaction. We assume that the reaction is first order with respect to the HCl reactant. Recalling Chapter 3, we can write the rate law for this first-order reaction as

$$\frac{dc_{HCl}}{dt} = -kc_{HCl} \tag{5.7}$$

In this case, the reaction takes place along a surface rather than homogeneously throughout the volume of a gas. Thus, the homogeneous reaction rate law given by Equation 5.7 should be transformed into a heterogeneous reaction rate law that describes the rate of consumption of the HCl gas per unit area of surface [e.g., mol/(cm²·s)]. This transformation is accomplished by employing a heterogeneous rate constant in Equation 5.7, k', which has units of length/time (whereas k has units of time^{-1}):

$$J_{HCl} = -k'c_{HCl}^{\circ} \tag{5.8}$$

where c_{HCl}° represents the concentration of HCl(g) that is supplied to the reaction. In this heterogeneous rate expression, the rate of consumption of HCl is now expressed per unit area of surface per unit time. Note that this area-specific consumption rate essentially represents the *flux* of HCl reacting on the surface of the solid.

During an active gas corrosion process, one often wants to know how quickly the solid surface is being etched away. Based on the rate at which the HCl(g) is consumed, we can determine the rate at which the Ti(s) is consumed using the reaction stoichiometry. We know that for every 4 mol of HCl(g) consumed, 1 mol of Ti(s) is consumed. Thus, the Ti is consumed at one-fourth the rate of the HCl:

$$J_{\text{Ti}} = \frac{1}{4} J_{\text{HCl}} = -\frac{k' c^{\circ}_{\text{HCl}}}{4} \tag{5.9}$$

This can be further converted into an etching rate (i.e., the decrease in the thickness of the Ti solid per unit time, dx/dt) via some simple algebraic transformations using the density and molecular weight of the Ti. Consider first that the number of moles of Ti (n_{Ti}) in a Ti plate of thickness x and area A can be calculated as

$$n_{\text{Ti}} = V_{\text{Ti}} \frac{\rho_{\text{Ti}}}{M_{\text{Ti}}} = Ax \frac{\rho_{\text{Ti}}}{M_{\text{Ti}}} \tag{5.10}$$

Using this relation and recognizing that J_{Ti} represents the moles of Ti consumed per unit area (A) of surface per unit time, J_{Ti} can be converted into an etching rate (dx/dt):

$$J_{\text{Ti}} = \frac{1}{A} \frac{dn_{\text{Ti}}}{dt} = \frac{1}{A} \frac{d[Ax(\rho_{\text{Ti}}/M_{\text{Ti}})]}{dt} = \frac{\rho_{\text{Ti}}}{M_{\text{Ti}}} \frac{dx}{dt}$$

$$\frac{dx}{dt} = \frac{M_{\text{Ti}}}{\rho_{\text{Ti}}} J_{\text{Ti}} \tag{5.11}$$

Finally, inserting the Ti reaction rate law (Equation 5.9) into this expression allows for the etching rate to be calculated as a function of the heterogeneous reaction rate constant and the HCl(g) concentration:

$$\frac{dx}{dt} = \frac{M_{\text{Ti}}}{\rho_{\text{Ti}}} J_{\text{Ti}}$$

$$= -\frac{M_{\text{Ti}}}{\rho_{\text{Ti}}} \frac{k' c^{\circ}_{\text{HCl}}}{4} \tag{5.12}$$

Typically, the HCl gas concentration will be specified as a partial pressure rather than a molar concentration. Recalling the ideal gas law, the HCl concentration can be converted to HCl pressure to obtain a final expression for the Ti etching rate as

$$\frac{dx}{dt} = -\frac{M_{\text{Ti}}}{\rho_{\text{Ti}}} \frac{k' P^{\circ}_{\text{HCl}}}{4RT} \tag{5.13}$$

This expression indicates that the Ti etching rate is constant; thus we would expect the thickness of a Ti plate undergoing active gas corrosion to decrease linearly with time. The expression also indicates that the etching rate will increase linearly with increasing HCl(g) pressure. Since HCl is the etchant, this makes sense! The effect of temperature is less obvious. While temperature appears directly in the denominator of this expression, recall that the rate constant k' is an exponentially temperature-activated quantity. Thus, the exponential increase in k' with increasing temperature dominates over the T^{-1} term in this expression; the overall effect is that the etching rate will increase rapidly with increasing temperature.

Example 5.2

Question: (a) Calculate the etching rate of a Ti surface undergoing active gas corrosion in HCl(g) assuming that the surface reaction is rate controlling. (b) Under these conditions, how much time would be required to etch 1 mm deep into a Ti plate? The following information is provided:

- $k'_0 = 1.2 \times 10^2$ cm/s
- $\Delta G_{act} = 50$ kJ/mol
- $P_{HCl} = 0.01$ atm
- $M_{Ti} = 47.9$ g/mol
- $\rho_{Ti} = 4.5$ g/cm^3
- $T = 1500$ K

Solution: (a) This is a fairly straightforward plug-and-chug exercise. First, we apply Equation 3.70 to determine the value of the reaction rate constant (k') at 1500 K:

$$k' = k'_0 \exp\left(-\frac{\Delta G_{act}}{RT}\right)$$

$$= (1.2 \text{ m/s}) \exp\left(-\frac{50,000 \text{ J/mol}}{8.314 \text{ J/(mol} \cdot \text{K)} \cdot 1500 \text{ K}}\right)$$

$$= 0.022 \text{ m/s}$$

Then, applying this value together with the other provided quantities to Equation 5.13 (paying careful attention to SI units!) gives

$$\frac{dx}{dt} = -\frac{M_{Ti}}{\rho_{Ti}} \frac{k' P^\circ_{HCl}}{4RT}$$

$$= -\frac{0.0479 \text{ kg/mol}}{4500 \text{ kg/m}^3} \cdot \frac{0.022 \text{ m/s} \cdot 0.01 \text{ atm} \frac{101,300 \text{ Pa}}{1 \text{ atm}}}{4 \cdot 8.314 \text{ J/(mol} \cdot \text{K)} \cdot 1500 \text{ K}}$$

$$= -4.8 \times 10^{-9} \text{ m/s} = -4.8 \times 10^{-7} \text{ cm/s} = -4.8 \text{ nm/s}$$

Where the negative sign indicates that material is being removed (etched away). This etching rate does not seem very fast. However, assuming a monolayer of atoms is ≈ 0.2 nm thick, this etching rate corresponds to about 20 atomic layers per second!

(b) Even with 20 atomic layers removed from the Ti surface every second, it will take a long time to etch 1 mm of Ti:

$$\Delta t = \frac{\Delta x}{dx/dt} = \frac{0.1 \text{ cm}}{4.8 \times 10^{-7} \text{ cm/s}} = 208,000 \text{ s} \approx 58 \text{ h}$$

Diffusion Control As was illustrated in Figure 5.3, depending on the conditions, either surface reaction or gas diffusion could be rate limiting for the overall active gas corrosion process. Having examined the scenario where the surface reaction was rate limiting in the previous section, we now examine the scenario where gas diffusion is rate limiting. While either diffusion of reactants to the surface or diffusion of products away from the surface could be rate limiting, we will assume that the diffusion of HCl(g) to the Ti surface is the rate-limiting step in this case. Because four times more HCl is required for reaction than $TiCl_4$ is produced, and because the other product, H_2, is much smaller (and hence faster diffusing) than HCl, this is likely a reasonable assumption. A rate limitation based on diffusion of products away from the surface could be treated in an analogous manner.

Under the situation where the rate of reactant species transport to the surface is much slower than the surface reaction rate, the concentration of the reacting species will be depleted to essentially zero at the surface (since the reacting species can be consumed at the surface faster than they can arrive). However, somewhere far away from the reacting surface, deep within the bulk gas phase, the reactant concentration will be maintained (by flowing gas convection) at the original (supplied) value (c°_{HCl}). Thus, a *diffusion zone* of thickness δ will form through which the reactant species must diffuse to reach the surface. This is shown schematically in Figure 5.4. The reactant concentration will vary from c°_{HCl} (in the bulk) to zero (at the surface) across this diffusion zone. Assuming the reactants are continuously supplied into the bulk gas phase, this diffusion zone will achieve a steady-state thickness with the reactant concentration varying in an approximately linear fashion across it.

Under these circumstances, the rate (flux) at which HCl is transported to the Ti surface by diffusion across the diffusion zone may be calculated using Fick's first

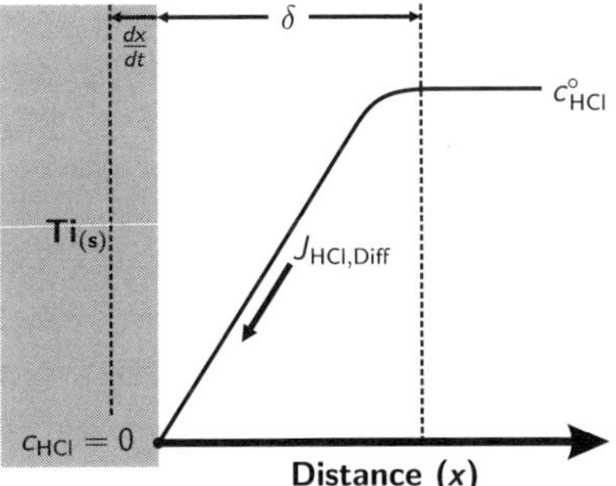

FIGURE 5.4 Schematic illustration of the active gas corrosion of Ti by HCl(g) when controlled by the diffusion of HCl(g) to the surface. Under diffusion control, the reaction rate is limited by the rate at which the HCl(g) reactant can diffuse across the diffusion zone (of thickness δ) to the surface. At steady state, the concentration profile across the diffusion zone is typically approximated as linear, enabling the diffusion flux to be calculated using a straightforward solution of Fick's first law.

law (Chapter 4) as

$$J_{HCl} = -D_{HCl}\frac{0 - c^{\circ}_{HCl}}{\delta} = D_{HCl}\frac{c^{\circ}_{HCl}}{\delta} \tag{5.14}$$

Because the surface reaction rate is much faster than the diffusion of HCl to the surface, every HCl that arrives at the surface will immediately react, consuming Ti. Thus, the consumption flux of Ti(s) can be directly related to the diffusion flux of HCl by the reaction stoichiometry:

$$J_{Ti} = -\frac{1}{4}J_{HCl} = -D_{HCl}\frac{c^{\circ}_{HCl}}{4\delta} \tag{5.15}$$

As in the previous section, this flux-based rate expression can then be converted into an expression that quantifies the etch rate (dx/dt):

$$\frac{dx}{dt} = \frac{M_{Ti}}{\rho_{Ti}}J_{Ti} = -\frac{M_{Ti}}{\rho_{Ti}}D_{HCl}\frac{c^{\circ}_{HCl}}{4\delta} \tag{5.16}$$

Converting the HCl concentration to a HCl pressure yields the final expression for the Ti etching rate as

$$\frac{dx}{dt} = -\frac{M_{Ti}}{\rho_{Ti}}D_{HCl}\frac{P^{\circ}_{HCl}}{4RT\delta} \tag{5.17}$$

This expression indicates that the Ti etching rate is constant; thus we would expect the thickness of a Ti plate undergoing active gas corrosion to decrease linearly with time. Because D_{HCl} depends on both temperature and pressure $[D = D_0(T^{3/2}/P)]$, this partially offsets the direct T and P terms appearing in the expression. In the end, the etching rate tends to increase very weakly (as approximately $T^{1/2}$) with increasing temperature.

> **Example 5.3**
>
> **Question:** (a) Calculate the etching rate of a Ti surface undergoing active gas corrosion in HCl(g) assuming that diffusion of HCl to the Ti surface is rate controlling. (b) Under these conditions, how much time would be required to etch 1 mm deep into a Ti plate? The following information is provided:
>
> - $D_{HCl(g)}|_{1500 \text{ K}} = 0.2 \text{ cm}^2/\text{s}$
> - $\delta = 1 \text{ mm}$
> - $P_{HCl} = 0.01 \text{ atm}$
> - $M_{Ti} = 47.9 \text{ g/mol}$
> - $\rho_{Ti} = 4.5 \text{ g/cm}^3$
> - $T = 1500 \text{ K}$
>
> **Solution:** (a) Inserting the provided quantities into Equation 5.13 (while once again paying careful attention to SI units!) gives

$$\frac{dx}{dt} = -\frac{M_{Ti}}{\rho_{Ti}} D_{HCl} \frac{P^{\circ}_{HCl}}{4RT\delta}$$

$$= -\frac{0.0479 \text{ kg/mol}}{4500 \text{ kg/m}^3} \cdot 2 \times 10^{-5} \text{ m}^2/\text{s}$$

$$\cdot \frac{(0.01 \text{ atm})(101{,}300 \text{ Pa}/1 \text{ atm})}{4 \cdot 8.314 \text{ J}/(\text{mol} \cdot \text{K}) \cdot 1500 \text{ K} \cdot 10^{-3} \text{ m}}$$

$$= -4.3 \times 10^{-9} \text{ m/s} = -4.3 \times 10^{-7} \text{ cm/s} = -4.3 \text{ nm/s}$$

Thus, the etching rate under the assumption that diffusion limits the kinetics is almost the same as that calculated assuming the surface reaction is rate limiting. As we will see in the next section, the similarity between the surface reaction and diffusion rates in this situation means that both must be considered simultaneously to accurately determine the overall total etching rate under these conditions.

(b) As before, we see that it will take more than 60 h to etch 1 mm of Ti. Specifically,

$$\Delta t = \frac{\Delta x}{dx/dt} = \frac{0.1 \text{ cm}}{4.3 \times 10^{-7} \text{ cm/s}} = 233{,}000 \text{ s} \approx 65 \text{ h}$$

Mixed Control   In the previous two sections we have examined the kinetics of active gas corrosion from the standpoint of two limiting scenarios: (1) surface reaction control and (2) diffusion control. Under many conditions, it is quite likely that one of these two processes will limit the overall rate of corrosion, and hence one of these two limiting models can be used to calculate the corrosion rate. Under certain conditions, however, the surface reaction and diffusion rates may be comparable, in which case both will influence the overall rate of corrosion. When two series processes both affect the overall rate, they essentially act as two series resistances. Like electronic resistors, these two kinetic resistances will add in series. However, it is important to keep in mind one key point: the resistance of each process is effectively given by the *inverse* of its rate; thus,

$$R_{tot} = R_{rxn} + R_{diff}$$

$$\left[\left(\frac{dx}{dt}\right)_{tot}\right]^{-1} = \left[\left(\frac{dx}{dt}\right)_{rxn}\right]^{-1} + \left[\left(\frac{dx}{dt}\right)_{diff}\right]^{-1}$$

$$\left[\left(\frac{dx}{dt}\right)_{tot}\right]^{-1} = -\left(\frac{M_{Ti}}{\rho_{Ti}} \frac{k' P^{\circ}_{HCl}}{4RT}\right)^{-1} - \left(\frac{M_{Ti}}{\rho_{Ti}} D_{HCl} \frac{P^{\circ}_{HCl}}{4RT\delta}\right)^{-1}$$

$$\left(\frac{dx}{dt}\right)_{tot} = -\frac{M_{Ti} P^{\circ}_{HCl}}{4RT\rho_{Ti}(\delta/D_{HCl} + 1/k')} \tag{5.18}$$

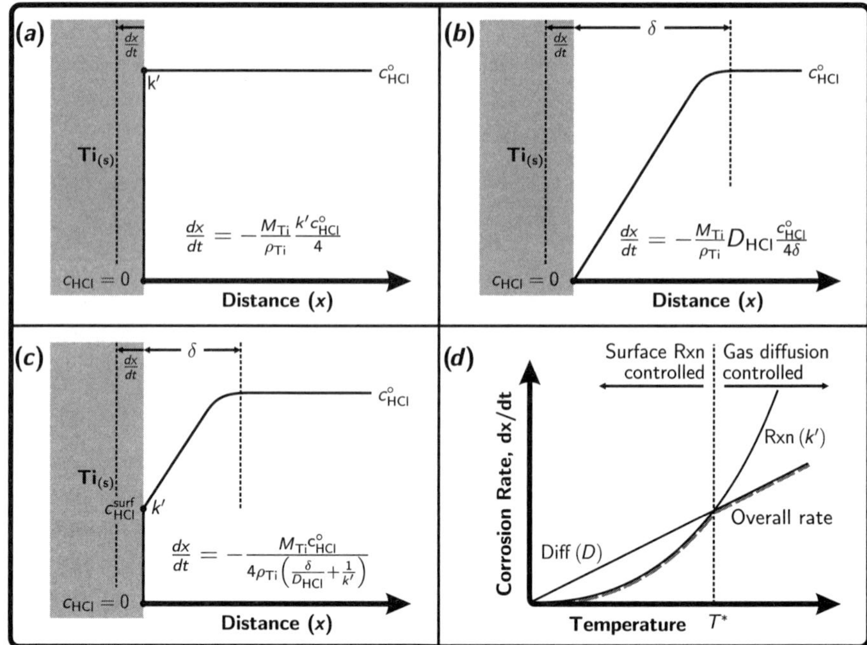

FIGURE 5.5 Summary of the key kinetic concepts associated with active gas corrosion under the surface reaction, diffusion, and mixed-control regimes. (*a*) Schematic illustration and corrosion rate equation for active gas corrosion under surface reaction control. (*b*) Schematic illustration and corrosion rate equation for active gas corrosion under reactant diffusion control. (*c*) Schematic illustration and corrosion rate equation for active gas corrosion under mixed control. (*d*) Illustration of the crossover from surface-reaction-controlled behavior to diffusion-controlled behavior with increasing temperature. The surface reaction rate constant (k') is exponentially temperature activated, and hence the surface reaction rate tends to increase rapidly with temperature. On the other hand, the diffusion rate increases only weakly with temperature. The slowest process determines the overall rate.

Examining the limits of this expression shows that when $k' \gg D_{HCl}/\delta$ (i.e., diffusion is slower and hence rate limiting), the equation reduces to the expression for diffusion control. In contrast, when $D_{HCl}/\delta \gg k'$ (i.e., the surface reaction is slower and hence rate limiting), the equation reduces to the expression for surface reaction control. The key concepts associated with active gas corrosion under the surface reaction, diffusion, and mixed-control regimes are summarized in Figure 5.5.

Example 5.4

Question: As we learned from Examples 5.2 and 5.3, for the active gas corrosion of Ti by HCl at $T = 1500$ K the surface reaction rate and diffusion rate are approximately equal. Considering both processes operating in series, calculate the actual overall etching rate for Ti under this situation.

Solution: Using Equation 5.18 and the etching rates calculated from the previous two examples, we have

$$\left[\left(\frac{dx}{dt}\right)_{\text{tot}}\right]^{-1} = \left[\left(\frac{dx}{dt}\right)_{\text{rxn}}\right]^{-1} + \left[\left(\frac{dx}{dt}\right)_{\text{diff}}\right]^{-1}$$

$$\left[\left(\frac{dx}{dt}\right)_{\text{tot}}\right]^{-1} = -\frac{1}{4.8 \text{ nm/s}} - \frac{1}{4.3 \text{ nm/s}}$$

$$\left(\frac{dx}{dt}\right)_{\text{tot}} = -2.3 \text{ nm/s}$$

Thus, the actual etching rate in this situation, taking into account both reaction and diffusion, is about half the rate calculated assuming that only one process controlled the kinetics!

When the reaction and diffusion rates are not so comparable, it is not necessary to consider them both in this manner. For example, if the surface reaction rate increases by 100× to 480 nm/s while the diffusion rate remains unchanged, the overall etching rates considering both reaction and diffusion would become

$$\left[\left(\frac{dx}{dt}\right)_{\text{tot}}\right]^{-1} = -\frac{1}{480 \text{ nm/s}} - \frac{1}{4.3 \text{ nm/s}}$$

$$\left(\frac{dx}{dt}\right)_{\text{tot}} = -4.26 \text{ nm/s} \approx \left(\frac{dx}{dt}\right)_{\text{diff}}$$

In this case, the diffusion rate is now 100× slower than the reaction rate, and hence it largely determines the overall etching rate.

Example 5.5

Question: Calculate the "crossover" temperature between the surface-reaction-controlled regime and the diffusion-controlled regime for the active gas corrosion of Ti(s) by HCl(g). In other words, calculate the temperature at which the rates of these two processes are equal. Use the same values provided in Examples 5.2 and 5.3 and assume that D_{HCl} is independent of temperature.

Solution: When $T = T^*$ (the crossover temperature), $(dx/dt)_{\text{rxn}} = (dx/dt)_{\text{diff}}$. Thus,

$$\frac{M_{\text{Ti}}}{\rho_{\text{Ti}}} \frac{k' P^{\circ}_{\text{HCl}}}{4RT^*} = \frac{M_{\text{Ti}}}{\rho_{\text{Ti}}} D_{\text{HCl}} \frac{P^{\circ}_{\text{HCl}}}{4RT^* \delta}$$

$$\frac{M_{\text{Ti}}}{\rho_{\text{Ti}}} \frac{k'_0 e^{-\Delta G_{\text{act}}/RT^*} P^{\circ}_{\text{HCl}}}{4RT^*} = \frac{M_{\text{Ti}}}{\rho_{\text{Ti}}} D_{\text{HCl}} \frac{P^{\circ}_{\text{HCl}}}{4RT^* \delta}$$

$$T^* = \frac{\Delta G_{\text{act}}}{R} \frac{1}{\ln\left(\dfrac{k'_0 \delta}{D_{\text{HCl}}}\right)}$$

$$= \left(\frac{50{,}000 \text{ J/mol}}{8.314 \text{ J/(mol} \cdot \text{K)}} \right) \left[\ln \left(\frac{(120 \text{ cm/s})(0.1 \text{ cm})}{0.2 \text{ cm}^2/\text{s}} \right) \right]^{-1}$$

$$= 1469 \text{ K}$$

This result makes sense, since at 1500 K we saw that the surface reaction rate was just a little higher than the diffusion rate. Since the surface reaction rate increases exponentially with temperature while the diffusion rate is only weakly temperature dependent, decreasing the temperature a little below 1500 K brings the two rates into equality.

5.3 CHEMICAL VAPOR DEPOSITION

In this section, we go from considering the *removal of material* from a solid surface to considering the *deposition of material* on a solid surface. One of the most popular ways to controllably deposit material on a solid surface is to introduce one or more gases that, when the pressure and temperature conditions are properly chosen, can react on a surface to deposit a solid film of material. Depending on the temperature, pressure, and growth requirements, there are many different gas chemistries available that can yield the deposition of many different types of solid films. These various gas–solid deposition reactions are collectively known as "chemical vapor deposition" (CVD) processes. CVD processes are used in the fabrication of integrated circuits, light-emitting diodes (LEDs), flat-panel displays, low-emissivity coatings for window glass, solar cells, and many other high-tech devices.

Frost: Natural Chemical Vapor Deposition Process

You are probably already familiar with at least one naturally occurring CVD process: the formation of frost (ice) on a cold humid morning in the winter. Frost is formed when the temperature is sufficiently cold that water vapor from the air can spontaneously nucleate and grow on solid surfaces as ice crystals. Frost will form, in preference to a snowfall, when the relative humidity and temperature conditions do not provide sufficient driving force to cause ice crystals to homogeneously form in the gas phase (snow); instead, only heterogeneous deposition on solid surfaces can occur. Heterogeneous nucleation on a solid surface is generally easier than homogeneous nucleation in the gas phase because of the reduction of surface energy afforded by heterogeneous nucleation on an existing surface. Thus, there is typically a thermodynamic "window" where conditions will favor the heterogeneous deposition of a film without the homogeneous nucleation of solid particles in the gas phase. Just as with frost formation, most CVD processes seek this magic thermodynamic window. We will learn more about homogeneous and heterogeneous nucleation and growth processes in Chapter 6.

Our kinetic treatment of CVD will be very similar to our treatment of active gas corrosion (except, of course, material is being deposited rather than removed). There will be one key difference, however. In our treatment of active gas corrosion, we made the implicit assumption that the reaction process went to completion. In the Ti corrosion example, as long as ANY HCl(g) was available to react, we assumed that it would do so completely. As you may recall from Chapter 3, the assumption that a reaction goes all the way to completion is valid in many cases. However, there are many other reactions that do not go all the way to completion. This is particularly true for many CVD reactions, whose thermodynamics are purposely tuned so that the driving forces for reaction are relatively small (and thus homogeneous nucleation is avoided).

Consider the CVD of Si from a trichlorosilane ($SiHCl_3$) gas precursor at 1300 K (Figure 5.6):

$$SiHCl_3(g) + H_2(g) \leftrightarrow Si(s) + 3HCl(g) \qquad \Delta G^\circ_{rxn}|_{T=1300 \text{ K}} = 8.46 \text{ kJ/mol}$$

In this case, ΔG°_{rxn} at 1300 K (a typical deposition temperature) is positive! Clearly, then, this reaction will not go to completion in the forward direction. In order to determine how far the reaction will proceed, we can calculate the equilibrium constant (K_{eq}) and use this information to determine the equilibrium pressures of the reactant and product species (recall Chapter 2, Example 2.3):

$$K_{eq} = \begin{cases} \exp\left(-\dfrac{\Delta G_{rxn}}{RT}\right) = \exp\left(-\dfrac{8460 \text{ J/mol}}{8.314 \text{ J/(mol} \cdot \text{K)} \cdot 1300 \text{ K}}\right) = 0.457 & (5.19) \\[2em] \dfrac{(P^{eq}_{HCl})^3 a_{Si}}{P^{eq}_{SiHCl_3} P^{eq}_{H_2}} = \dfrac{(P^{eq}_{HCl})^3}{P^{eq}_{SiHCl_3} P^{eq}_{H_2}} & (5.20) \end{cases}$$

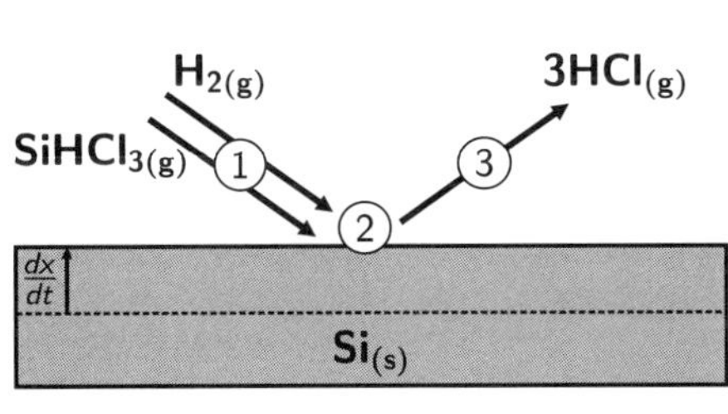

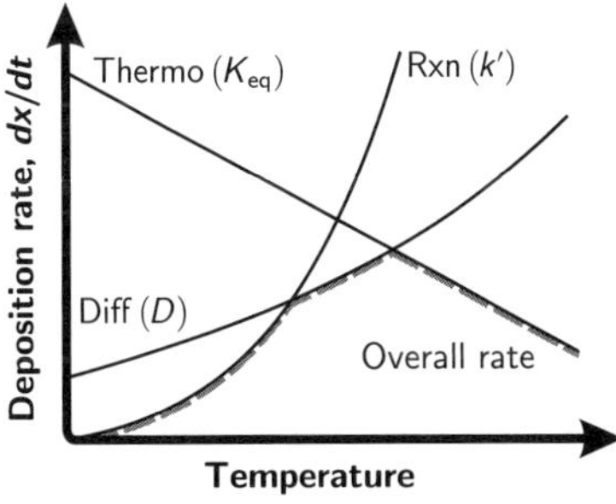

FIGURE 5.6   Schematic illustration of the chemical vapor deposition of a Si film from a $SiHCl_3$ gas precursor. This process involves three sequential steps: (1) transport of the $SiHCl_3$ and H_2 gas reactants to the surface; (2) reaction between $SiHCl_3$ and H_2 on the surface, creating Si(s) and HCl gas; and (3) transport of the HCl gas product away from the surface. As shown in the accompanying graph, the rate-limiting step can change depending on the conditions.

For sake of illustration, consider a CVD reaction that starts with the following partial pressures for the reactant gases (and initially no product gases):

- $P^{\circ}_{\mathrm{SiHCl_3}} = 0.10$ atm
- $P^{\circ}_{\mathrm{H_2}} = 0.90$ atm

The equilibrium pressures of the reactant and product gases will be

$$0.457 = \frac{(P^{\mathrm{eq}}_{\mathrm{HCl}})^3}{\left(0.1 - \frac{1}{3}P^{\mathrm{eq}}_{\mathrm{HCl}}\right)\left(0.9 - \frac{1}{3}P^{\mathrm{eq}}_{\mathrm{HCl}}\right)} \tag{5.21}$$

Therefore

$$P^{\mathrm{eq}}_{\mathrm{HCl}} = 0.20 \text{ atm}$$

$$P^{\mathrm{eq}}_{\mathrm{SiHCl_3}} = 0.1 - \frac{1}{3}0.2 = 0.033 \text{ atm}$$

$$P^{\mathrm{eq}}_{\mathrm{H_2}} = 0.9 - \frac{1}{3}0.2 = 0.833 \text{ atm}$$

Thus, with respect to the initial $\mathrm{SiHCl_3}$ partial pressure, the reaction will proceed only about two-thirds of the way to completion; one-third of the $\mathrm{SiHCl_3}$ gas will remain unreacted at equilibrium.

Based on this understanding of the incomplete nature of the CVD reaction, the kinetic behavior of this system under surface reaction, diffusion, and mixed control can now be developed. The results will be very similar to the active gas corrosion example with only minor changes due to the incomplete nature of the reaction and the different reaction stoichiometry of this example.

Surface Reaction Control   We will first examine the behavior of this system under the assumption that the surface reaction is the slowest step and hence controls the overall deposition rate. To start, we must write the rate law for this surface reaction. We assume that the reaction is first order with respect to the $\mathrm{SiHCl_3}$ reactant. However, we must modify the rate law for this first-order reaction to take into account the fact that the reaction does not go to completion. This yields

$$\frac{dc_{\mathrm{SiHCl_3}}}{dt} = -k(c_{\mathrm{SiHCl_3}} - c^{\mathrm{eq}}_{\mathrm{SiHCl_3}}) \tag{5.22}$$

where $c^{\mathrm{eq}}_{\mathrm{SiHCl_3}}$ represents the equilibrium concentration of $\mathrm{SiHCl_3(g)}$. This rate law correctly indicates that the net reaction rate will go to zero when the $\mathrm{SiHCl_3(g)}$ concentration reaches its equilibrium value.

Transformation to a heterogeneous (surface) reaction where the rate is expressed as a flux then yields

$$J_{\mathrm{SiHCl_3}} = -k'(c^{\circ}_{\mathrm{SiHCl_3}} - c^{\mathrm{eq}}_{\mathrm{SiHCl_3}}) \tag{5.23}$$

where $c^{\circ}_{\mathrm{SiHCl_3}}$ represents the concentration of $\mathrm{SiHCl_3(g)}$ that is supplied to the reaction.

Based on the rate at which the $SiHCl_3(g)$ is being consumed, we can determine the rate at which $Si(s)$ is being produced using the reaction stoichiometry. In this case, for every mole of $SiHCl_3(g)$ consumed, 1 mol of $Si(s)$ is produced. Thus,

$$J_{Si} = -J_{SiHCl_3} = k'(c^{\circ}_{SiHCl_3} - c^{eq}_{SiHCl_3}) \tag{5.24}$$

As in the active gas corrosion example, the flux of Si can be transformed into a deposition rate (i.e., the increase in the thickness of the Si film per unit time, dx/dt) via some simple algebraic transformations using the density and molecular weight of the Si:

$$J_{Si} = \frac{1}{A}\frac{dn_{Si}}{dt} = \frac{1}{A}\frac{Ax(\rho_{Si}/M_{Si})}{dt} = \frac{\rho_{Si}}{M_{Si}}\frac{dx}{dt}$$

$$\frac{dx}{dt} = \frac{M_{Si}}{\rho_{Si}}J_{Si} \tag{5.25}$$

Finally, inserting the Si reaction rate law (Equation 5.24) into this expression allows for the deposition rate to be calculated as a function of the heterogeneous reaction rate constant and the $SiHCl_3(g)$ concentration:

$$\frac{dx}{dt} = \frac{M_{Si}}{\rho_{Si}}J_{Si} = \frac{M_{Si}}{\rho_{Si}}k'(c^{\circ}_{SiHCl_3} - c^{eq}_{SiHCl_3}) \tag{5.26}$$

which can then be converted into a final expression involving gas partial pressures using the ideal gas law:

$$\frac{dx}{dt} = \frac{M_{Si}}{RT\rho_{Si}}k'(P^{\circ}_{SiHCl_3} - P^{eq}_{SiHCl_3}) \tag{5.27}$$

This expression indicates that the Si deposition rate is constant; thus we would expect the thickness of a Si film grown by this CVD process to increase linearly with time. The expression also indicates that the deposition rate will increase linearly with increasing $SiHCl_3(g)$ pressure. Since $SiHCl_3$ is the reactant, this makes sense! The effect of temperature is less obvious. While temperature appears directly in the denominator of this expression, recall that the rate constant k' is an exponentially temperature-activated quantity. In addition, $P^{eq}_{SiHCl_3}$ also depends on temperature ($P^{eq}_{SiHCl_3}$ increases with increasing temperature due to the fact that ΔS for this reaction is negative). The overall effect of these factors, however, is generally that deposition rate will increase with increasing temperature until the temperature is increased to the point where the reaction can no longer occur in the forward direction at all.

Example 5.6

Question: The deposition rate for the CVD of $Si(s)$ from $SiHCl_3(g)$ is 10 nm/s when the following reactant gas partial pressures are supplied to the CVD chamber at $T = 1300$ K:

- $P^{\circ}_{SiHCl_3} = 0.10$ atm
- $P^{\circ}_{H_2} = 0.90$ atm

The initial product gas concentrations are zero. Calculate the heterogeneous rate constant for the surface reaction process at 1300 K assuming that the surface reaction process is rate limiting, $M_{Si} = 28.1$ g/mol, and $\rho_{Si} = 2.65$ g/cm^3.

Solution: From the text (Equation 5.21) we know that $P^{eq}_{SiHCl_3} = 0.033$ atm for the reactant partial pressures supplied to the chamber in this problem at $T = 1300$ K. Applying this value along with the other information provided in the problem statement to Equation 5.27 and solving for k' yields

$$\frac{dx}{dt} = \frac{M_{Si}}{RT\rho_{Si}} k'(P^{\circ}_{SiHCl_3} - P^{eq}_{SiHCl_3})$$

$$k' = \frac{RT\rho_{Si}}{M_{Si}(P^{\circ}_{SiHCl_3} - P^{eq}_{SiHCl_3})}\left(\frac{dx}{dt}\right)$$

$$= \frac{8.314 \text{ J/(mol} \cdot \text{K)} \cdot 1300 \text{ K} \cdot 2650 \text{ kg/m}^3}{0.0281 \text{ kg/mol} \cdot (0.10 \text{ atm} - 0.0333 \text{ atm}) \cdot (101{,}300 \text{ Pa/1 atm})} 10$$

$$\times 10^{-9} \text{ m/s}$$

$$= 1.5 \times 10^{-3} \text{ m/s} = 0.15 \text{ cm/s}$$

Diffusion Control   In the diffusion control regime, either diffusion of $SiHCl_3$ or H_2 to the surface or diffusion of HCl away from the surface could be rate limiting. Here, we will assume that the diffusion of $SiHCl_3(g)$ to the surface is the rate-limiting step. Because $SiHCl_3$ is a far larger molecule (and hence slower diffusing) than the other two gas species, this is likely a reasonable assumption. A rate limitation based on the diffusion of one of the other species could be modeled in a nearly identical manner.

In this case, because the reaction does not go to completion, the $SiHCl_3$ concentration at the surface cannot fall below $c^{eq}_{SiHCl_3}$ even when the transport of $SiHCl_3$ to the surface is much slower than the surface reaction rate. Thus, as illustrated in Figure 5.7, the reactant concentration will vary across the diffusion zone from $c^{\circ}_{SiHCl_3}$ (in the bulk) to $c^{eq}_{SiHCl_3}$ (at the surface).

As before and assuming again a linear concentration gradient across the diffusion zone, the rate (flux) at which $SiHCl_3$ is transported to the surface by diffusion may be calculated using Fick's first law as

$$J_{SiHCl_3} = -D_{SiHCl_3}\frac{c^{eq}_{SiHCl_3} - c^{\circ}_{SiHCl_3}}{\delta} \tag{5.28}$$

By applying the reaction stoichiometry, the corresponding Si flux is then

$$J_{Si} = J_{SiHCl_3} = -D_{SiHCl_3}\frac{c^{eq}_{SiHCl_3} - c^{\circ}_{SiHCl_3}}{\delta}$$

$$= D_{SiHCl_3}\frac{c^{\circ}_{SiHCl_3} - c^{eq}_{SiHCl_3}}{\delta} \tag{5.29}$$

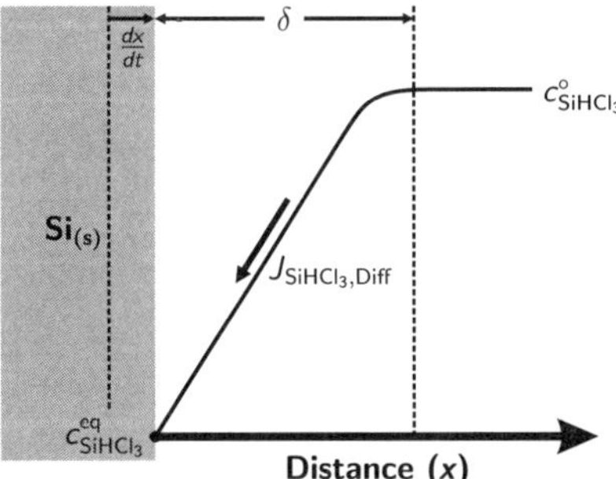

FIGURE 5.7 Schematic illustration of the CVD growth of a Si film when controlled by the diffusion of $SiHCl_3(g)$ to the surface. Under diffusion control, the reaction rate is limited by the rate at which the $SiHCl_3(g)$ reactant can diffuse across the diffusion zone (of thickness δ) to the surface. At steady state, the concentration profile across the diffusion zone is typically approximated as linear, enabling the diffusion flux to be calculated using a straightforward solution of Fick's first law.

As in the previous section, this flux-based rate expression can then be converted into an expression that quantifies the deposition rate (dt/dx):

$$\frac{dx}{dt} = \frac{M_{Si}}{\rho_{Si}} J_{Si}$$

$$= \frac{M_{Si}}{\rho_{Si}} D_{SiHCl_3} \frac{c^o_{SiHCl_3} - c^{eq}_{SiHCl_3}}{\delta} \tag{5.30}$$

Converting the $SiHCl_3$ concentration to a $SiHCl_3$ pressure yields the final expression for the Si deposition rate as

$$\frac{dx}{dt} = \frac{M_{Si}}{RT\rho_{Si}} D_{SiHCl_3} \frac{P^o_{SiHCl_3} - P^{eq}_{SiHCl_3}}{\delta} \tag{5.31}$$

This expression indicates that the Si deposition rate is constant; thus we would expect the thickness of a Si film to increase linearly with time. Because D_{SiHCl_3} depends on both temperature and pressure $[D = D_0(T^{3/2}/P)]$, this partially offsets the direct T and P terms appearing in the expression. In addition, $P^{eq}_{SiHCl_3}$ also depends on temperature and on the partial pressures of the other gases in the system. In general, the deposition rate tends to increase very weakly (as approximately $T^{1/2}$) with increasing temperature (until the temperature is increased to the point where the CVD reaction can no longer occur in the forward direction at all).

Mixed Control In the mixed-control regime, the surface reaction and diffusion rates are comparable, and thus both influence the overall rate of deposition. Analogous to the active gas corrosion example, the growth rate under mixed control for this

CVD process can be determined by adding these two kinetic resistances together in series:

$$R_{\text{tot}} = R_{\text{rxn}} + R_{\text{diff}}$$

$$\left[\left(\frac{dx}{dt}\right)_{\text{tot}}\right]^{-1} = \left[\left(\frac{dx}{dt}\right)_{\text{rxn}}\right]^{-1} + \left[\left(\frac{dx}{dt}\right)_{\text{diff}}\right]^{-1}$$

$$\left[\left(\frac{dx}{dt}\right)_{\text{tot}}\right]^{-1} = \left(\frac{M_{\text{Si}}}{\rho_{\text{Si}}}\frac{k'(P^{\circ}_{\text{SiHCl}_3} - P^{\text{eq}}_{\text{SiHCl}_3})}{RT}\right)^{-1} + \left(\frac{M_{\text{Si}}}{\rho_{\text{Si}}}D_{\text{SiHCl}_3}\frac{P^{\circ}_{\text{SiHCl}_3} - P^{\text{eq}}_{\text{SiHCl}_3}}{RT\delta}\right)^{-1}$$

$$\left(\frac{dx}{dt}\right)_{\text{tot}} = \frac{M_{\text{Si}}(P^{\circ}_{\text{SiHCl}_3} - P^{\text{eq}}_{\text{SiHCl}_3})}{RT\rho_{\text{Si}}(\delta/D_{\text{SiHCl}_3} + 1/k')} \tag{5.32}$$

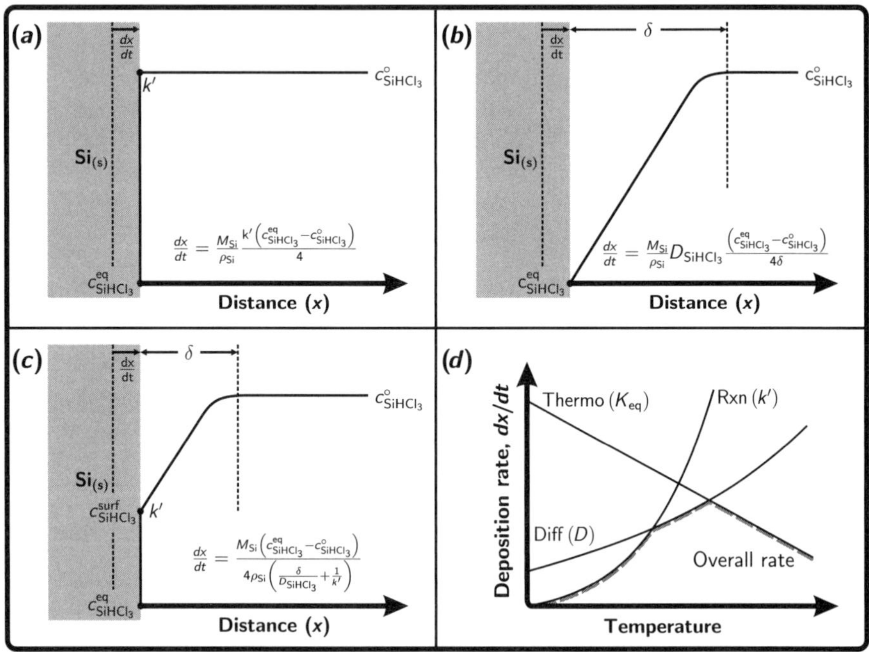

FIGURE 5.8 Summary of the key kinetic concepts associated with CVD under the surface reaction, diffusion, and mixed-control regimes. (*a*) Schematic illustration and deposition rate equation for CVD under surface reaction control. (*b*) Schematic illustration and deposition rate equation for CVD under reactant diffusion control. (*c*) Schematic illustration and deposition rate equation for CVD under mixed control. (*d*) Illustration of the crossover from surface-reaction-controlled behavior to diffusion-controlled behavior with increasing temperature. The surface reaction rate constant (k') is exponentially temperature activated, and hence the surface reaction rate tends to increase rapidly with temperature. On the other hand, the diffusion rate increases only weakly with temperature. For CVD processes where the reactions become less thermodynamically favorable with increasing temperature (common), the rate will eventually fall at higher temperatures as the CVD process becomes unfavorable thermodynamically. The slowest process determines the overall rate.

Examining the limits of this expression shows that when $k' \gg D_{\text{SiHCl}_3}/\delta$ (i.e., diffusion is slower and hence rate limiting), the equation reduces to the expression for diffusion control. In contrast, when $D_{\text{SiHCl}_3}/\delta \gg k'$ (i.e., the surface reaction is slower and hence rate limiting), the equation reduces to the expression for surface reaction control. The key concepts associated with CVD under the surface reaction, diffusion, and mixed-control regimes are summarized in Figure 5.8.

Halogen Light Bulbs: "Self-Healing" Gas Corrosion/Chemical Vapor Deposition Equilibrium

The halogen light bulb provides a fascinating example of a kinetic system involving active gas corrosion and chemical vapor deposition processes. The reason halogen light bulbs are much brighter than standard incandescent light bulbs is because they operate at higher filament temperatures, which enables the tungsten filament inside the bulb to emit light at much higher intensity. Standard light bulbs cannot operate at such high filament temperatures because the tungsten filament is susceptible to evaporation. As electrical current flows through the filament and heats it to the operating temperature by resistive heating, narrow spots in the filament can lead to localized areas of higher resistive heating, which leads to further loss of tungsten from these areas and hence greater constriction of the filament (as illustrated in Figure 5.9a). This, in turn, leads to even higher localized resistive

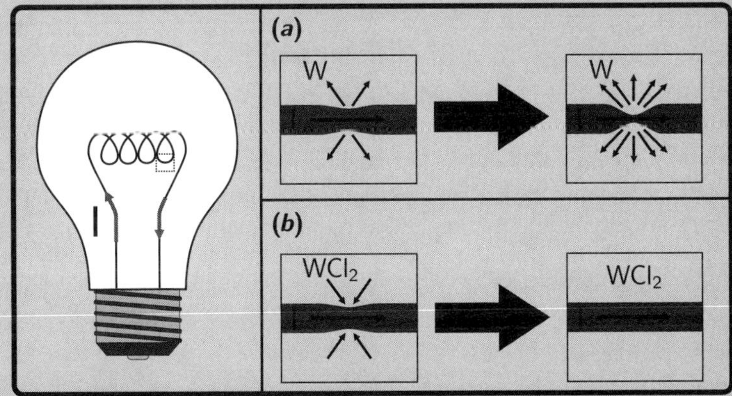

FIGURE 5.9   (a) The operating principle of a standard incandescent light bulb. Current is passed through the tungsten filament, generating resistive heating that leads to light emission. A positive-feedback cycle can occur when narrow spots in the filament incur higher resistive heating, leading to further loss of tungsten from these areas and hence greater constriction of the filament. This, in turn, leads to even higher localized resistive heating and to further tungsten loss from these areas, eventually leading to catastrophic failure. (b) The operating principle of a halogen light bulb. The interplay between W(s), HCl(g), and WCl$_2$(g) leads to a negative-feedback cycle where the increased resistive heating at narrow spots in the filament induces CVD of W(s) from WCl$_2$(g) in these regions, leading to *self-healing* of the filament and enabling operation at higher temperatures than standard incandescents.

heating and to further tungsten loss from these areas, eventually causing catastrophic failure. This accelerating cycle of increased heating leading to increased evaporation leading to failure is an example of a *positive-feedback* cycle.

The halogen light bulb solves this issue by exploiting the unique active gas corrosion/CVD chemistry of the $W(s)/HCl(g)/WCl_2(g)$ "halogen" system. Halogen light bulbs are filled with $HCl(g)$, which at operating temperatures leads to the following reaction equilibrium:

$$W(s) + HCl(g) \leftrightarrow WCl_2(g) + H_2(g) \tag{5.33}$$

The kinetics of this system are such that moderate operating temperatures favor the forward reaction direction, while high operating temperatures favor the reverse reaction direction. Thus, as the bulb heats up, a small amount of tungsten is actively corroded from the surface by the $HCl(g)$, producing an overpressure of $WCl_2(g)$ in the bulb. However, at higher temperatures, when localized "hot spots" develop due to constrictions in the tungsten filament, the reverse reaction becomes favored in these regions, and tungsten metal is deposited, leading to "self-healing" of the filament (as illustrated in Figure 5.9*b*). This self-healing phenomenon is an example of *negative feedback* and enables the bulb to be operated at higher temperatures than standard incandescents.

Science of Avalanches: Kinetics of Snowpack Evolution

Did you know that the kinetic principles of active gas corrosion and CVD processes can be used to understand some of the factors leading to dangerous avalanche conditions in mountain areas? From 1950 to 2012, avalanches have killed nearly 1000 people in the United States. Intriguingly, Colorado has had far and away the most avalanche deaths over this period (> 250). In comparison, California has sustained only about 60 avalanche deaths during this same period. This may come as a surprise, since the California Sierra Nevada mountains typically receive more than twice as much snow as the Colorado Rocky Mountains and California has 7× more people than Colorado! Can kinetic factors be contributing to this startling discrepancy? The answer is "yes."

Figure 5.10*a* illustrates one of the primary mechanisms responsible for most avalanches: a weak layer of "rotten" snow deep in the snowpack. A trigger (such as a skier, snowboarder, or hiker) can stress the snowpack, inducing fracture that propagates along the weak layer. The weak layer acts as a glide surface (the rotten snow can even act like ball bearings), enabling the overlying snowpack to slide down the mountain in a catastrophic avalanche. The frequency of avalanches thus correlates very strongly with the presence of weak layers in the snowpack.

Colorado is located in the interior of the United States, while California is on the coast. The resulting differences in climate lead to several important differences in the mountain snowpack in these two regions. California tends to have a thick, wet snowpack with relatively warm and wet winter air temperatures. In contrast, Colorado tends to have a thinner snowpack, while the winter air temperature tends to be colder and drier. As illustrated in Figure 5.10, this leads to significant differences in the temperature and moisture (P_{H_2O}) gradients in the California (Figure 5.10b) versus Colorado (Figure 5.10c) snowpacks. The larger temperature/moisture gradients in the Colorado snowpack induce significant active gas corrosion of ice crystals from the bottom layers of the snowpack and their subsequent redeposition (CVD) in the top layers of the snowpack according to the following reaction equilibrium:

$$\begin{array}{ccc} H_2O(s) & & H_2O(g) \\ Lower\ T & \leftrightarrow & Higher\ T \end{array}$$

Thus, in the Colorado snowpack, this process tends to eat away snow from the bottom layers of the snowpack and redeposit it the upper layers of the snowpack (Figure 5.10d). This process can lead to a weak, rotten bottom layer in the snowpack overlaid by a strong, dense, coherent top "slab" layer: the prefect conditions for dangerous avalanches.

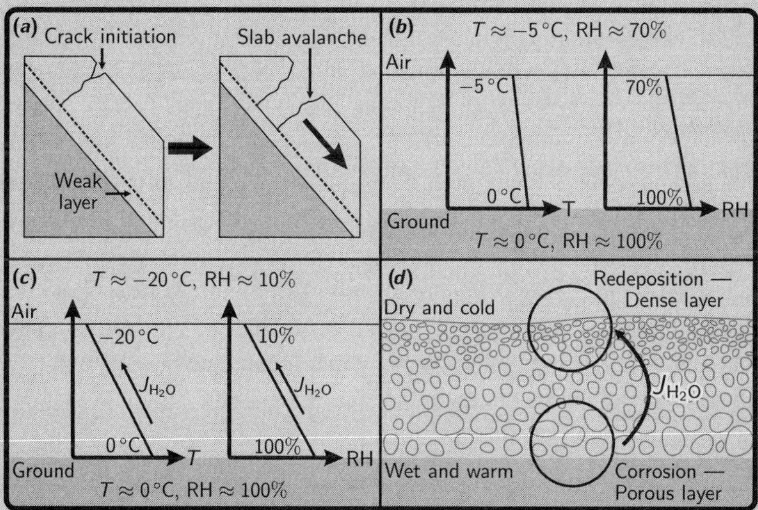

FIGURE 5.10   (a) One of the chief causes of avalanches, a weak buried layer in the snowpack can easily fracture and act as a glide plane enabling the overlying coherent snowpack to slide down the mountain in a catastrophic avalanche. (b) California tends to have a thick, wet snowpack with relatively warm and wet winter air temperatures. This leads to an "equilibrated" snowpack with only small temperature/moisture gradients. (c) Colorado tends to have a thinner snowpack, while the winter air temperature tends to be colder and drier. This leads to an unstable snowpack with large temperature/moisture gradients. (d) The large temperature/moisture gradients in the thin Colorado snowpack drive the formation of a "rotten" layer at the base of the snowpack due to active gas corrosion while redeposition of this moisture in the upper regions of the snowpack lead to a dense overlying "slab" of coherent snow. This provides a perfect setup for killer avalanches.

5.4 ATOMIC LAYER DEPOSITION

One issue with chemical vapor deposition is that it is often difficult to tune the growth conditions to achieve an optimal deposition. Sometimes the deposition of thin, highly conformal coatings over extremely intricate or high-aspect-ratio structures is desired. (See Figure 5.11 for examples.) In such cases, the CVD conditions should be tuned to operate in the surface-reaction-limited regime. By ensuring that the surface reaction is much slower than the diffusion rate, this provides a better chance that all surfaces (even those far from the bulk gas supply stream) will be evenly coated with material. This variation on CVD is sometimes known as "chemical vapor infiltration." Even using this principle, however, perfectly conformal coatings are often impossible to achieve. In order to overcome this issue, a new type of deposition scheme, known as "atomic layer deposition" (ALD) was developed.

ALD is a cyclic, self-limiting gas–solid surface reaction process that allows films to be grown one atomic layer at a time. A classical ALD reaction scheme for the deposition of Al_2O_3 is shown in Figure 5.12. The deposition process consists of two distinct reaction steps that are repeatedly cycled to build up the Al_2O_3 film layer by layer:

$$\text{Step 1: } AlOH(s) + Al(CH_3)_3(g) \rightarrow Al_2O(CH_3)_2(s) + CH_4(g)$$

$$\text{Step 2: } Al_2O(CH_3)_2(s) + 2H_2O(g) \rightarrow Al_2O_2(OH)(s) + 2CH_4(g) + \tfrac{1}{2}H_2(g)$$

(Repeat)

An important point is that both of the reaction steps are *self-limiting*. In reaction step 1, once all available $AlOH(s)$ surface sites have reacted with the $Al(CH_3)_3(g)$ precursor, the reaction comes to a stop because the CH_3 groups that now terminate the surface are unreactive toward $Al(CH_3)_3(g)$. Thus, only a single atomic layer of

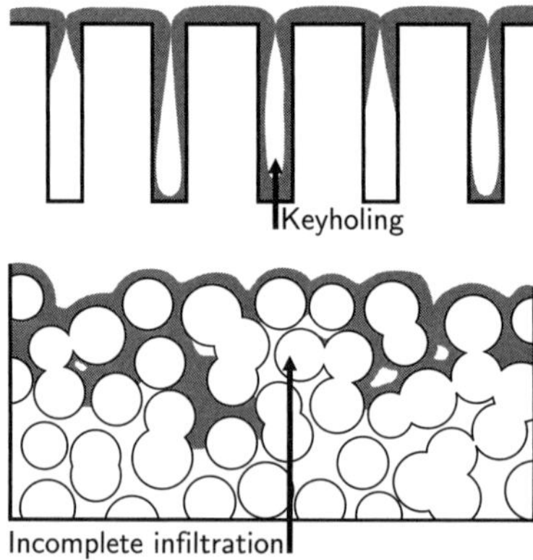

FIGURE 5.11 Conformal deposition over high-aspect-ratio structures can be difficult to achieve by CVD due to "keyhole" effects and other difficulties posed by the long diffusion paths required to access the inner regions of such complex structures.

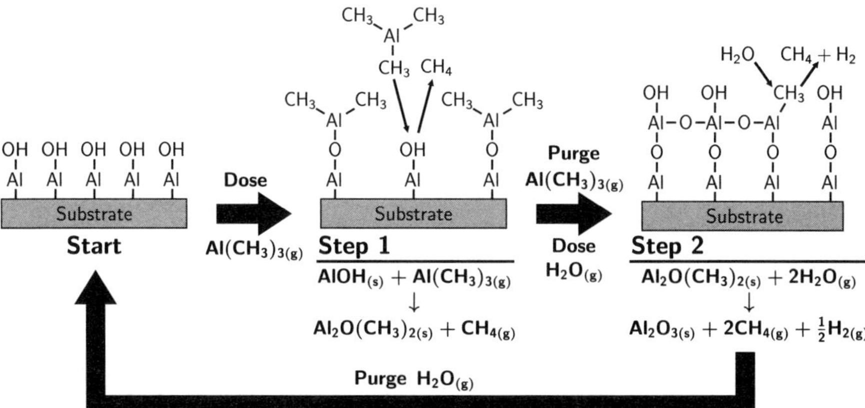

FIGURE 5.12 Schematic illustration of the ALD reaction scheme for the deposition of Al_2O_3. The deposition process involves two distinct reaction steps (step 1, step 2) followed by purge steps (purge 1, purge 2) that are repeatedly cycled to build up the Al_2O_3 film layer by layer.

$Al_2O(CH_3)_2(s)$ is formed. The reaction can be held for a sufficiently long period of time to ensure that the entire surface of the sample is reacted without worrying that parts of the surface will continue to react to form a second layer of material. The chamber is then purged of any excess $Al(CH_3)_3(g)$ and the second reaction step proceeds. In the second step, water vapor is introduced into the chamber and reacts with the $CH_3(s)$ groups terminating the surface and replaces them with terminating OH groups. Once all available $CH_3(s)$ surface sites have reacted with the water vapor precursor, the reaction again comes to a stop because the OH(s) groups now terminating the surface are unreactive toward H_2O. However, because the surface has now been "reactivated" with a fresh set of AlOH(s) surface sites, the excess water vapor can be purged and reaction step 1 can now be repeated, enabling a fresh atomic layer of $Al_2O(CH_3)_2(s)$ to be formed.

ALD can lead to beautiful, ultrathin, highly conformal coatings, even on extremely high aspect ratio structures. Figure 5.13 provides a couple of amazing examples. One shortcoming of the ALD technique is that it is extremely slow. In addition, the precursor gas chemicals can be very expensive and there is significant precursor waste (due to the repeated precursor fill/purge cycles). Thus, it is generally only used to deposit relatively thin (<100-nm) films for high-value applications (e.g., integrated circuits). Example 5.7 calculates the growth rate of a typical ALD coating process.

Example 5.7

Question: In the atomic layer deposition of alumina (Al_2O_3), the average deposition rate is approximately 0.5 ML (ML = monolayer) per cycle. A single cycle consists of two purge steps and two reaction steps. Even with careful attention to valve, chamber, and gas flow design, typically at least a few seconds is required to purge the chamber, introduce the precursor gas, and then ensure that all surfaces undergo complete reaction for each step. For the purposes of this question, assume that one complete cycle takes 10 s. Based on this assumption, determine the time it takes to deposit a 1-μm-thick film of Al_2O_3 by ALD. Assume that 1 ML of Al_2O_3 is 4 Å thick.

Solution: The problem statement indicates that 10 s is required for one complete ALD cycle, which results in the deposition of 0.5 ML of Al_2O_3. Thus, we can approximate the growth rate as

$$\frac{dx}{dt} = \frac{0.5 \text{ ML}}{10 \text{ s}} = \frac{0.5 \text{ ML}(4 \text{ Å}/1 \text{ ML})}{10 \text{ s}} = 0.2 \text{ Å/s} = 72 \text{ nm/h}$$

The time required to deposit 1 μm of Al_2O_3 is

$$\Delta t = \frac{\Delta x}{dx/dt} = \frac{1000 \text{ nm}}{72 \text{ nm/h}} \approx 14 \text{ h}$$

As this example illustrates, ALD is an extremely slow process and thus is generally only used when the deposition of highly conformal and extremely thin films (<100 nm) is needed.

It is important to note that a single ALD cycle generally does not lead to the deposition of a complete monolayer of material. For instance, in this example, the problem statement indicated that each ALD cycle resulted in the deposition of 0.5 ML of Al_2O_3. The sub-monolayer growth rate per cycle is due to steric factors—the bulky ligands associated with the $Al_2O(CH_3)_2(s)$ surface groups restrict access to surface sites, meaning that only a fraction of surface sites can react during each cycle. This is common with most ALD processes.

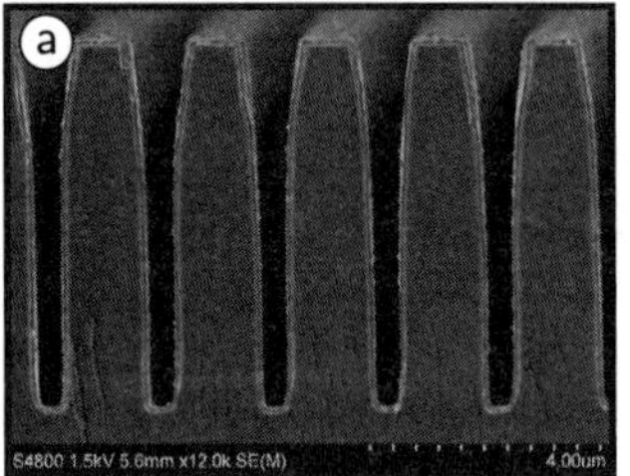
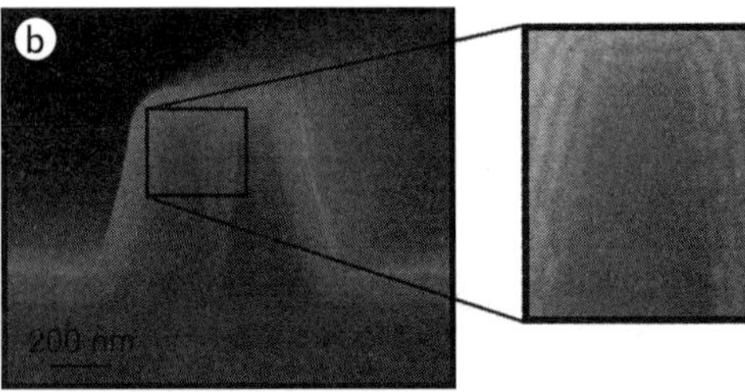

FIGURE 5.13   Examples of thin-film coatings obtained via ALD. (*a*) Germanium antimony telluride film deposited on a high-aspect-ratio trench. Reprinted with permission from V. Pore, T. Hatanpää, M. Ritala, and M. Leskelä, "Atomic layer deposition of metal tellurides and selenides using alkylsilyl compounds of tellurium and selenium," *Journal of the American Chemical Society*, vol. 131, no. 10, pp. 3478–3480, 2009. Copyright 2009 American Chemical Society. (*b*) A nanolaminate of alternating layers of aluminum oxide and titanium oxide. Reprinted with permission from A. Säynätjoki, T. Alasaarela, A. Khanna, L. Karvonen, P. Stenberg, M. Kuittinen, A. Tervonen, and S. Honkanen, "Angled sidewalls in silicon slot waveguides: Conformal filling and mode properties," *Optical express*, vol. 17, no. 23, pp. 21 066–21 076, Nov. 2009. Copyright 2009 The Optical Society.

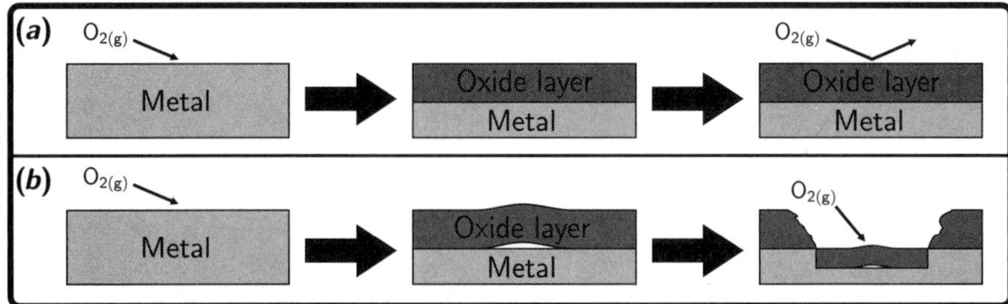

FIGURE 5.14   Schematic illustration of (*a*) passive oxidation and (*b*) active oxidation.

5.5 PASSIVE OXIDATION

In this section, we consider another important gas–solid kinetic process: oxidation. Many metals and semiconductors will spontaneously form a thin oxide coating on their surface when exposed to oxygen (air). As illustrated schematically in Figure 5.14, there are two main categories of oxidation: *passive* and *active*. Passive oxidation occurs when the oxide layer that forms on the surface provides a protective coating that inhibits further oxide growth. Passive oxidation is therefore a *self-limiting* process that gradually slows down (or even completely stops) as the oxide grows thicker. Active oxidation occurs when the oxide layer that forms on the surface does not protect against further oxidation. For example, the oxide may spall or flake off the surface, thereby continually exposing fresh surface for further oxidation. The formation of rust on iron is an example of an active oxidation process. During active oxidation, the oxidation process is not self-limiting and therefore oxidation can continue until the entire volume of a material is completely oxidized. Whether a material forms a passive or active oxide layer depends, among other factors, on the relative molar volume and mechanical properties of the oxide and the underlying material. Oxides that do not have large molar volume changes relative to their parent material and adhere strongly and uniformly to the underlying material tend to form passive oxide coatings, while oxides that undergo a large molar volume change relative to the parent material and/or adhere poorly to the underlying material tend to facilitate active oxidation.

Both passive and active oxidation processes have significant commercial implications and much research has been invested in understanding the kinetics of these processes. In this section, we will examine one of these processes as an example: the passive oxidation of Si.

Si/SiO$_2$ and the CMOS Revolution

It has been claimed that the passive oxidation of Si is one of the most commercially significant oxidation processes in the world. Without a detailed understanding of this process, none of the myriad electronic products we enjoy today would have ever been possible. Interestingly, the reason that Si has become the most widely used semiconductor material on Earth is not because Si is the best semi-conductor. In fact, germanium (Ge) or galium–arsenide (GaAs) are both known to be "better" semiconductors in many respects. The reason Si is the semiconductor of choice is because it happens to form one of the "best" passivating oxides (SiO_2). The ability to precisely grow and control the thickness of the passivating oxide film that forms on Si enabled the development of the complimentary metal–oxide–semiconductor (CMOS) transistor. CMOS transistors form the heart of most integrated circuits. As the name implies, they are fabricated from a sandwichlike structure consisting of a metal, an oxide, and a semiconductor layer (Figure 5.15). The thinner the oxide layer, the better the performance of the transistor. The ability to grow a thin (yet defect-free) passive oxide directly on top of a Si wafer has enabled Si semiconductor technology to dominate over all other semiconductor alternatives.

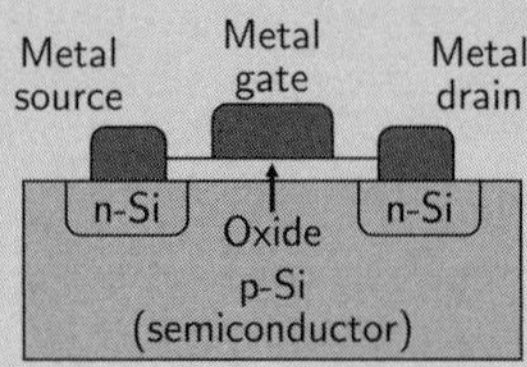

FIGURE 5.15 Schematic illustration of a silicon-based CMOS device. The oxide layer is grown directly on top of the Si wafer substrate using a passive thermal oxidation process.

The oxidation of a Si surface can be described by the following reaction:

$$Si(s) + O_2(g) \rightarrow SiO_2(s) \tag{5.34}$$

As illustrated in Figure 5.16, this process involves four sequential steps:

1. Transport of O_2 gas reactant to the (growing) SiO_2 surface
2. Decomposition of the O_2 on the SiO_2 surface to O atoms
3. Diffusion of the O atoms through the (growing) SiO_2 layer to the SiO_2/Si interface[1]
4. Reaction between the O atoms and the Si to form SiO_2

The slowest of these four series steps will control the overall rate of oxidation. In general, steps 3 and 4 tend to be the slowest. Because the speed of step 3 depends on the thickness of the oxide, an interesting crossover behavior is observed as the

---

[1] Alternatively, one can imagine that Si atoms could diffuse through the SiO_2 layer and react with O atoms at the SiO_2/air interface. However, Si diffusion through SiO_2 turns out to be much slower than O diffusion through SiO_2. Because these two diffusion pathways occur *in parallel*, the *faster* one dominates.

oxide grows thicker. During the initial stages of oxidation, when the oxide coating is extremely thin, the interfacial reaction (step 4) dominates the kinetics. However, as the oxide grows thicker, diffusion (step 3) becomes slower and slower and eventually dominates the kinetics. As a result, the oxide growth rate continually decreases as the oxide grows thicker.

Just as we did with active gas corrosion and chemical vapor deposition, the kinetics of passive oxidation can be quantitatively modeled using the reaction and diffusion principles we learned in Chapters 3 and 4 of this textbook. In developing our model, we will follow in the footsteps of Bruce Deal and Andy Grove. Their "Deal–Grove" model for the passive oxidation of Si paved the way for the precise growth of oxide layers for CMOS transistors. After developing the Deal–Grove model, Andy Grove went on to cofound Intel and served as its CEO from 1987 to 1998—proving just how rewarding a firm understanding of kinetics can be!

Interfacial Reaction Control   We will assume that the overall rate of Si oxidation is controlled by either the reaction between oxygen and Si at the SiO_2/Si interface (step 4) or the diffusion of oxygen through the SiO_2 film (step 3). In the interfacial reaction control regime, the kinetics of silicon oxidation are determined by the reaction between oxygen atoms and Si to form SiO_2. This reaction can be modeled as first order with respect to the oxygen concentration in the film at the SiO_2/Si interface (this concentration is designated in Figure 5.16 as c_3):

$$J_{rxn} = -k' c_3 \tag{5.35}$$

Diffusion Control   In the diffusion control regime, the kinetics of silicon oxidation are controlled by the rate of oxygen diffusion through the SiO_2 film to the SiO_2/Si

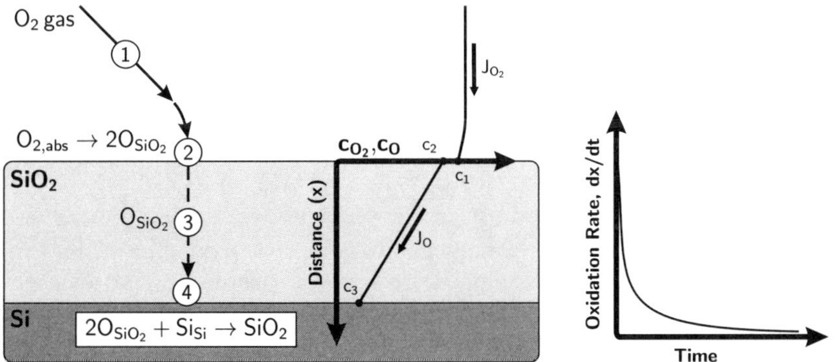

FIGURE 5.16   Schematic illustration of the passive oxidation of Si. This process involves four sequential steps: As shown in the accompanying graph, the oxidation rate becomes slower and slower as the oxide grows thicker and thicker.

interface. Using Fick's first law and assuming a linear concentration gradient of oxygen through the SiO_2 layer, the rate of oxygen diffusion is given by

$$J_{\text{diff}} = -D_{O,SiO_2} \frac{c_3 - c_2}{x} \tag{5.36}$$

where x is the thickness of the SiO_2 film and c_2 is the oxygen concentration in the oxide film at the SiO_2/gas interface.[2]

Mixed Control The interfacial reaction and diffusion processes in the passive oxidation of Si are tightly coupled. Thus, it is better not to consider them in isolation, but instead to consider them together. Since these reactions occur in series, their rates must be equal. In other words,

$$J_{\text{tot}} = J_{\text{diff}} = -J_{\text{rxn}} \tag{5.37}$$

where J_{tot} is the overall rate of the oxidation process and the negative sign appearing in front of J_{rxn} reflects the fact that the delivery of oxygen to the interface by diffusion is balanced by the *consumption* of oxygen at the interface by reaction. Inserting Equations 5.35 and 5.36 into Equation 5.37 allows for concentration c_3 to be determined:

$$-J_{\text{rxn}} = J_{\text{diff}}$$

$$k' c_3 = -D_{O,SiO_2} \frac{c_3 - c_2}{x}$$

$$c_3 = \frac{D_{O,SiO_2} c_2}{k'x + D_{O,SiO_2}} = \frac{c_2}{k'x/D_{O,SiO_2} + 1} \tag{5.38}$$

Since $J_{\text{tot}} = -J_{\text{rxn}} = J_{\text{diff}}$, this result for c_3 can then be reinserted into the expression for either J_{rxn} or J_{diff} to create an expression for the overall rate of oxidation as a function of c_2, D_{O,SiO_2}, and k':

$$J_{\text{tot}} = -J_{\text{rxn}}$$

$$= k' \frac{c_2}{k'x/D_{O,SiO_2} + 1} = \frac{c_2}{x/D_{O,SiO_2} + \frac{1}{k'}} \tag{5.39}$$

This expression should look vaguely familiar! In fact, it is quite similar to the mixed-control equations we developed for active gas corrosion and for chemical vapor deposition (Equations 5.18 and 5.32).

Equation 5.39 quantifies the oxidation rate in terms of the oxygen flux to the SiO_2/Si interface. Using the reaction stochiometry and the relative molar volumes

---

[2]Note that c_2 is distinct from c_1, which is the oxygen concentration in the gas phase at the SiO_2/gas interface. The terms c_1 and c_2 can be related to one another by the solubility of oxygen in the SiO_2 film, or alternatively by describing the kinetics of the incorporation reaction that leads to the dissolution of oxygen into the SiO_2 film. Because this reaction tends to be much faster than the other kinetics processes, a detailed treatment of it is generally not needed.

of Si and SiO_2, it is possible to convert this flux into an oxide growth rate (dx/dt). This conversion is somewhat more complicated than the previous cases for active gas corrosion and CVD since material is not simply being removed or deposited. Instead, one material (Si) is being converted into another material (SiO_2). For sake of simplicity, however, the overall effects of this conversion can be incorporated into a simple growth thickness constant G, in which case

$$\frac{dx}{dt} = \frac{1}{G}J_{tot} = \frac{1}{G}\frac{c_2}{x/D_{O,SiO_2} + 1/k'} \tag{5.40}$$

Passive Oxide Thickness as Function of Time   An important point associated with the passive oxide growth rate, as expressed by Equation 5.40, is that the growth rate dx/dt *depends on the oxide thickness x.* Thus, the oxide growth rate is constantly changing as the oxide grows thicker! In order to understand exactly how the oxide thickness increases with time, it is necessary to integrate Equation 5.40. Integrating over the dummy variables x' and t' from $x' = 0$ to $x' = x$ and from $t' = 0$ to $t' = t$ yields

$$\int_0^x \left(\frac{x'}{D_{O,SiO_2}} + \frac{1}{k'} \right) \, dx = \frac{c_2}{G} \int_0^t dt'$$

$$\frac{x^2}{2D_{O,SiO_2}} + \frac{x}{k'} = \frac{c_2}{G}t \tag{5.41}$$

Solving for t yields the well-known "parabolic oxidation law" for passive oxidation processes:

$$t = \frac{G}{2c_2 D_{O,SiO_2}}x^2 + \frac{G}{c_2 k'}x \tag{5.42}$$

As can be seen from Equation 5.42, the parabolic oxidation law is so named because the time–thickness dependence follows a quadratic (parabolic) equation. The first term in Equation 5.42 (the x^2 term) dictates the growth limit under the diffusion-controlled regime, while the second term in Equation 5.42 dictates the growth limit under the reaction-controlled regime. Figure 5.17 plots Equation 5.42 with (*a*) thickness as the dependent axis and (*b*) time as the dependent axis. The figure shows how the two limiting terms in this equation combine to determine the overall growth rate. As can be seen by studying these plots, there is a crossover from a reaction-limited linear growth rate when x (or t) is small to a diffusion-limited square-root growth rate when x (or t) becomes large. The critical thickness (x_{crit}) at which this crossover occurs can be determined by setting the two terms in Equation 5.42 equal to one another:

$$\frac{G}{2c_2 D_{O,SiO_2}}x_{crit}^2 = \frac{G}{c_2 k'}x_{crit}$$

$$x = \frac{2D_{O,SiO_2}}{k'} \tag{5.43}$$

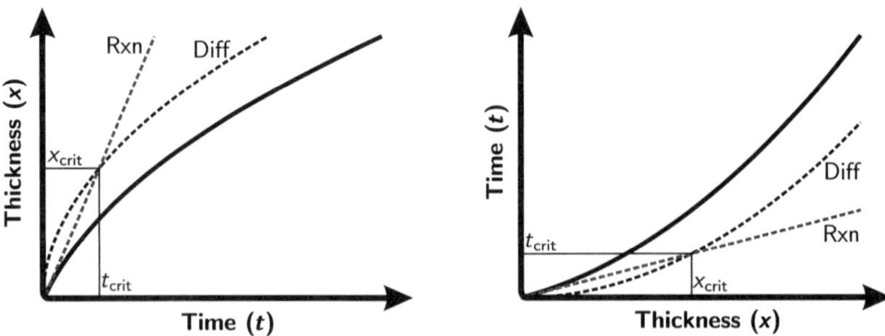

FIGURE 5.17   The parabolic oxidation law plotted in terms of (*a*) thickness as the dependent axis and (*b*) time as the dependent axis. A crossover from a reaction-limited linear growth rate to a diffusion-limited square-root growth rate occurs at a critical oxide thickness x_{crit} when the two limiting rate processes are equal.

Thus:

- when $x \ll 2D_{O,SiO_2}/k'$, interfacial reaction controls the oxidation rate. Growth is *linear* (i.e., thickness increases linearly with time).
- when $x \gg 2D_{O,SiO_2}/k'$, diffusion through the oxide controls the oxidation rate. Growth is *parabolic* (i.e., thickness increases with the square root of time).

5.6 CHAPTER SUMMARY

This chapter examined gas–solid kinetic processes. We saw how to apply the basic tools we learned in calculating thermodynamic driving forces (Chapter 2), reaction rates (Chapter 3), and mass diffusion (Chapter 4) to understand and model a number of important gas–solid kinetic processes including adsorption/desorption, active gas corrosion, chemical vapor deposition, and passive oxidation. The main points introduced in this chapter include:

- Gas–solid kinetic processes are fundamentally *heterogeneous* as they involve both a gas phase and a solid phase.
- The impingement rate, or impingement flux, J'_s, quantifies the maximum rate at which atoms can move from the gas phase to the solid phase or vice-versa in the absence of other limiting kinetic factors. The impingement rate is given by

$$J'_s = \frac{\alpha P}{\sqrt{2\pi MRT}}$$

where the sticking coefficient α is a number between 0 and 1 that quantifies the fraction of impinging gas atoms that stick to the surface.

- The impingement rate characterizes the maximum intrinsic rate at which gas molecules immediately above a solid can strike the surface of a solid. The impingement rate does not take concentration gradients into effect, and thus it represents the maximum rate of transport for a gas-phase species to a surface when diffusion is unimportant. When calculating the rate of transport of a gas-phase species to a solid surface, the decision to use the impingement rate versus the diffusion equation depends on the length scale of the transport relative to the mean free path of the transporting gas molecules. Under vacuum pressures and for small distances (where the mean free path is smaller than the distance of travel), direct line-of-sight impingement can determine the transport rate and hence the impingement rate can be used to quantify the flux of a species to a surface. However, at higher pressures, where molecules undergo many collisions along their journey and their concentration varies spatially, transport is instead determined by diffusion and Fick's laws should be used to quantify the flux.

- When atoms are deposited onto or removed from the surface of a solid, this process often preferentially occurs at higher energy sites on the surface of the solid, such as at step edges or kink sites. Atoms at these sites have more dangling/unsatisfied bonds and thus are more susceptible to reaction.

- Active gas corrosion is a gas–solid kinetic process involving etching (removal) of a solid surface by a corrosive gas species. The rate of this corrosion process depends on both the rate of transport of gases to/from the solid surface and the rate of the corrosion reaction on the solid surface. Depending on the temperature and pressure conditions, either the gas diffusion or the surface reaction process can limit the overall corrosion rate. An overall corrosion rate can be derived which takes into account both processes according to

$$\left[\left(\frac{dx}{dt}\right)_{\text{tot}}\right]^{-1} = \left[\left(\frac{dx}{dt}\right)_{\text{rxn}}\right]^{-1} + \left[\left(\frac{dx}{dt}\right)_{\text{diff}}\right]^{-1}$$

  In general, the diffusion rate term $(dx/dt)_{\text{diff}}$ is proportional to D_i/δ (where D_i is the diffusivity of the rate-controlling gas-phase species and δ is the diffusion layer thickness), while the reaction rate term $(dx/dt)_{\text{rxn}}$ is proportional to k' (the heterogeneous reaction rate constant for the surface corrosion reaction). Thus, when $k' \gg D_i/\delta$, the diffusion term controls the overall rate, while for $D_i/\delta \gg k'$, the reaction term controls the overall rate. Because reaction rates tend to increase exponentially with temperature while gas-phase diffusion is only weakly temperature dependent, higher temperatures tend to lead to diffusion control (diffusion is slow relative to reaction), while lower temperatures can lead to reaction control (reaction is slow relative to diffusion).

- Chemical vapor deposition (CVD) is a process involving the heterogeneous reaction of one or more gas-phase species on a solid surface, resulting in the deposition of a solid film of material. Depending on the temperature, pressure, and growth requirements, there are many different gas chemistries available that can yield the deposition of many different types of solid films. CVD processes are used in the fabrication of many important devices, including integrated circuits, LEDs, flat-panel displays, low-emissivity coatings for window glass, and solar cells.

- Just as the rate of *material removal* during active gas corrosion can depend on either gas transport to/from the surface or the reaction process on the surface, the overall rate of *material deposition* during a CVD process also depends on the rate of transport of gases to/from the solid surface as well as the rate of the deposition reaction on the solid surface. Depending on the temperature and pressure conditions, either the gas diffusion or the surface reaction process can limit the overall deposition rate. An overall deposition rate can be derived that takes into account both processes according to

$$\left[\left(\frac{dx}{dt}\right)_{\text{tot}}\right]^{-1} = \left[\left(\frac{dx}{dt}\right)_{\text{rxn}}\right]^{-1} + \left[\left(\frac{dx}{dt}\right)_{\text{diff}}\right]^{-1}$$

  In general, the diffusion rate term $(dx/dt)_{\text{diff}}$ is proportional to D_i/δ (where D_i is the diffusivity of the rate-controlling gas-phase species and δ is the diffusion layer thickness), while the reaction rate term $(dx/dt)_{\text{rxn}}$ is proportional to k' (the heterogeneous reaction rate constant for the surface deposition reaction). Thus, when $k' \gg D_i/\delta$, the diffusion term controls the overall rate, while for $D_i/\delta \gg k'$, the reaction term controls the overall rate. Because reaction rates tend to increase exponentially with temperature while gas-phase diffusion is only weakly temperature dependent, higher temperatures tend to lead to diffusion control (diffusion is slow relative to reaction), while lower temperatures can lead to reaction control (reaction is slow relative to diffusion).

- Atomic layer deposition (ALD) is a cyclic, self-limiting gas–solid surface reaction process that allows films to be grown one atomic layer at a time. The unique precursor chemistries used in ALD ensure self-limiting reaction steps, so that only one layer of chemical species can react with the surface per cycle. A series of alternating reaction and purge cycles are used to grow up the film in a highly controlled step-by-step manner. This control can result in extremely uniform and highly conformal film coatings but also results in very slow deposition rates. As a result, it is generally only used to deposit relatively thin (<100-nm) films for high-value applications (such as integrated circuits or electronic devices).

- *Oxidation* is another example of a gas–solid kinetic process. In an oxidation process, oxygen molecules from the gas phase oxidize (react with) the surface of a solid (typically a metal). Oxidation processes can be either *active* or *passive*.

- *Active oxidation* occurs when the oxidation of a solid surface does not result in a self-protective coating, and thus the oxidation can proceed indefinitely. For example, in the active oxidation of iron ("rust"), the oxide layer that forms easily spalls or flakes of the surface of the iron, thereby continually exposing fresh surface for further oxidation. Since active oxidation is not self-limiting, the oxidation process can continue until the entire volume of the material is completely oxidized.

- *Passive oxidation* occurs when the oxide layer that forms on the surface provides a protective coating that inhibits further oxide growth. Passive oxidation is therefore a self-limiting process that gradually slows down (or eventually completely stops) as the oxide grows thicker.

- The growth expression for passive oxidation is remarkably similar to those derived for active gas corrosion and for chemical vapor deposition. The oxide growth rate depends on the rate of solid-state diffusion through the growing oxide layer as well as the rate of the oxidation reaction at the interface. When the oxide first begins to form, the growth rate is determined by the kinetics of the oxidation reaction. However, as the oxide grows thicker, diffusion through the oxide layer becomes rate limiting. The growth rate is often modeled using a "parabolic oxidation law" of the form

$$t = Ax^2 + Bx$$

where t is the time elapsed, x is the thickness of the oxide, and A and B are constants that depend, among other things, on the solid-state diffusivity (D) and the reaction rate constant (k'), respectively. The first term (Ax^2) dictates the growth limit under the diffusion-controlled regime, while the second term (Bx) dictates the growth limit under the reaction-controlled regime. The crossover from reaction-controlled to diffusion-controlled oxide growth occurs when $Ax^2 = Bx$, that is, when $x = B/A$. When the oxide is thinner than B/A, reaction determines the oxide growth rate, and the thickness increases linearly with time. When the oxide thickness exceeds B/A, diffusion dominates the oxide growth rate, and the thickness increases with the square root of time.

5.7 CHAPTER EXERCISES

Review Questions

Problem 5.1. Define/explain the following. Use schematic illustrations/diagrams as appropriate.

(a) Sticking coefficient

(b) Impingement flux

(c) Kink site

(d) Diffusion layer thickness

(e) Atomic layer deposition

Problem 5.2. Consider the chemical vapor deposition of silicon nitride (Si_3N_4) by the following reaction:

$$3SiH_4(g) + 4NH_3(g) \leftrightarrow Si_3N_4(s) + 12H_2(g)$$

(a) On the same graph, make a schematic plot of the natural logarithm of the Si_3N_4 growth rate versus $1/T$ (K^{-1}) for each of the following three growth processes and identify each on the plot:

1. A surface-reaction-controlled process

2. A gas diffusion-controlled process

3. The actual rate-limiting rate when both are occurring in series

(b) Explain why each plot has the temperature dependence shown.

(c) Make a plot of the SiH_4 gas pressure as a function of distance away from the growing silicon nitride film surface for the following three cases:

   1. Surface reaction control

   2. Gas diffusion control

   3. When both processes have about the same rate

   Clearly identify each case and indicate important distances and pressures. Assume that the gases are flowing over a flat substrate on which the deposition is being made.

Problem 5.3. Niobium (Nb) is being oxidized to Nb_2O_5 in 1 atm air at $1300\,^\circ C$.

(a) Write the equation for this reaction.

(b) If the diffusion coefficient of oxygen in Nb_2O_5 is $10^{-13}\,cm^2/s$ and that of niobium is $10^{-8}\,cm^2/s$, give the species, Nb or O, that controls the rate of oxidation. Explain why this atom or ion controls the rate.

(c) If this oxidation process forms a passive oxide layer on the surface of the Nb, make a plot of the oxide thickness of a function of time for two temperatures. Indicate the higher of the two temperatures.

Problem 5.4. Starting from the parabolic oxidation rate law described by Equation 5.42, determine an expression for the critical time t_c at which the oxidation behavior transitions from reaction to diffusion-limited growth.

Calculation Questions

Problem 5.5. A vacuum reactor is used to synthesize WO_3 nanoparticles for use in an electrochromic window device. To create the WO_3 nanoparticles, a tungsten (W) metal filament 1.0 mm in diameter and 10 cm long is heated to $2200\,^\circ C$ in a vacuum with a trace pressure of oxygen gas. Assuming all of the W evaporating from the surface of the filament is rapidly and completely oxidized to WO_3, determine the maximum rate of production of WO_3 from this reactor (g/s). The equilibrium vapor pressure of W at $2200\,^\circ C$ is approximately 10^{-10} atm.

Problem 5.6. A metallic calcium thin film has been freshly deposited by evaporation in a vacuum chamber containing an oxygen partial pressure of 10^{-4} torr for use as a bottom contact in an organic solar cell. Since metallic calcium oxidizes extremely easily, it is critical that the subsequent layer in the solar cell is deposited before the surface of the calcium is "contaminated" by a significant amount of adsorbed oxygen. In order to estimate this time, calculate how long it takes for the surface of the calcium film to be coated by a monolayer of oxygen assuming that the surface of the calcium film is initially completely clean. Assume a sticking coefficient of unity for the oxygen gas and a chamber temperature of 300 K. Calcium forms a face-centered-cubic (FCC) structure with a lattice constant $a = 5.6\,Å$.

Problem 5.7. Titanium undergoes active gas corrosion in HCl according to the reaction

$$Ti(s) + 4HCl(g) \leftrightarrow TiCl_4(g) + 2H_2(g)$$

The rate of etching of the Ti surface is found to be 2.88 µm/s. The following additional information is provided:

- $k_0 = 1.2 \times 10^2$ cm/s
- $P_{HCl} = 0.10$ atm
- $M_{Ti} = 48$ g/mol
- $\rho_{Ti} = 4.5$ g/cm^3
- $T = 1327\,°C$

(a) Calculate the rate constant for the reaction.

(b) Calculate the activation energy ΔG_{act} for the reaction.

(c) What is the time required to remove 5 mm of Ti?

(d) What is the corrosion rate after a time $t = 20$ h?

Problem 5.8. A silicon wafer is being oxidized in an atmosphere of pure oxygen at 1200 K. The thickness of the oxide, x (in µm), as a function of time t (in seconds) is given by the parabolic oxidation law:

$$x^2 + Bx = Ct$$

where $B = 0.040$ µm and $C = 0.045$ µm^2/s.

(a) On a separate sheet, make a computer-drawn plot of the oxide thickness x as a function of time t for oxide thickness between 0 and 20 µm.

(b) What is the rate of change of the oxide thickness (µm/s) when the oxide is 10 µm thick?

(c) When the oxide is 10 µm thick, is the oxidation reaction controlled by diffusion through the oxide film or by the reaction at the Si/SiO$_2$ interface? Justify your answer with an equation.

(d) If air is used in the reaction chamber instead of pure oxygen, do you expect the oxidation rate to change when $x = 10$ µm? If so, will it increase or decrease? Fully explain your answer.

(e) If the temperature in the reaction chamber is decreased to 1000 K, do you expect the oxidation rate to change when $x = 10$ µm? If so, will it increase or decrease? Fully explain your answer.

CHAPTER 6

LIQUID–SOLID AND SOLID–SOLID PHASE TRANSFORMATIONS

6.1 WHAT IS A PHASE TRANSFORMATION?

Liquid–solid and solid–solid phase transformations are also known as *condensed-matter* phase transformations. Condensed-matter phase transformations, like other kinetic processes, are driven by thermodynamics. When a region of matter can *lower its total free energy* by changing its composition, structure, symmetry, density, or any other phase-defining aspect, a phase transformation can occur.

In most condensed-matter phase transformations, pressure is typically not a chief controlling variable (although it is important in certain circumstances). This is in distinct contrast to the gas–solid processes discussed in the previous chapter, where gas-phase partial pressures typically played a central role.

Condensed-matter phase transformations can be broadly divided into two main categories: *diffusional* transformations and *diffusionless* (or "fluxless") transformations. Figure 6.1 schematically illustrates the difference between diffusional and diffusionless phase transformations. Diffusionless phase transformations do not require the net transport of atoms across a phase boundary. For example, phase transformations involving a change in spin or magnetic moment do not require the movement (i.e., diffusion) of atoms. Thus, they are *diffusionless* transformations. Likewise, certain changes in crystal structure or symmetry do not require diffusional fluxes—they can be accomplished by the collective shearing movement of atoms. Examples of such processes include the martensitic transformation in steel, or certain cubic-to-tetragonal phase transformations.

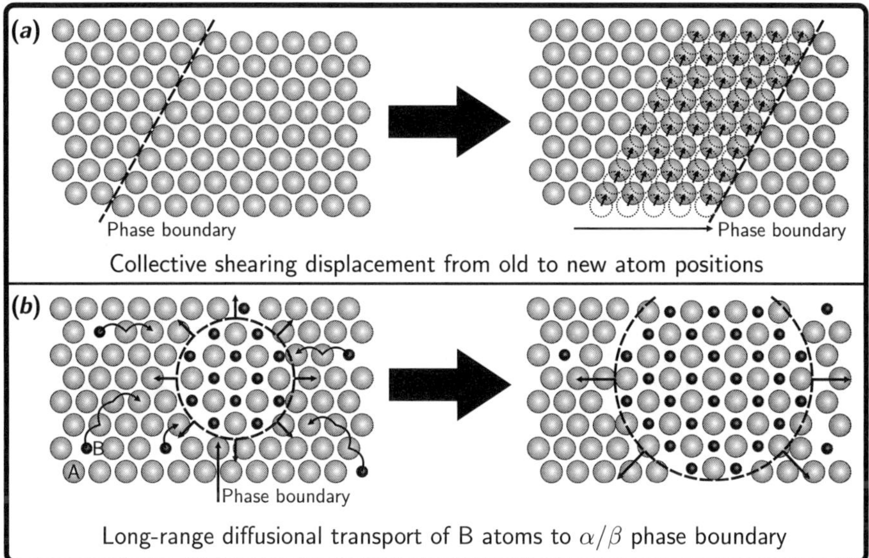

FIGURE 6.1   (*a*) Illustration of a diffusionless phase transformation involving a displacive martensitic transformation. This process does not require the net transport of atoms across a phase boundary. (*b*) Illustration of a diffusional phase transformation involving nucleation and growth, which requires the transport of atoms across a phase boundary.

Our focus in this text will be on diffusional transformations. Diffusional transformations can be further subdivided into two main types: *continuous* and *discontinuous*. Gibbs (of the Gibbs phase rule and the Gibbs free energy) articulated the difference between continuous and discontinuous phase transformations as follows:

- *Continuous phase transformations* are initially small in degree but large in extent. Spinodal decomposition is a classic example of a continuous phase transformation. In a spinodal transformation, a single-phase material gradually separates into two phases via gradual changes in local composition (small in degree). However, the process occurs more or less homogeneously throughout the entire material system (large in extent).

- *Discontinuous phase transformations* are large in degree but initially small in extent. Nucleation and growth represents the classic example of a discontinuous phase transformation. In a nucleation-and-growth process, the new phase possesses distinctly and abruptly different properties from its parent (large in degree) but its creation is a highly localized event (small in extent). The spatial extent of the phase transformation is subsequently increased by growth of the new phase.

Figure 6.2 schematically illustrates the differences between a *continuous* spinodal phase transformation and a *discontinuous* nucleation and growth transformation.

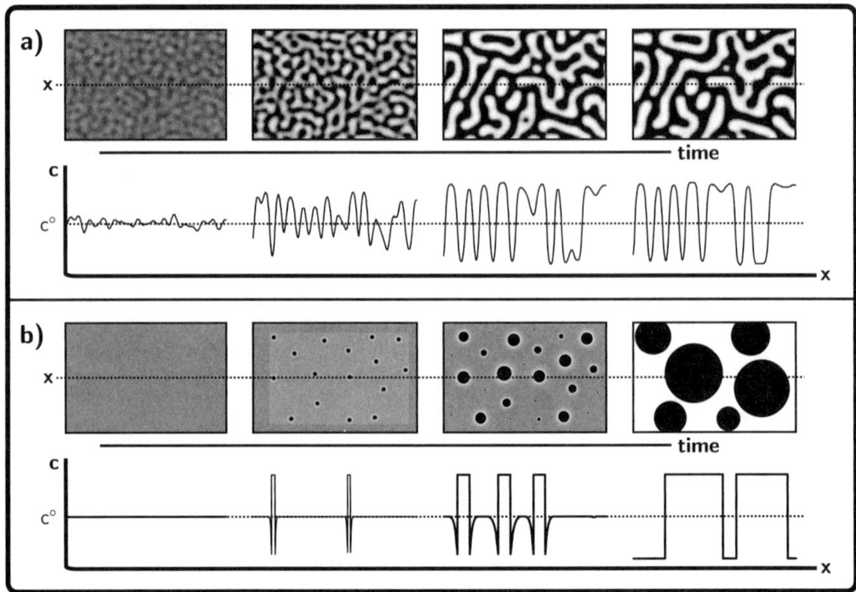

FIGURE 6.2   (*a*) A spinodal phase transformation is an example of a *continuous* phase transformation. Phase transformation gradually occurs everywhere (small in degree, large in extent). (*b*) The nucleation-and-growth process is an example of a *discontinuous* phase transformation). Phase transformation abruptly occurs in a few places (large in degree, small in extent).

One final point about condensed-matter phase transformations is that, like most other materials kinetic processes, they are heterogeneous and involve phase boundaries or interfaces. Thus, the properties of the interface (i.e., the interfacial energy, curvature, etc.) play a crucial role in phase transformation kinetics. Interfaces also play an important role in other kinetic processes that do not involve phase transformations. For example, grain growth or sintering of single-phase materials is driven by interfacial energy considerations and involves net transport of atoms, just as in a phase transformation (although no new phases are being created). Many of the concepts developed in our treatment of phase transformation kinetics can thus be applied to these single-phase microstructural evolution processes. These processes, which include sintering, grain growth, and coarsening, will be covered in Chapter 7.

6.2 DRIVING FORCES FOR TRANSFORMATION: TEMPERATURE AND COMPOSITION

In order for a phase transformation to occur, a driving force must be present. For most of the condensed-matter phase transformations discussed in this chapter, the driving force is supplied by a change in temperature or composition. Temperature and composition are two of the primary processing "knobs" that materials engineers have at their disposal to manipulate the structure and property of materials for various applications.

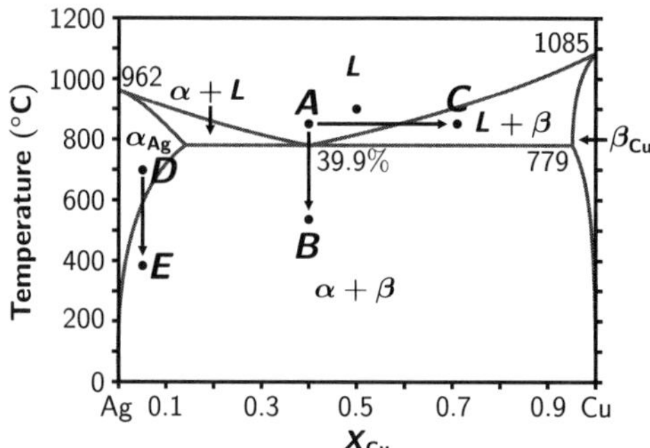

FIGURE 6.3   Example temperature and composition-induced phase transformations in the Cu–Ag binary eutectic alloy system.

Phase diagrams map out the effect of temperature and composition changes on the equilibrium phase behavior of a materials system. Thus, these diagrams are crucial for understanding phase transformations. As an example, consider the binary eutectic phase diagram of the Cu–Ag system shown in Figure 6.3. As the figure illustrates, changes in the temperature or composition can induce a variety of different phase transformations. For example, decreasing the temperature of the liquid Cu–Ag melt from point A to point B will cause the melt to solidify into a mixture of solid α and β phases. Similarly, adding a sufficient amount of pure Cu to shift the composition of the melt from point A to point C will cause the liquid melt to partially solidify into the solid β phase, even though the temperature is not decreased. As another example, mixing equal amounts of pure *solid* Cu at 900 °C with pure *solid* Ag at 900 °C will result in a *liquid* melt, although the temperature does not change! Solid–solid phase transformations also occur on this phase diagram—for example, decreasing the temperature of pure solid α phase initially at point D on the phase diagram to point E will result in the solid-state precipitation of a second phase β.

For the purposes of this chapter, it is assumed that students are familiar with binary phase diagrams and basic calculations involving phase diagrams, including the lever rule. For a review of phase diagrams, the reader is advised to consult *Phase Equilibria, Phase Diagrams and Phase Transformations: Their Thermodynamic Basis* [8], *Thermodynamics of Materials* [9], or *Introduction to Phase Equilibria in Ceramic Systems* [10]. Phase diagrams will serve as our maps for predicting and understanding phase transformations, so ensure that you can read these maps before proceeding!

6.2.1   Calculating ΔG_V

The driving force (ΔG) for transformation due to a change in temperature can be estimated using relatively simple thermodynamic relations. For example, consider the transformation from the pure α phase to the mixture of α and β phases by changing the temperature from point D to E in Figure 6.3. This reaction can be schematically written as

$$\alpha(s) \rightarrow \alpha(s) + \beta(s) \qquad \Delta G_V \tag{6.1}$$

where ΔG_V quantifies the free energy change *per unit volume* of material. The volumetric free-energy change ΔG_V can be related to the more common molar free-energy change ΔG via knowledge of the *molar volume* (V_m) of the material:

$$\Delta G_V = \frac{\Delta G}{V_m} = \frac{\Delta G}{M/\rho} \tag{6.2}$$

where the molar volume V_m (m^3/mol) is given by the molecular weight of the material, M (kg/mol), divided by the density of the material, ρ (kg/m^3).

As with all reaction free-energy changes, this ΔG_V can be expressed in terms of ΔH_V and ΔS_V as

$$\Delta G_V = \Delta H_V - T\,\Delta S_V \tag{6.3}$$

At the equilibrium temperature (indicated on Figure 6.3 as T_E), we know that ΔG_V must be zero, and so we have

$$\Delta G_V|_{T=T_E} = 0 = \Delta H_V - T_E\,\Delta S_V$$

$$\Delta S_V = \frac{\Delta H_V}{T_E} \tag{6.4}$$

or, alternatively,

$$\Delta H_V = \Delta S_V\,T_E \tag{6.5}$$

Since ΔH_V and ΔS_V do not vary much with temperature, we can approximate the free energy change for the transformation at any arbitrary temperature $T \neq T_E$ as

$$\begin{aligned}
\Delta G_V(T) &= \Delta H_V - T\,\Delta S_V \\
&\approx \Delta H_V - T\frac{\Delta H_V}{T_E} \\
&= \Delta H_V\frac{T_E - T}{T_E} \\
&= \Delta H_V\frac{\Delta T}{T_E}
\end{aligned} \tag{6.6}$$

or, alternatively,

$$\Delta G_V(T) \approx \Delta S_V\,T_E - \Delta S_V\,T = \Delta S_V(T_E - T) = \Delta S_V\,\Delta T \tag{6.7}$$

where $\Delta T = T_E - T$ is known as the amount of *supercooling* below (or *superheating* above) the equilibrium transformation temperature. If ΔH_V is negative (e.g., heat is released, as occurs in solidification), then decreasing the temperature below the equilibrium temperature will drive the transformation (i.e., ΔG_V will be negative). Similarly, melting (which absorbs heat) is favored by increasing the temperature

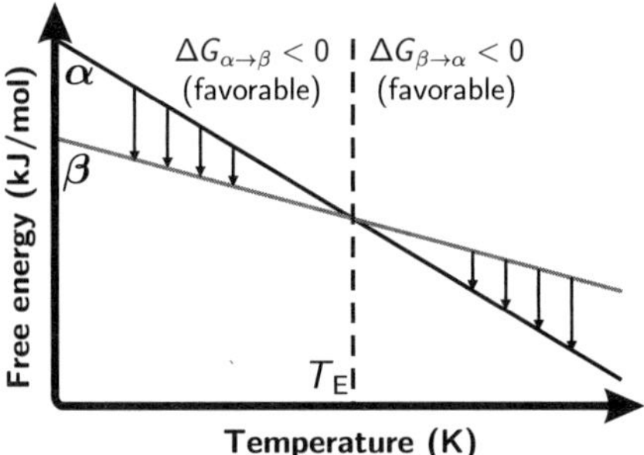

FIGURE 6.4 Free-energy curves illustrating the driving force for transformation between two phases (α and β) as a function of temperature. At $T = T_E$, $G_\alpha = G_\beta$ and hence $\Delta G_V = 0$. As the temperature decreases below T_E, the driving force for the transformation reaction $\alpha \to \beta$ increases. Similarly, as the temperature increases above T_E, the driving force for the transformation reaction $\beta \to \alpha$ increases.

above the equilibrium temperature. Equation 6.6 tell us that the size of the driving force for transformation increases linearly with increasing ΔT. Thus, the larger the deviation from the equilibrium temperature, the larger the driving force for transformation. (This should make sense!). See Figure 6.4 for further illustration.

Similar although somewhat more complex analyses can be applied to estimate ΔG_V when a large change in composition (at constant temperature) triggers a phase transformation. In such cases, the analysis depends on the functional dependence of the Gibbs free energy of mixing between the components involved in the system. Such analyses are beyond the scope of the present text.

Because solids and liquids are fairly incompressible, the external gas pressure typically has very little influence on the thermodynamics or kinetics of condensed-matter phase transformations. However, because condensed-matter phase transformations can be accompanied by large volume changes (since reactant and product phases may have different densities), the resulting internal stresses/strains can have significant impacts on the thermodynamics and kinetics of the phase transformation. These stress or strain energy effects are often incorporated as an additional term in ΔG_V. As an example, consider the nucleation of spherical particles of a new phase in a parent matrix with isotropic elastic properties, where a volume change (and hence a strain) occurs upon nucleation. In this case, the additional strain energy term ΔG_{strain} is given as

$$\Delta G_{\text{strain}} = 4\mathbf{G}\epsilon^2 \tag{6.8}$$

where $\mathbf{G}$ is the shear modulus of the parent phase and ϵ is the dilational strain due to transformation (which can be calculated from the specific volume difference between the two phases). In this introductory textbook, we will not consider the effects of stress/strain on phase transformations in more detail—however, for an example of a materials system where such effects have important practical repercussions, read the dialog box below on transformation-toughened zirconia.

Example 6.1

Question: Calculate ΔG_V for the condensation of water vapor to liquid water (i.e., "rain formation") at 298 K and atmospheric pressure assuming supersaturation of the water vapor such that $P_{H_2O(g)} = 0.1$ atm. You are given the following information about this reaction:

$$H_2O(g) \rightarrow H2O(l)$$

- $\Delta H° = -44.0$ kJ/mol
- $\Delta S° = -118.9$ J/(mol $\cdot$ K)

Solution: Based on $\Delta H°$ and $\Delta S°$, we can calculate $\Delta G°$ for this reaction as

$$\Delta G° = \Delta H° - T\Delta S°$$

$$= -44{,}000 \text{ J/mol} - 298 \text{ K} \cdot -118.9 \text{ J/(mol} \cdot \text{K)}$$

$$= -8568 \text{ J/mol}$$

Now, $\Delta G°$ gives the Gibbs free energy for this reaction at STP, in other words at $T = 298$ K and $P_{H_2O(g)} = 1$ atm. However, we need to calculate ΔG for this reaction at $T = 298$ K and $P_{H_2O(g)} = 0.1$ atm. In order to adjust for the difference in pressure, we must recall Equation 2.16 from Chapter 2:

$$\Delta G = \Delta G° + RT \ln Q$$

$$= \Delta G° + RT \ln \left[\frac{a_{H_2O(l)}}{a_{H_2O(g)}} \right] = \Delta G° + RT \ln \left[\frac{1}{P_{H_2O(g)}/P°} \right]$$

$$= -8568 \text{ J/mol} + 8.314 \text{ J/(mol} \cdot \text{K)} \cdot 298 \text{ K} \cdot \ln \left[\frac{1}{0.1 \text{ atm}/1 \text{ atm}} \right]$$

$$= -2863 \text{ J/mol}$$

Now, to convert this molar free-energy change into a volumetric free-energy change, we need to divide by the molar volume of liquid water:

$$\Delta G_V = \frac{\Delta G}{V_m} = \frac{\Delta G}{\frac{M}{\rho}}$$

$$= \frac{-2863 \text{ J/mol}}{(18.0 \text{ g/mol})/(1 \text{ g/cm}^3)}$$

$$= -159 \text{ J/cm}^3 = -159 \times 10^{-6} \text{ J/m}^3$$

Transformation-Induced Toughening: Exploiting a Pressure-Induced Phase Transformation to Make Tough Ceramics

Stress and strain effects can have an important effect on solid–solid phase transformations. Transformation toughening in ceramics is a classic example. In certain ceramic systems, such as *partially stabilized zirconia*, a pressure-induced phase transformation is exploited to dramatically increase the toughness of the ceramic—providing the ability to create amazingly resilient ceramic hammers, nails, knives, and other unique objects.

The basic idea behind transformational toughening is illustrated in Figure 6.5 using zirconia as an example. During fabrication, the zirconia ceramic is partially stabilized in its metastable (nonequilibrium) tetragonal phase by the addition of a small amount of yttrium dopant (Figure 6.5*a*). During use, the stress concentration associated with any crack tip that develops in the ceramic induces the metastable tetragonal phase to locally transform to the equilibrium monoclinic phase (Figure 6.5*b*). The volume expansion associated with this phase transformation puts the crack into compression, retarding its growth and thereby significantly enhancing the fracture toughness (Figure 6.5*c*). Similar to the halogen light bulb discussion in Chapter 5, this phenomenon is an example of *negative feedback*, and it can be used to significantly increase the lifetime and reliability of critical ceramic components.

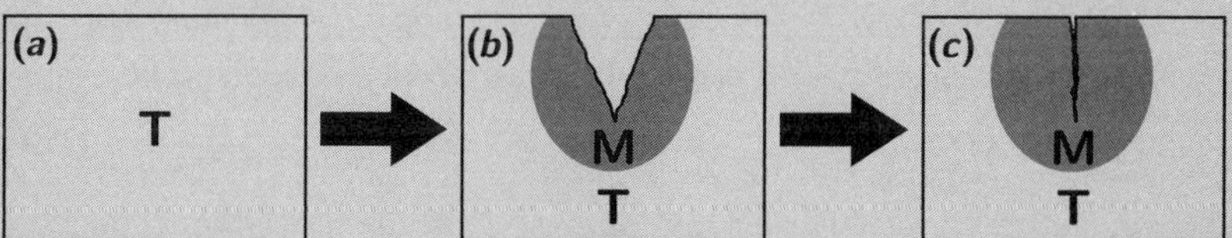

FIGURE 6.5 Mechanism of transformation toughening in zirconia. (*a*) The zirconia ceramic is synthesized in the metastable tetragonal phase (T) by the addition of a small amount of yttrium dopant during fabrication. (*b*) During use, the stress concentration associated with any crack tip that develops in the ceramic induces the metastable tetragonal phase to locally transform to the equilibrium monoclinic phase (M). (*c*) The volume expansion associated with this phase transformation puts the crack into compression, retarding its growth and thereby significantly enhancing the fracture toughness.

6.3 SPINODAL DECOMPOSITION: A CONTINUOUS PHASE TRANSFORMATION

The thermodynamic differences between spinodal decomposition (a continuous phase transformation) and nucleation and growth (a discontinuous phase transformation) are illustrated in Figure 6.6. At a given temperature, the volume free

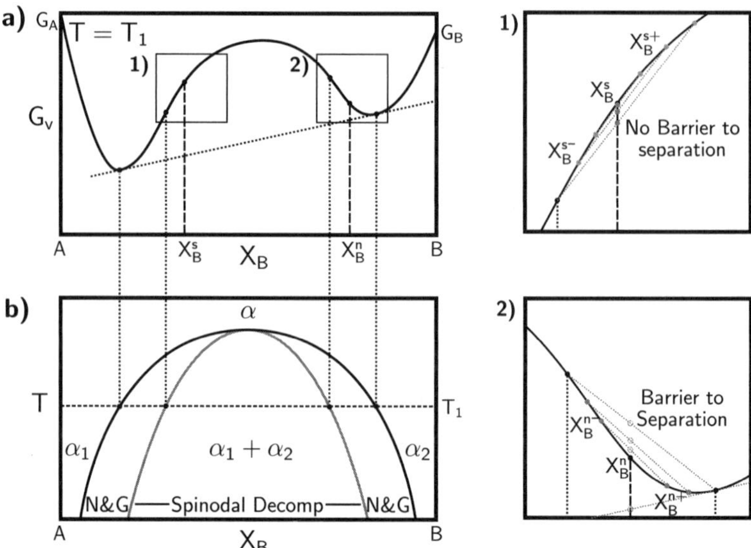

FIGURE 6.6   (*a*) Volume free energy (G_V) versus composition for a system exhibiting a large positive heat of mixing. (*b*) Resulting phase diagram for this system, exhibiting a spinodal region.

energy–composition curve for this system can be divided into two side regions where the curvature is positive (concave up) separated by a middle region where the curvature is negative (concave down). The region of negative curvature corresponds to compositions where spinodal decomposition can occur, while the regions of positive curvature correspond to compositions where the phase transformation will instead proceed via nucleation and growth.

In order to understand the reason for this distinction, consider what happens during phase transformation for a system with overall composition X_s, which is inside the spinodal region, as compared to a system with overall composition X_n, which is outside the spinodal region. As shown in the figure, for compositions inside the spinodal region, gradual separation into two distinct compositions can freely occur with an immediate and monotonic decrease in the overall Gibbs free energy of the system. There is no barrier to overcome, and thus the reaction can take place over a wide region starting with small and gradual changes in composition (small in degree, large in extent). In comparison, for compositions outside the spinodal region, initiating decomposition requires an increase in the overall Gibbs free energy of the system. Although the system will eventually separate into two distinct phases that sum to give an overall smaller Gibbs free energy, the initial local composition changes are associated with a temporary increase in the system free energy, and thus this represents an energy barrier that must be overcome for decomposition to occur. Because of this barrier, we must depend on random thermal fluctuations to create sporadic favorable instances of sufficiently large local compositional separation (large in degree, small in extent), which can then stabilize and grow larger to accomplish the phase

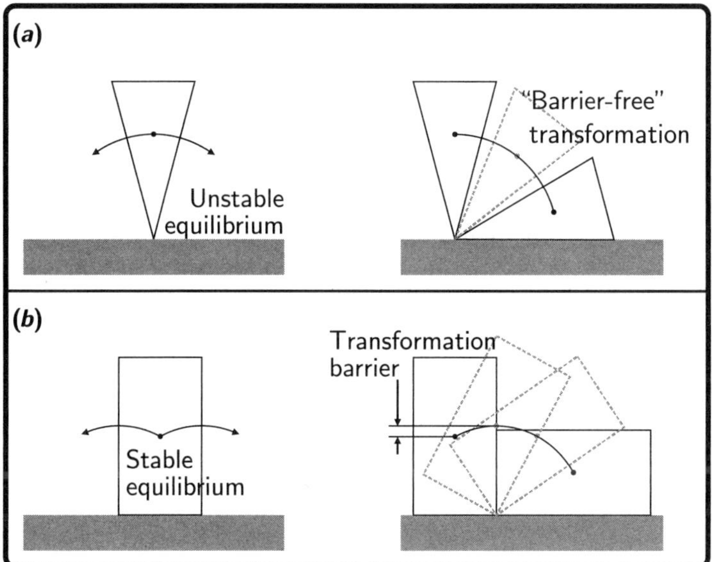

FIGURE 6.7 (*a*) Mechanical analogy for spinodal decomposition process. (*b*) Mechnical analogy for nucleation-and-growth process [11].

transformation (i.e., nucleation and growth). This has important implications on the kinetics (speed) of the transformation process.

The difference between spinodal decomposition and nucleation and growth can be captured using a handy mechanical analogy, as illustrated in Figure 6.7.

6.4 SURFACES AND INTERFACES

An important truth of nature has significant implications on the kinetics of condensed-matter phase transformations:

Surfaces and interfaces of materials have greater free energy than the bulk.

Surfaces/interfaces generally have higher energy than the bulk because atoms at the surface/interface have missing or dangling bonds and/or possess altered structural arrangement. The "extra" energy associated with a surface or interface compared to the bulk is quantified by a surface energy term, γ:

$$\gamma = \left(\frac{\partial G}{\partial A} \right)_{T,P} \tag{6.9}$$

Thus, γ quantifies the amount of excess (compared to the bulk) free energy per unit area of surface or interface.

Because surface and interfaces have higher energy than the bulk and because a phase transformation necessarily creates new interfaces, these interfaces represent an important energetic "cost" that must be paid in order to create the new phase. A crucial difference between continuous phase transformations (such as spinodal decomposition) and discontinuous phase transformations (such as nucleation and growth) is in how this energy cost is paid:

- *During spinodal decomposition*, the energy cost associated with creating the new phase interfaces is gradually and continuously paid as the phase transformation proceeds. As illustrated in Figure 6.2, the interfaces between the two phases are initially quite diffuse and only sharpen gradually with time; thus the additional energy required for interface formation must also only be gradually paid for. This energy can be easily offset by the decrease in volume free energy associated with the decomposition into two lower energy phases. Thus, the total system free energy (volume free energy + interfacial free energy) can gradually and continuously decrease as the two phases separate, and there is no "up-front" energy barrier impeding the transformation process.
- *During nucleation and growth*, the energy cost associated with creating the new phase interfaces must be paid up front immediately upon nucleation of the new phase. As Figure 6.2 shows, the nucleation process creates immediate and abrupt interfaces between the two phases, so the additional energy required for interface formation must be paid at the onset of phase separation. This energy can only be offset by the decrease in volume energy associated with the decomposition into two lower energy phases *if the nucleating particle is sufficiently large*. Thus, the total system free energy (volume free energy + interfacial free energy) goes through an intermediate maximum, which represents an up-front energy barrier impeding the transformation process.

6.4.1 Estimating Surface Energies

In order to understand the origin and principles of surface energies in greater detail, let us consider a few illustrative examples.

First, recall a favorite childhood pastime: blowing soap bubbles. As illustrated in Figure 6.8, work is done when a soap bubble is made bigger. The work required to expand the soap bubble is directly related to the amount of new surface area created:

Work done to expand bubble = Energy stored in creating new bubble surface area

$$F\,dr = 2\gamma(dA)$$

$$\approx 2\gamma(8\pi r\,dr)$$

$$\gamma \approx \frac{F}{16\pi r} \tag{6.10}$$

The factor of 2 accounts for the fact that two surfaces are created (inside surface and outside surface). The units for γ are force/distance, which converts to energy/area.

Now consider the formation of new surfaces in a solid material. Imagine that a solid block of a crystalline material is cut in half as shown in Figure 6.9. This process creates two sets of fresh surface but involves the breaking of atomic bonds as the two

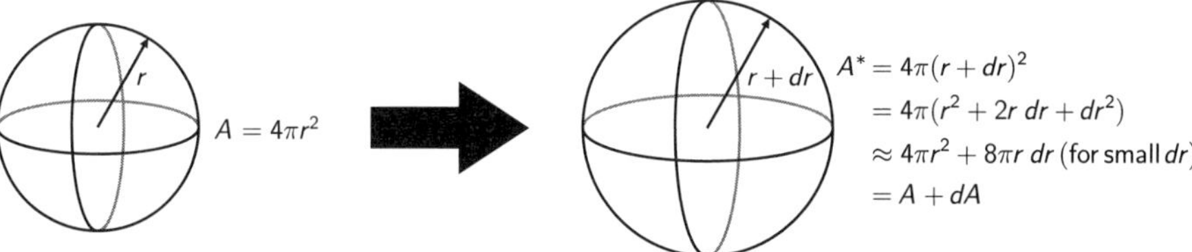

$$A = 4\pi r^2$$

$$A^* = 4\pi(r + dr)^2$$
$$= 4\pi(r^2 + 2r\,dr + dr^2)$$
$$\approx 4\pi r^2 + 8\pi r\,dr \text{ (for small } dr)$$
$$= A + dA$$

FIGURE 6.8 The work required to expand a soap bubble is directly related to the creation of additional bubble surface area.

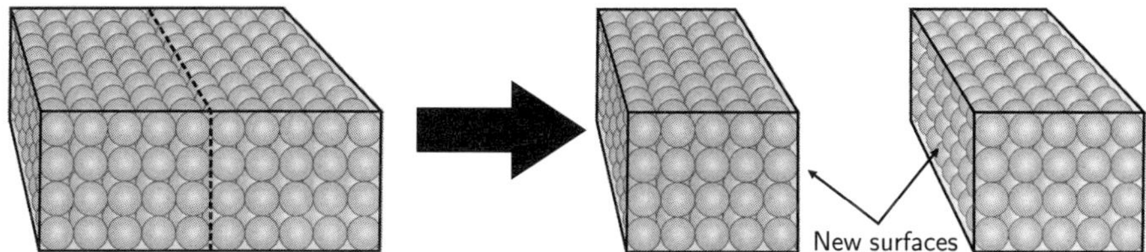

FIGURE 6.9 Cutting a block of solid material in half creates two fresh surfaces. The number and energy of the bonds that are broken per unit area of surface can be used to estimate the surface energy.

halves of the solid are separated from one another. The number of broken bonds per unit area of surface created (n_b) can be estimated as

$$n_{\mathrm{b}} \approx \frac{C}{3}\frac{N}{A} \tag{6.11}$$

where N/A is the number of atoms per unit area of surface and C is the *coordination number* for the atoms. The symbol C describes how many neighboring atoms each atom is bonded, or *coordinated to* in the solid and depends on the atomic arrangement and typically varies from as low as 4 to as high as 12. The factor of $\frac{1}{3}$ accounts for the fact that approximately one-third of the bonds between a fresh surface atom and its neighbors are broken when the surface is created (bonds with "below-plane" and "in-plane" neighbors are preserved; it is only "above-plane" bonds that are broken).

In order to calculate the surface energy, it is next necessary to calculate how much energy is associated with each broken bond. The molar heat of sublimation (ΔH_s) gives the enthalpy required to sublimate 1 mol of a substance from the solid to the gas phase; thus it provides a useful estimate of the amount of energy required to break all of the bonds in 1 mol of a substance. Assuming again that each atom in a solid is bonded to C nearest neighbors, the energy per bond (ϵ) can be estimated as

$$\epsilon = \frac{\Delta H_s}{0.5 C N_A} \tag{6.12}$$

where Avogadro's number (N_A) converts ΔH_s from energy/mole to energy/atom and the factor of 0.5 in the denominator accounts for the fact that every bond is shared by two atoms.

Combining Equations 6.11 and 6.12, the surface energy can be estimated:

$$\gamma = \frac{1}{2}n_b\epsilon \tag{6.13}$$

where the factor of $\frac{1}{2}$ accounts for the fact that cleaving the solid creates two sets of surface (just as two sets of surface were created in the soap bubble example). Evaluating this expression gives

$$\gamma = \frac{1}{2}n_b\epsilon$$

$$= \frac{1}{2}\frac{C}{3}\frac{N}{A}\frac{\Delta H_s}{0.5CN_A}$$

$$= \frac{N}{A}\frac{\Delta H_s}{3N_A} \tag{6.14}$$

This expression indicates that a material with a higher molar heat of sublimation (high ΔH_s) will tend to have a higher surface energy. As this expression suggests, surface energy in a solid is also surface orientation and surface termination dependent. Thus, different crystallographic surface orientations will be more or less favored depending on their surface energies. This factor often helps explain the preference for certain geometric shapes evidenced in a wide variety of free crystals and materials microstructures.

Example 6.2

Question: Silver has the FCC crystal structure with a unit cell parameter $a_0 = 4.09$ Å. Each atom is coordinated to 12 neighbors. Given that the molar heat of sublimation for Ag is 284 kJ/mol, estimate the surface energy (in J/cm^2) of the Ag (100) surface.

Solution: Using Equation 6.14 as a point of departure, we must first determine the atomic surface density, (N/A). We can do this by using our knowledge of the FCC structure and the lattice constant value given in the problem statement. The packing of the (100) plane in the FCC structure is given in Figure 6.10. Based on this diagram, we can determine that two atoms are packed into an area of a_0^2. Thus,

$$\text{For FCC (100):} \qquad \frac{N}{A} = \frac{2}{a_0^2} \tag{6.15}$$

Inserting this result along with the other values provided in the problem statement into Equation 6.14 yields

$$\gamma_{Ag,(100)} = \frac{2}{a_0^2}\frac{\Delta H_{s,Ag}}{3N_A}$$

$$= \frac{2}{(4.09\times10^{-10}\text{ m})^2}\cdot\frac{284{,}000\text{ J/mol}}{3\times6.022\times10^{23}/\text{mol}}$$

$$= 1.88\text{ J/m}^2 = 1.88\times10^{-4}\text{ J/cm}^2 \tag{6.16}$$

For comparison, the experimentally measured value is $\approx 1.8 \times 10^{-4}$ J/cm^2.

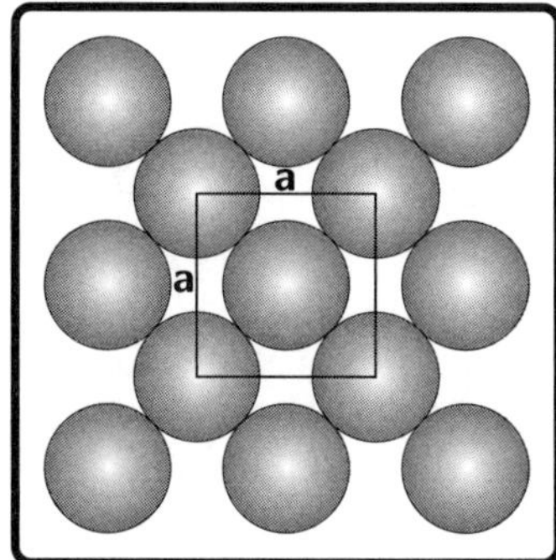

FIGURE 6.10 Atomic arrangement of the (100) plane in the FCC structure.

6.4.2 Interfacial Energy Balances

The relative interfacial energies between different phases influence how they interact with each other. As an example, consider the wetting behavior of a liquid water droplet on the surface of a solid, as illustrated in Figure 6.11. The contact angle between the water droplet and the solid surface (θ) is determined by the balance between three different interfacial energy terms:

1. The interfacial energy of the liquid–solid interface, γ_{ls}
2. The interfacial energy of the liquid–vapor interface, γ_{lv}
3. The interfacial energy of the solid–vapor interface, γ_{sv}

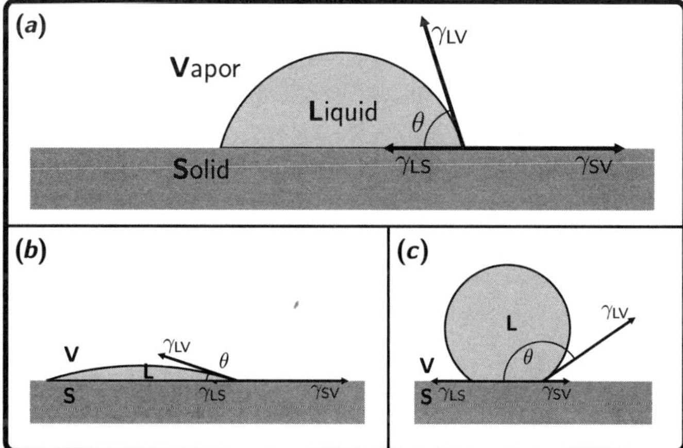

FIGURE 6.11 Interfacial energy balances determine the wetting behavior of a liquid droplet in contact with a solid surface. (*a*) The balance of the *x*-direction components of the interfacial energy forces determines the contact angle θ. (*b*) If $\gamma_{ls} \ll \gamma_{sv}$, $\theta \to 0°$ and completely hydrophilic (spreading) behavior occurs. (*c*) If $\gamma_{ls} \gg \gamma_{sv}$, $\theta \to 180°$ and completely hydrophobic behavior occurs.

Equilibrium occurs when a contact angle (θ) is established that exactly balances the x-direction components of the interfacial energy forces[1]:

$$\gamma_{ls} = \gamma_{sv} + \gamma_{lv}\cos(180 - \theta) = \gamma_{sv} - \gamma_{lv}\cos(\theta) \tag{6.17}$$

Depending on the relative magnitude of γ_{ls} versus γ_{sv}, the water droplet can exhibit wetting (hydrophilic) or nonwetting (hydrophobic) behavior:

- If $\gamma_{ls} \ll \gamma_{sv}$, $\theta \to 0°$ and completely hydrophilic (spreading) behavior occurs. Systems exhibiting $\theta < 90°$ are considered wetting. Thus, if the liquid phase "likes" the solid phase (i.e., has lower interfacial energy) more than the vapor phase likes the solid phase, wetting will occur.
- If $\gamma_{ls} \gg \gamma_{sv}$, $\theta \to 180°$ and completely hydrophobic behavior occurs. Systems exhibiting $\theta > 90°$ are considered nonwetting. Thus, if the vapor phase likes the solid phase (i.e., has lower interfacial energy) more than the liquid phase likes the solid phase, nonwetting behavior will occur.

The interfacial energy balance associated with grain boundaries explains why a dihedral angle of $120°$ is frequently observed at grain triple points, as illustrated in Figure 6.12a. Assuming an isotropic material where all grain boundaries possess approximately the same interfacial energy, irrespective of orientation, the interfacial energy force balance at grain triple points requires that the three grain boundaries converge with equal $120°$ angles.

Figure 6.12b illustrates the force balances associated with the nucleation of a new second phase (phase β) along the grain boundary of a parent

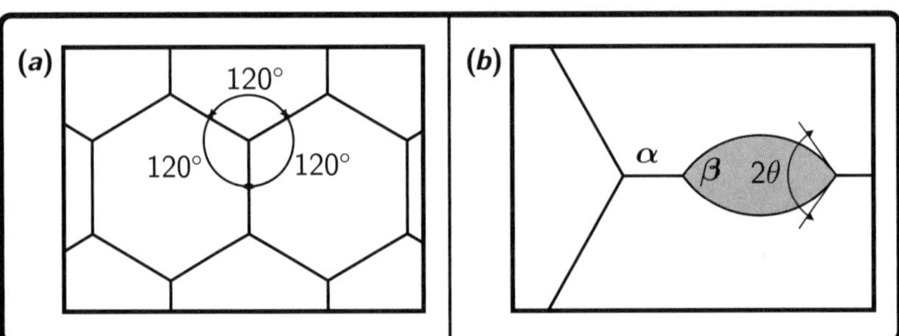

FIGURE 6.12 (a) Isotropic grain boundary energies result in a characteristic $120°$ angle of convergence at grain boundary triple points. (b) Nucleation of a new phase often proceeds at heterogeneous nucleation sites associated with surfaces or interfaces. The wetting angle of the new phase nucleating along a grain boundary interface of the old phase is determined by the relative surface energies of the interfaces.

---

[1]Although it appears from this figure that the z-direction component of the liquid–vapor interfacial energy force is uncompensated, it is in fact offset by the force of gravity pulling on the droplet (not shown).

phase (phase α). In this case, the wetting angle between the two phases is given by

$$\gamma_{\alpha\alpha} = 2\gamma_{\alpha\beta} \cos(\theta) \tag{6.18}$$

where $\gamma_{\alpha\alpha}$ is the interfacial grain boundary energy of phase α and $\gamma_{\alpha\beta}$ is the interfacial energy of the phase α/phase β contact. If the two phases prefer each other much more than phase α prefers itself (i.e., $2\gamma_{\alpha\beta} \ll \gamma_{\alpha\alpha}$), then $\theta \to 0°$ and nucleation of phase β along the phase α grain boundaries is favored. However, if phase α prefers itself much more than it prefers phase β (i.e., $\gamma_{\alpha\alpha} \ll 2\gamma_{\alpha\beta}$), then $2\theta \to 180°$ and nucleation of phase β along the phase α grain boundaries is not favored.

6.4.3 Overview of Important Surface/Interface Energy Effects

The existence of surface and interface energies has wide implications in materials kinetics. Here are some of the most important implications:

- Systems seek to minimize their total free energy G. Since surfaces and interfaces carry "excess" free energy, systems will seek to minimize the amount of surface/interface area per unit volume.
- Smaller particles have higher surface/volume ratios. Thus, smaller particles tend to be less stable, be more reactive, have lower melting point (relative to the bulk), and so on.
- The desire to decrease interface/volume ratio provides the driving force for many aspects of microstructural evolution, including coarsening or ripening (Figure 6.13a), grain growth (Figure 6.13b), and sintering (Figure 6.13c). These issues will be explored in Chapter 7.
- The energy cost to create surfaces/interfaces leads to a *nucleation barrier* in condensed-matter phase transformations. Therefore, nucleation-based phase transformations can only occur if *the energy released by creating the new volume of the second phase sufficiently offsets the energy expended in creating the new interfacial area*. This leads to a minimum viable nucleation size and thus helps determine the speed at which nucleation can proceed. These issues will be discussed in the next section!

6.5 NUCLEATION

As we discussed in Section 6.2, a *thermodynamic driving force*, quantified by ΔG_V, must be present in order for a phase transformation to occur. In general, the magnitude of this driving force increases linearly with the degree of deviation from equilibrium, as expressed, for example, via

$$\Delta G_V = \Delta H_V \frac{\Delta T}{T_E} = \Delta S_V \Delta T \tag{6.19}$$

Thus, the larger the degree of supercooling below (or superheating above) the equilibrium temperature, the larger the driving force for phase transformation.

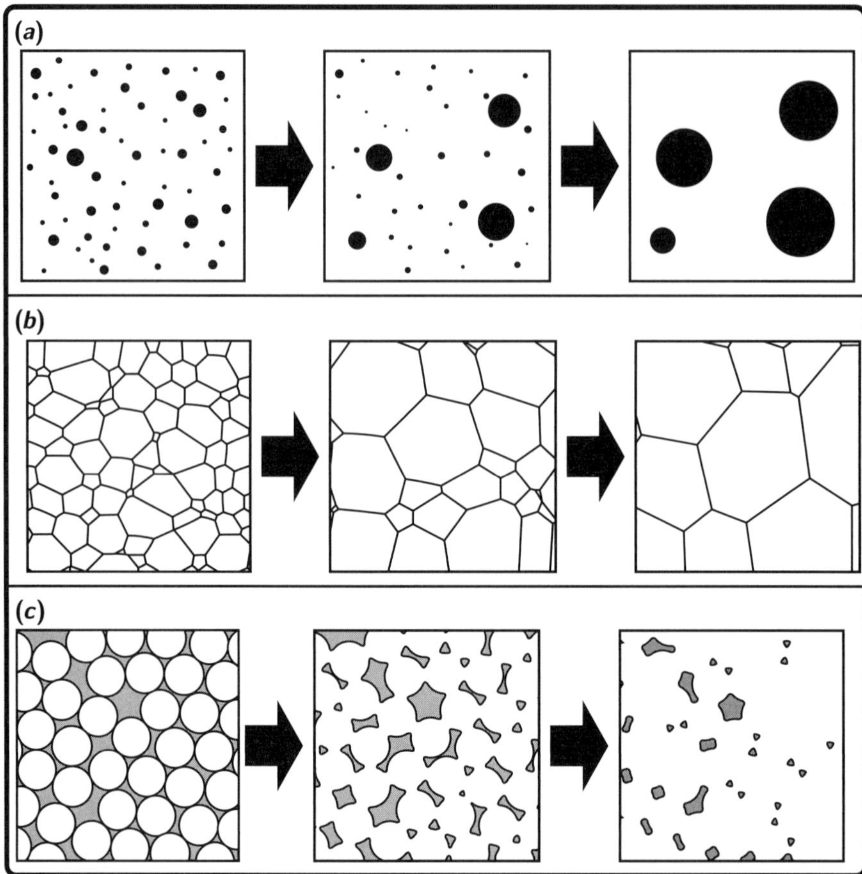

FIGURE 6.13   (*a*) The coalescence of many small second-phase precipitates into a few larger precipitates is driven by the reduction in interfacial area/energy. (*b*) Grain growth is likewise driven by the reduction in grain boundary area per unit volume. (*c*) Sintering is driven by the reduction in surface energy as well as a reduction in surface curvature, since highly curved surfaces possess higher energy (per unit area) than low-curvature surfaces. These phenomena, which are crucial to understanding and predicting the kinetics of microstructure evolution, will be examined in detail in Chapter 7.

While Equation 6.19 provides a way to estimate the driving force for a phase transformation, it says nothing about the *speed* of the transformation process. For phase transformations governed by nucleation and growth, the speed of transformation depends, in part, on the rate at which viable nuclei of the new phase can form. This is quantified by the nucleation rate $\dot{N}$.

6.5.1 Homogeneous Nucleation

Consider the homogeneous nucleation of β particles in an α matrix as illustrated in Figure 6.14. The nucleation process illustrated in Figure 6.14 is referred to as

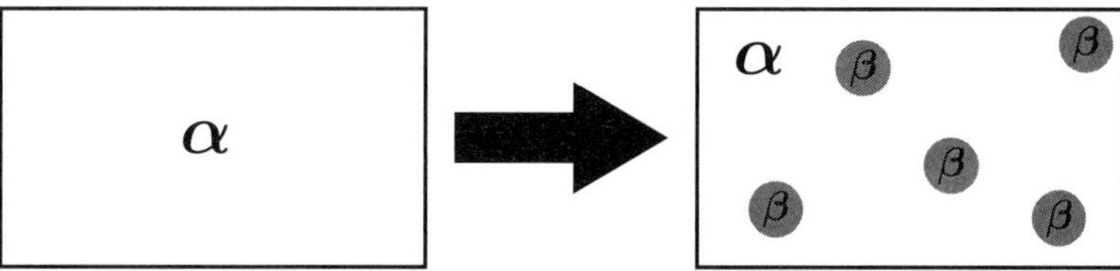

FIGURE 6.14   Homogeneous nucleation of β nuclei in an α-phase matrix.

homogeneous nucleation because the nucleation is envisioned to proceed without specific locational preference anywhere inside the volume of the α phase. This is in contrast to *heterogeneous nucleation*, to be discussed later, which preferentially occurs at specific favored locations such as along surfaces, walls, or grain boundaries of the parent phase.

At the atomic level, the formation of a β nucleus requires local atomic rearrangement and replacement, as A and B atoms must move in order to establish a composition consistent with the β phase and the lattice must reorganize itself so that the atomic positions are consistent with the pattern of the β lattice. The probability that a sufficient number of atoms will simultaneously align themselves into the β phase to create a viable nucleus is very small, and since the rearrangement requires diffusion, this probability is highly temperature dependent. Furthermore, creation of a β nucleus creates new interfacial area, which has an associated energy cost. Therefore, nucleation-based phase transformations can only occur if the energy released by creating the new volume of the second phase sufficiently offsets the energy expended in creating the new interfacial area. This leads to a minimum viable nucleation size.

Calculating Minimum Viable Nucleus Size (r*) and Nucleation Activation Energy (ΔG^*)   Since a spherical nucleus minimizes the amount of interfacial area per unit volume, it represents the most likely nucleus shape.[2] The total free-energy change, ΔG_{tot}, involved in forming a spherical nucleus of radius r is therefore

$$\Delta G_{\text{tot}} = \Delta G_{\text{volume}} + \Delta G_{\text{interface}}$$

$$= \frac{4}{3}\pi r^3 \, \Delta G_V + 4\pi r^2 \gamma \tag{6.20}$$

where ΔG_V is the free energy per unit volume released upon creating the new second-phase particle (in order for a phase transformation to occur, this term *must* be negative), and γ is the interfacial energy per unit area associated with the creation of the new interfacial area associated with the particle. The γ term is always positive, since energy is expended in making an interface. Figure 6.15 plots the three terms in Equation 6.20 as a function of the nucleus radius r. As the figure indicates, there is a critical nucleus size r^* beyond which the total free energy of the system begins to

---

[2]Certain factors such as anisotropic strain energy contributions can lead to a preference for other shapes, but this is beyond the scope of the present text.

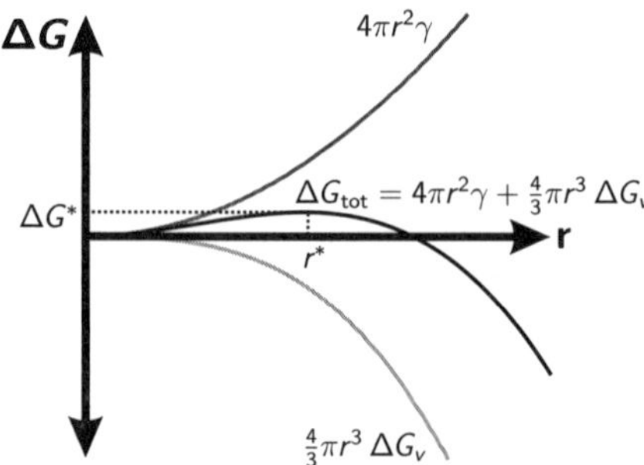

FIGURE 6.15   Volume, surface, and total free-energy changes associated with the homogeneous nucleation of a spherical particle of radius r.

decrease with further increases in the nucleus size. Thus, r^* represents the *minimum viable nucleus size*. If nuclei are formed with $r < r^*$, it is energetically unfavorable for them to continue growing, and they will likely spontaneously decompose back to the parent phase. However, if nuclei are formed with $r > r^*$, it is energetically favorable for them to persist and continue to grow.

The critical nucleus size r^* can be calculated by taking the derivative of Equation 6.20 and setting it equal to zero:

$$r = r^* \quad \text{when} \quad \frac{\partial \Delta G_{\text{tot}}}{\partial r} = 0 = \frac{\partial \frac{4}{3}\pi r^3}{\partial r}\Delta G_V + \frac{\partial 4\pi r^2}{\partial r}\gamma$$

$$0 = 4\pi (r^*)^2 \Delta G_V + 8\pi r^* \gamma \tag{6.21}$$

Therefore:

$$r^* = -\frac{2\gamma}{\Delta G_V}$$

Incorporating Equation 6.19 into Equation 6.21 explicitly captures how r^* depends on the degree of undercooling (or superheating):

$$r^* = -\frac{2\gamma T_{\text{E}}}{\Delta H_V \, \Delta T} = -\frac{2\gamma}{\Delta S_V \, \Delta T} \tag{6.22}$$

Thus, r^* decreases with increasing ΔT. This effect is shown in Figure 6.16.

As shown in Figure 6.15, there is an energy barrier of size ΔG^* associated with the critical nucleus size r^*; ΔG^* represents the activation energy required to form a viable nucleus. It can be determined by inserting the solution for r^* back into Equation 6.20:

$$\Delta G^* = \frac{4}{3}\pi (r^*)^3 \Delta G_V + 4\pi (r^*)^2 \gamma$$

$$= \frac{4}{3}\pi \left(-\frac{2\gamma}{\Delta G_V}\right)^3 \Delta G_V + 4\pi \left(-\frac{2\gamma}{\Delta G_V}\right)^2 \gamma$$

$$= \frac{16\pi \gamma^3}{3(\Delta G_V)^2} \tag{6.23}$$

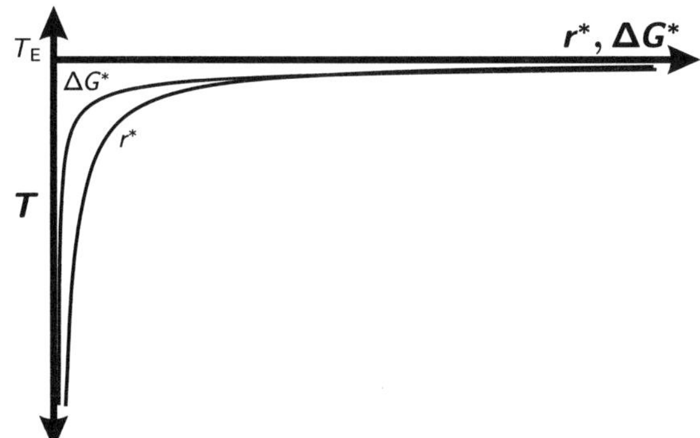

FIGURE 6.16 r^* and ΔG^* decrease with increasing undercooling.

Incorporating Equation 6.19 into Equation 6.23 explicitly captures how ΔG^* depends on the degree of undercooling (or superheating):

$$\Delta G^* = \frac{16\pi\gamma^3(T_E)^2}{3(\Delta H_V)^2(\Delta T)^2} = \frac{16\pi\gamma^3}{3(\Delta S_V)^2(\Delta T)^2} \tag{6.24}$$

Thus, ΔG^* decreases with increasing ΔT. This effect is shown in Figure 6.16.

As shown in the following example, both r^* and ΔG^* tend to be small numbers.

Example 6.3

Question: Calculate r^* and ΔG^* for the homogeneous nucleation of rain (liquid water) from water vapor at 298 K and atmospheric pressure assuming supersaturation of the water vapor such that $P_{H_2O(g)} = 0.1$ atm. The surface energy of liquid water in contact with humid (saturated) air at $T = 298$ K is $\gamma_{lv} = 7.2 \times 10^{-6}$ J/cm^2.

Solution: To answer this question, we must first calculate ΔG_V for the condensation of liquid water under the conditions given. Fortunately, this was the subject of Example 6.1, which yielded $\Delta G_V = -159$ J/cm^3 at $T = 298$ K and $P_{H_2O(g)} = 0.1$ atm. Armed with this result, we can then apply Equation 6.21 to calculate r^*:

$$r^* = -\frac{2\gamma}{\Delta G_V}$$

$$= -\frac{2 \cdot 7.2 \times 10^{-6} \text{ J/cm}^2}{-159 \text{ J/cm}^3}$$

$$= 9.1 \times 10^{-8} \text{ cm} = 9.1 \text{ Å}$$

The radius of a water molecule is about 1.5 Å. Thus this critical nucleus size represents a cluster of about 210 water molecules packed together!

Next, we apply Equation 6.23 to calculate ΔG^*:

$$\Delta G^* = \frac{16\pi\gamma^3}{3(\Delta G_V)^2}$$

$$= \frac{16\pi(7.2 \times 10^{-6} \text{ J/cm}^2)^3}{3(-159 \text{ J/cm}^3)^2}$$

$$= 2.5 \times 10^{-19} \text{ J}$$

Typically ΔG^* values will be on the order of 10^{-18}–10^{-21} J. While these seem like extremely small numbers, they represent the activation energy on a *per-nuclei* basis. When translated into an activation energy *per mole* of nuclei (by multiplying by N_A), the resulting activation energies are approximately 1000–100,000 J/mol.

Calculating Concentration of Viable Nuclei (n*)   Now that we have determined the activation energy for nucleation, it is possible, using statistical thermodynamics concepts, to calculate the concentration of viable nuclei (n^*) that will form at a given temperature T. As with other activated processes, the probability that a viable nucleus can form is exponentially temperature dependent:

$$n^* = n_0 e^{-N_A \Delta G^*/(RT)} \tag{6.25}$$

In this equation, n_0 is the molar site density ($n_0 = 1/V_m$, where $V_m = M/\rho$ is molar volume). Avogadro's number (N_A) is needed to transform ΔG^*, which has units of Joules, into a molar energy, with units of J/mol.[3] The term n^* will have the same units as n_0, that is, moles/vol. The variation of n^* with temperature for a phase transformation that occurs on cooling a system below T_E is shown in Figure 6.17a. The concentration of viable nuclei reaches a peak at an intermediate temperature somewhere well below T_E. This is due to two competing phenomena. First, the activation energy for nucleation, as expressed by ΔG^*, decreases with decreasing temperature below T_E (refer to Figure 6.16). This helps increase n^* with decreasing temperature since the numerator in Equation 6.25 decreases. At the same time, however, because nucleus formation is a temperature-activated process, as T decreases, this decreases the denominator in Equation 6.25, which has an opposing effect. For a nucleation-based phase transformation induced by heating above T_E, the nucleation activation energy decreases and the temperature-activated probability of nucleation increases, resulting in a monotonically increasing nucleation concentration n^* with increasing temperature, as shown in Figure 6.17b.

---

[3] Alternatively, we could replace R in denominator of the exponential with k and keep the units on ΔG^* in joules.

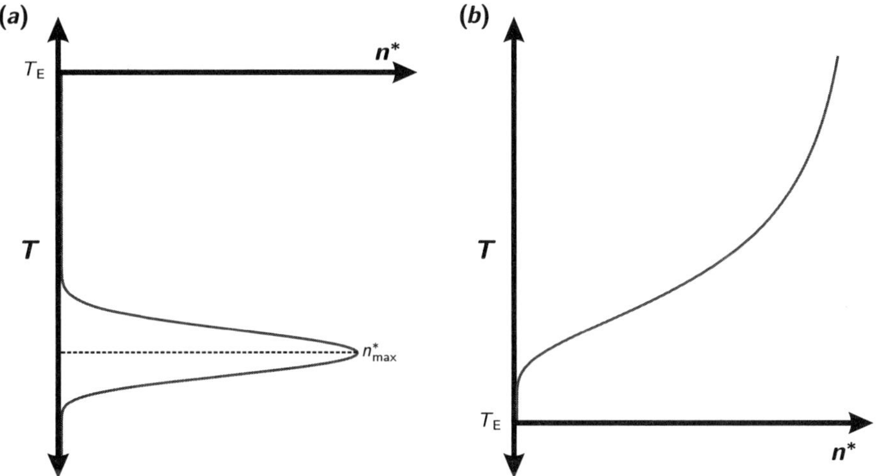

FIGURE 6.17 (*a*) For a phase transformation that occurs on cooling a system below T_E, the concentration of viable nuclei (n^*) reaches a peak at an intermediate temperature somewhere well below T_E. Although decreasing the temperature increases the driving force for transformation and thus decreases ΔG^*, nucleation is a temperature-activated process and thus becomes less favored as the temperature decreases. (*b*) For a nucleation-based phase transformation induced by heating above T_E, the nucleation activation energy decreases and the temperature-activated probability of nucleation increases, resulting in a monotonically increasing nucleation concentration n^* with increasing temperature.

Example 6.4

Question: Calculate n^* for the homogeneous nucleation of rain (liquid water) from water vapor at 298 K and atmospheric pressure assuming supersaturation of the water vapor such that $P_{H_2O(g)} = 0.1$ atm. In order to estimate the nucleation site density (n_0) assume that the liquid water is nucleating in air at atmospheric pressure, so assume a site density value consistent with the molar concentration (mol/vol) of water in air at $P_{H_2O(g)} = 0.1$ atm.

Solution: To answer this question, we must first calculate n_0. To estimate the nucleation site density, we can calculate the molar concentration of water molecules in air using the ideal gas law as

$$PV = nRT$$

$$c = \frac{n}{V} = \frac{P}{RT}$$

$$= \frac{0.1 \text{ atm} \cdot 101,300 \text{ Pa/atm}}{8.314 \text{ J/(mol} \cdot \text{K)} \cdot 298 \text{ K}}$$

$$= 4.09 \text{ mol/m}^3$$

Inserting this result and the value we calculated for ΔG^* from Example 6.3 into Equation 6.25 gives

$$
\begin{aligned}
n^* &= n_0 \exp\left(-\frac{N_A \Delta G^*}{RT}\right) \\
&= (4.09 \text{ mol/m}^3) \ \exp\left[-\left(\frac{6.022 \times 10^{23}/\text{mol} \cdot 2.47 \times 10^{-19} \text{ J}}{8.314 \text{ J}/(\text{mol} \cdot \text{K}) \cdot 298 \text{ K}}\right)\right] \\
&= 3.45 \times 10^{-26} \text{ mol/m}^3
\end{aligned}
$$

Multiplying by N_A to calculate the number of viable nuclei per cubic meter yields

$$
n^* = 3.45 \times 10^{-26} \text{ mol/m}^3 \cdot 6.022 \times 10^{23}/\text{mol} = 2.08 \times 10^{-2} \text{ nuclei/m}^3
$$

Thus, on average there will be about one viable homogeneous raindrop nucleus for every 50 m^3 of atmosphere. This is a small number and indicates how difficult homogeneous nucleation can be! Because homogeneous nucleation is so improbable, and because there are plenty of heterogeneous nucleation sites for raindrop formation in the atmosphere (e.g., dust particles), raindrop formation (and many other phase transformations) typically occurs by *heterogeneous nucleation* rather than homogeneous nucleation. We will discuss heterogeneous nucleation in the following section.

6.5.2 Heterogeneous Nucleation

So far, we have only considered the ideal case of homogeneous nucleation. However, in reality, nucleation rarely occurs homogeneously. It is far more common for nucleation to occur *heterogeneously* at specific sites, such as on surfaces, interfaces, or grain boundaries. The reason for this effect is shown in Figure 6.18 for the heterogeneous nucleation of a solid phase from a molten liquid along a mold wall. If $\gamma_{s/wall}$ is lower than γ_{sl}, it can be shown that the overall surface energy of the nucleating solid is

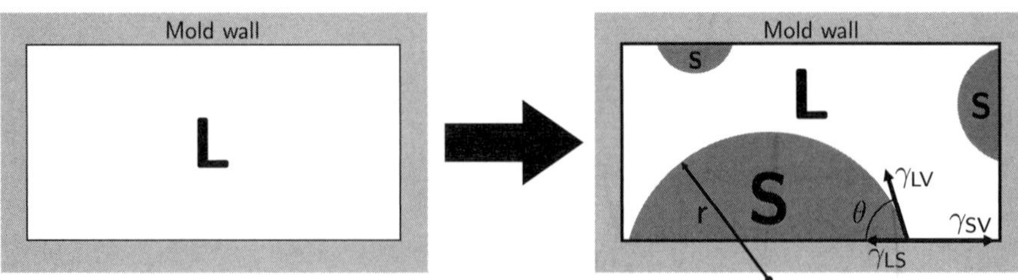

FIGURE 6.18   Heterogeneous nucleation of a solid phase from a molten liquid at a mold wall.

lower if it forms heterogeneously in contact with the wall rather than homogeneously in the melt. This is true even though the nucleus no longer has a spherical shape that minimizes its surface-to-volume ratio.

Considering the geometry of the nucleating phase depicted in Figure 6.18, both the volume free-energy and interfacial free-energy terms are modified compared to the homogeneous nucleation case due to the change in shape and surface energy. If θ is the contact angle that the nucleating solid phase forms with the mold wall, then the volume of this "spherical cap" can be described by

$$V(\theta) = \tfrac{4}{3}\pi r^3 f(\theta) \tag{6.26}$$

where $f(\theta)$ describes the volume of the spherical cap of radius r relative to that of a full sphere of the same radius:

$$f(\theta) = \frac{2 - 3\cos\theta + \cos^3\theta}{4} \tag{6.27}$$

If $\theta = 90°$ (a hemisphere), $f(\theta) = 0.5$. Thus, the overall volume free-energy term becomes

$$\Delta G_{\text{volume}} = \tfrac{4}{3}\pi r^3 \,\Delta G_V f(\theta) \tag{6.28}$$

The interfacial free-energy term must likewise be modified to account for the different surface area and interfacial energy contributions associated with the nucleating spherical cap-shaped particle:

$$\begin{aligned}
\Delta G_{\text{interface}} &= A_{\text{sl}}\gamma_{\text{sl}} + A_{\text{s/wall}}\gamma_{\text{s/wall}} - A_{\text{wall/l}}\gamma_{\text{wall/l}} \\
&= 2\pi r^2(1 - \cos\theta)\gamma_{\text{sl}} + \pi r^2(1 - \cos^2\theta)(\gamma_{\text{s/wall}} - \gamma_{\text{wall/l}})
\end{aligned} \tag{6.29}$$

Note that the third term in this expression, which is associated with the wall/liquid interfacial energy, is *subtracted* from the other two terms because this interface is *replaced* by the solid/wall interface upon nucleation of the particle. From the interfacial energy force balance derived in Equation 6.17, we can relate the interfacial energy terms to the contact angle θ via

$$\gamma_{\text{wall/l}} = \gamma_{\text{s/wall}} + \gamma_{\text{sl}}\cos\theta \tag{6.30}$$

And, so by substituting this into Equation 6.29, we have

$$\begin{aligned}
\Delta G_{\text{interface}} &= 2\pi r^2(1 - \cos\theta)\gamma_{\text{sl}} + \pi r^2(1 - \cos^2\theta)(-\gamma_{\text{sl}}\cos\theta) \\
&= 2\pi r^2\gamma_{\text{sl}} - 3\pi r^2(\cos\theta)\gamma_{\text{sl}} + \pi r^2(\cos^3\theta)\gamma_{\text{sl}} \\
&= \pi r^2\gamma_{\text{sl}}(2 - 3\cos\theta + \cos^3\theta) \\
&= \pi r^2\gamma_{\text{sl}}[4f(\theta)] = 4\pi r^2\gamma_{\text{sl}}f(\theta)
\end{aligned} \tag{6.31}$$

The total free energy associated with the nucleating phase then becomes

$$\Delta G_{tot} = \Delta G_{volume} + \Delta G_{interface}$$

$$= \tfrac{4}{3}\pi r^3 \,\Delta G_V f(\theta) + 4\pi r^2 \gamma f(\theta) \tag{6.32}$$

Thus, compared to the case for homogeneous nucleation, we find that in heterogeneous nucleation both the volume and surface energy terms are simply modified by the $f(\theta)$ factor, which essentially accounts for the geometry of the cap versus the sphere. Since $f(\theta)$ decreases with decreasing θ, the more that the solid phase likes to wet the mold wall, the greater the benefit associated with heterogeneous nucleation over homogeneous nucleation. This effect is shown in Figure 6.19, which compares the total free energy as a function of nucleus size (r) for homogeneous nucleation versus heterogeneous nucleation at several different values of θ.

As was done for homogeneous nucleation, the critical nucleus size r^*_{het} for heterogeneous nucleation can be calculated by taking the derivative of Equation 6.32 and setting it equal to zero. This exercise gives

$$r = r^*_{het} \quad \text{when} \quad \frac{\partial \Delta G_{tot}}{\partial r} = 0 = \frac{\partial \tfrac{4}{3}\pi r^3}{\partial r}\Delta G_V f(\theta) + \frac{\partial 4\pi r^2}{\partial r}\gamma f(\theta)$$

$$0 = 4\pi (r^*_{het})^2 \,\Delta G_V f(\theta) + 8\pi r^*_{het}\gamma f(\theta) \tag{6.33}$$

$$r^*_{het} = -\frac{2\gamma}{\Delta G_V}$$

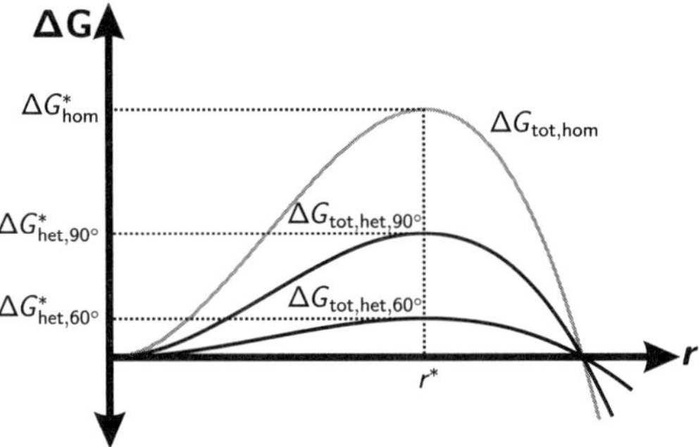

FIGURE 6.19　Comparison of the total free energy as a function of nucleus size (r) for homogeneous nucleation versus heterogeneous nucleation at $\theta = 90°$ and $\theta = 60°$. Heterogeneous nucleation significantly reduces the nucleation activation barrier, ΔG^*, with smaller θ (better wetting) yielding a larger reduction. It is important to note that although heterogeneous nucleation decreases ΔG^*, it *does not alter the critical nucleus size r^*.*

Thus, the critical nucleus size for heterogeneous nucleation *does not change* compared to the value for homogeneous nucleation. This can be seen in Figure 6.19. On the other hand, as shown in Figure 6.19, the heterogeneous nucleation activation barrier, ΔG^*_{het}, does change compared to the case for homogeneous nucleation. ΔG^*_{het} can be determined by inserting the solution for r^*_{het} back into Equation 6.32:

$$\Delta G^*_{het} = \tfrac{4}{3}\pi (r^*)^3 \Delta G_V f(\theta) + 4\pi (r^*)^2 \gamma f(\theta)$$

$$= \tfrac{4}{3}\pi \left(-\frac{2\gamma}{\Delta G_V}\right)^3 \Delta G_V f(\theta) + 4\pi \left(-\frac{2\gamma}{\Delta G_V}\right)^2 \gamma f(\theta)$$

$$= \frac{16\pi\gamma^3}{3(\Delta G_V)^2} f(\theta) \tag{6.34}$$

Finally, the concentration of viable nuclei (n^*) that can form at any given temperature T is likewise modified due to the change in ΔG^*:

$$n^*_{het} = n_0 e^{-N_A \Delta G^*_{het}/(RT)} \tag{6.35}$$

where ΔG^*_{het} decreases strongly with decreasing θ, and thus n^*_{het} increases exponentially with decreasing θ. As we will see in the following example, compared to ΔG^* for homogeneous nucleation, ΔG^*_{het} for heterogeneous nucleation is reduced to $1/2\Delta G^*$ for $\theta = 90°$ but is reduced to $\approx 1/6\Delta G^*$ for $\theta = 60°$ and to $\approx 1/75\Delta G^*$ for $\theta = 30°$. This can result in many orders of magnitude increase in the concentration of viable heterogeneous nuclei versus homogeneous nuclei, as shown in Figure 6.20.

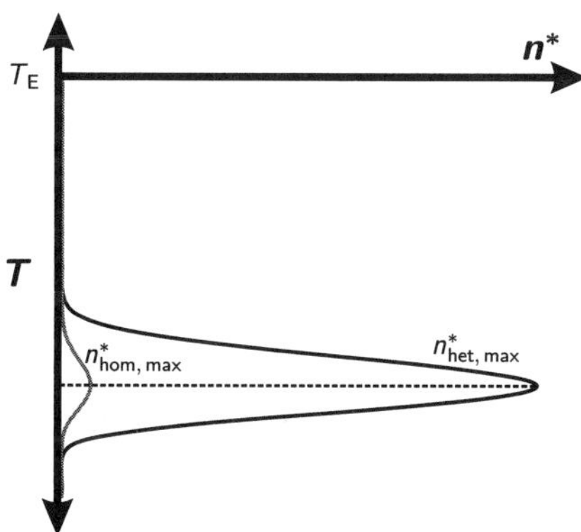

FIGURE 6.20 For a heterogeneous phase transformation that occurs on cooling a system below T_E, the concentration of viable heterogeneous nuclei (n^*_{het}) is much higher than the concentration of viable homogeneous nuclei (n^*) at all temperatures, explaining why heterogeneous nucleation tends to dominate transformation processes upon cooling.

Example 6.5

Question: If the contact angle for the nucleation of solid Au from molten Au on a mold wall is 50°, by what factor is ΔG^*_{het} reduced versus ΔG^*_{hom}?

Solution: Using Equation 6.34 and Equation 6.23, we can compute the ratio of $\Delta G^*_{\text{het}}/\Delta G^*_{\text{hom}}$ as

$$
\frac{\Delta G^*_{\text{het}}}{\Delta G^*_{\text{hom}}} = \frac{\frac{16\pi\gamma^3}{3(\Delta G_V)^2}f(\theta)}{\frac{16\pi\gamma^3}{3(\Delta G_V)^2}} = f(\theta)
$$

$$
= \frac{2 - 3\cos\theta + \cos^3\theta}{4}
$$

$$
= \frac{2 - 3\cos(50°) + \cos^3(50°)}{4} = 0.084
$$

Thus, ΔG^*_{het} is reduced by about a factor of 12 compared to ΔG^*_{hom}.

Cloud Seeding

Cloud seeding, an intriguing real-world example of heterogeneous nucleation, has been examined for more than 100 years as a way to induce or increase precipitation (rain or snow) from clouds. While there are a number of different approaches, the most common technique is to inject silver iodide particles into a cloud via either aircraft or surface-based ordinance. Silver iodide provides an excellent lattice constant match to ice, thereby providing a seed surface with low interfacial energy that facilitates heterogeneous nucleation under supercooled conditions. The greatly increased heterogeneous nucleation rate inside the cloud due to the AgI crystals therefore leads to enhanced precipitation. Although cloud seeding is currently practiced worldwide by numerous public and private agencies, its effectiveness is still somewhat controversial. Much of this controversy stems from the fact that it is difficult to determine how much precipitation would have fallen in a given storm if cloud seeding was not used. Current consensus suggests that cloud seeding of winter storms in mountain terrain likely leads to enhanced local snowfall, and the technique is currently used by a number of major ski resorts. Warm-weather cloud seeding is less clearcut; most long-term studies have been inconclusive. In warm-weather clouds, water vapor content tends to be higher and the nucleation of liquid droplets, rather than ice crystals, is more typical. Under these conditions, the benefit of silver-iodide-based heterogeneous nucleation is likely minimized. Interestingly, recent research suggests that dry ice or salt-based cloud seeding techniques may provide better efficacy in these warm-cloud situations.

Heterogeneous Nucleation at Grain Boundaries

In condensed-matter systems, heterogeneous nucleation can also occur at microstructural features such as grain boundaries or dislocations. Consider heterogeneous nucleation at a grain boundary as illustrated in Figure 6.21. The lenticular (lenslike) shape of the nucleating β-phase particle is a consequence of the minimization of total interfacial energy while satisfying the equilibrium contact angle θ established by the interfacial energy force balance. The lenticular particle can be viewed as the combination of two back-to-back symmetric spherical caps of radius r. The volume of this particle is thus twice the volume of a spherical cap of the same radius:

$$V(\theta) = \tfrac{4}{3}\pi r^3 [2f(\theta)] \tag{6.36}$$

and so the overall volume free-energy term becomes

$$\Delta G_{\text{volume}} = \tfrac{4}{3}\pi r^3 \, \Delta G_V [2f(\theta)] \tag{6.37}$$

The interfacial free-energy term becomes

$$\Delta G_{\text{interface}} = 2A_{\alpha\beta}\gamma_{\alpha\beta} - A_{\alpha\alpha}\gamma_{\alpha\alpha}$$
$$= 4\pi r^2 (1 - \cos\theta)\gamma_{\alpha\beta} - \pi r^2 (1 - \cos^2\theta)(\gamma_{\alpha\alpha}) \tag{6.38}$$

where $A_{\alpha\beta}$ is the interfacial contact area for one side of the lenticular particle. From the interfacial energy force balance for grain boundary nucleation as derived in Equation 6.18, we can relate the interfacial energy terms to the contact angle θ via

$$\gamma_{\alpha\alpha} = 2\gamma_{\alpha\beta}\cos\theta \tag{6.39}$$

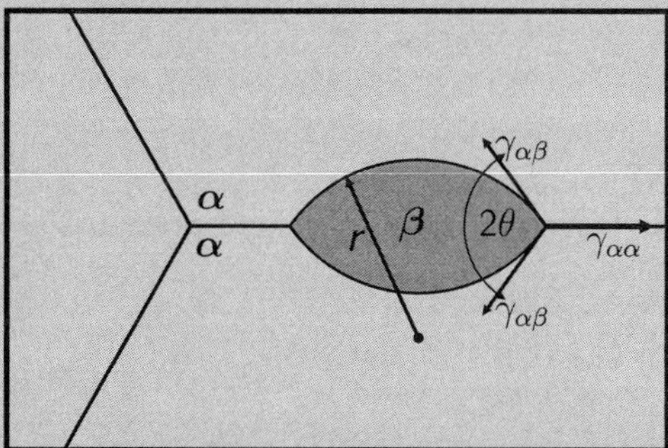

FIGURE 6.21 Heterogeneous nucleation of a *lenticular* β-phase particle at a grain boundary in the parent α-phase matrix. The lenticular (lenslike) shape of the nucleating β-phase particle is a consequence of the minimization of total interfacial energy while satisfying the equilibrium contact angle θ established by the interfacial energy force balance. The lenticular particle can be viewed as the combination of two back-to-back symmetric spherical caps of radius r.

And so by substituting this into Equation 6.38, we have:

$$\Delta G_{\text{interface}} = 4\pi r^2 (1 - \cos\theta)\gamma_{\alpha\beta} - \pi r^2 (1 - \cos^2\theta)(2\gamma_{\alpha\beta}\cos\theta)$$

$$= 2\pi r^2 \gamma_{\alpha\beta}(2 - 3\cos\theta + \cos^3\theta)$$

$$= 2\pi r^2 \gamma_{\text{sl}}[4f(\theta)] = 4\pi r^2 \gamma_{\text{sl}}[2f(\theta)] \tag{6.40}$$

The total free energy associated with the nucleating phase then becomes

$$\Delta G_{\text{tot}} = \Delta G_{\text{volume}} + \Delta G_{\text{interface}}$$

$$= \tfrac{4}{3}\pi r^3 \Delta G_V[2f(\theta)] + 4\pi r^2 \gamma[2f(\theta)] \tag{6.41}$$

Thus, compared to the case for homogeneous nucleation, we find that for heterogeneous nucleation at a grain boundary, both the volume and surface energy terms are modified by a factor of $2f(\theta)$, which essentially accounts for the geometry of the lenticular shape (which is like two spherical caps) versus a sphere. Since $f(\theta)$ decreases with decreasing θ, the more that the β phase likes to wet the grain boundary, the greater the benefit associated with heterogeneous grain boundary nucleation (just as with heterogeneous nucleation on a surface). Since the modification factor for grain boundary nucleation is $2f(\theta)$ versus $f(\theta)$ for surface nucleation, it is generally expected that heterogeneous grain boundary nucleation is not quite as favorable as heterogeneous surface nucleation (although both are more favorable than homogeneous nucleation). However, as there are often many more grain boundary nucleation sites available in a material compared to free-surface nucleation sites, the overall *amount* of grain boundary nucleation may often be higher.

6.5.3 Nucleation Rate

So far, we have only determined the number of viable nuclei for homogeneous and heterogeneous nucleation processes (n^*, n^*_{het}) as a function of the temperature T. What we would really like to determine is the *rate* of nucleation, $\dot{N}$, that is, the concentration of viable nuclei that are created *per unit time* during a transformation process.

A nucleus that has obtained exactly the critical viable nucleus size r^* is perched in a tenuous position. Two things can happen with essentially equal probability: (1) the nucleus can lose a few atoms, in which case it ceases to be viable and "falls" back down the activation energy hill, *or* (2) the nucleus can gain a few more atoms, at which point it begins to accelerate down the other side of the activation energy hill, where it becomes highly energetically favorable to continue growing. Thus, the rate of nucleation is determined by the concentration of viable nuclei and the rate at which they are "activated" through the addition of a few more atoms to ensure their continued existence (and growth). Expressed mathematically,

$$\dot{N} = v n^* \tag{6.42}$$

where v is the frequency (1/s) with which fresh molecules or atoms collide with the nucleus. Since the units for n^* are typically mol/m^3 (or mol/cm^3), the units for $\dot{N}$ are $mol/(m^3 \cdot s)$ [or $mol/(cm^3 \cdot s)$].

For the case of nucleation from the vapor phase, the impingement rate (Equation 5.3 from Chapter 5) multiplied by the surface area of the nucleus governs the frequency with which atoms collide with the nucleus. Thus, we have for $\dot{N}$

$$\dot{N}_{(\text{from vapor})} = \left[\frac{\alpha P}{(2\pi M R T)^{1/2}} A_{\text{m}}^* \right] [n_0 e^{-\Delta G^*/RT}] \tag{6.43}$$

where A_{m}^* is the molar area (area/mol) of the critically sized nucleus.[4]

For the case of nucleation in a condensed-matter phase, diffusion controls the rate at which fresh atoms can arrive at the surface of the nucleus. In this case, the nucleation rate may be expressed as

$$\dot{N}_{(\text{condensed phase})} = [v_0 e^{(-\Delta G_m/RT)}][n_0 e^{-\Delta G^*/RT}] \tag{6.44}$$

where v_0 is the jump frequency (1/s) of atoms at the surface of the nucleus and ΔG_{m} describes the activation energy for the diffusion (J/mol) of atoms to the surface of the nucleus.

For a condensed-matter transformation that occurs on cooling below an equilibrium temperature T_{E}, the overall nucleation rate as a function of temperature is shown in Figure 6.22. The overall nucleation rate is a result of two factors: the concentration of critical nuclei n^*, which obtains a peak at intermediate temperatures below T_{E}, and the diffusion term $v = v_0 e^{(-\Delta G_m/RT)}$, which increases exponentially with increasing temperature. As a result, the peak for $\dot{N}$ is skewed toward higher temperatures than the peak for n^*. However, as T approaches T_{E}, both n^* and $\dot{N}$ fall to zero. Similarly, at very low temperature, both n^* and $\dot{N}$ approach zero. Thus, an *intermediate* undercooling temperature results in the greatest rate of nucleation.

The peak in $\dot{N}$ at intermediate temperatures explains how glasses can be formed, even from materials systems that do not tend to form into glasses (e.g., metallic glasses). The key to forming glasses in such systems is to *rapidly quench them* from the melt, so that they solidify without having time for crystalline nuclei to form. In this way, the random amorphous structure of the liquid melt is "frozen" in place before the material has time to nucleate and grow organized crystalline domains.

---

[4]For a spherical nucleus of critical size r^*, the volume and area of the critically sized nucleus are $V^* = 4/3\pi(r^*)^3$ and $A^* = 4\pi(r^*)^2$, respectively. The molar content of one critically sized nucleus, M^*, is given by $M^* = V^*/V_{\text{m}}$, where V_{m} is the molar volume ($V_{\text{m}} = M/\rho$). The molar area of a critically sized nucleus is therefore $A_{\text{m}}^* = A^*/M^* = 4\pi(r^*)^2/[4/3\pi(r^*)^3(\rho/M)] = (3M)/(r^*\rho)$.

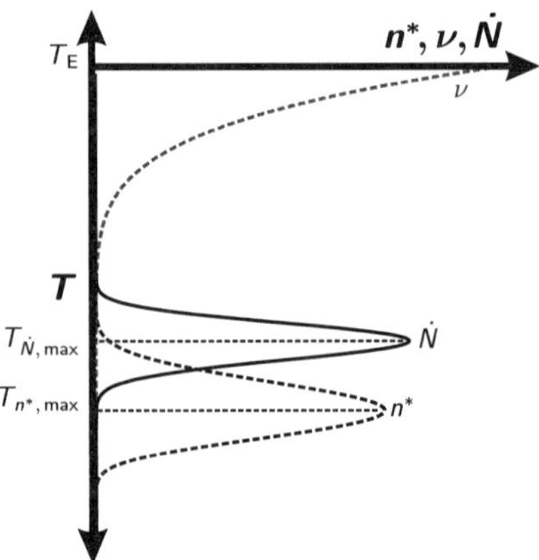

FIGURE 6.22   For a phase transformation that occurs on cooling a system below T_E, the overall nucleation rate $\dot{N}$ is a result of two factors: the concentration of critical nuclei n^*, which obtains a peak at intermediate temperatures below T_E, and the diffusion term $\nu = \nu_0 e^{(-\Delta G_\mathrm{m}/RT)}$, which increases exponentially with increasing temperature. As a result, the peak for $\dot{N}$ is skewed toward higher temperatures than the peak for n^*. However, as T approaches T_E, both n^* and $\dot{N}$ fall to zero. Similarly, at very low temperature, both n^* and $\dot{N}$ approach zero. Thus, there is an intermediate cooling temperature that results in the greatest rate of nucleation.

Example 6.6

Question: Have you ever heard the expression that it is "too cold to snow?" This expression is based on the observation that precipitation drops off rather abruptly when it is extremely cold (i.e., at temperatures well below the freezing point). This seemingly counterintuitive observation is actually underpinned by solid theory. Your professor gives you two possible reasons for the phenomenon: (1) Very cold air holds less moisture than warmer air, and so there is less water vapor available for precipitation. (2) The nucleation rate decreases at very low temperatures, leading to fewer, smaller snowflakes. Which explanation is more likely?

Solution: To address this question, we will consider the likelihood of second explanation—the suppression of the nucleation rate. Specifically, we will first try to determine the temperature at which $\dot{N}$ reaches a maximum ($T_{\dot{N}_{\mathrm{max}}}$). At temperatures below $T_{\dot{N}_{\mathrm{max}}}$, $\dot{N}$ should begin to decrease. For the formation of snow nuclei from the vapor phase, we start with Equation 6.43:

$$\dot{N}_{\text{(from vapor)}} = \left[\frac{\alpha P}{(2\pi MRT)^{1/2}} A_\mathrm{m}^* \right] [n_0 e^{-\Delta G^*/RT}]$$

Temperature appears both in the preexponential factor (as a $T^{-1/2}$ term) and in the exponential. Because the $T^{-1/2}$ dependence in the preexponential factor is extremely weak compared to the exponential temperature-dependent term, we will ignore it, simplifying our calculation significantly. We seek to determine the temperature where $\dot{N}$ attains a maximum, in other words, where

$$\frac{d[e^{-\Delta G^*/RT}]}{dT} = 0 \tag{6.45}$$

Recall that ΔG^* itself has a strong temperature dependence built into it as described by the degree of undercooling (Equation 6.24):

$$\Delta G^* = \frac{16\pi\gamma^3}{3(\Delta S_V)^2(\Delta T)^2} = \frac{16\pi\gamma^3}{3(\Delta S_V)^2(T_E - T)^2} = \frac{C}{(T_E - T)^2} \tag{6.46}$$

where we've lumped all the temperature independent terms into the constant C. Inserting Equation 6.46 into Equation 6.45 and taking the derivative with respect to temperature, we have

$$\frac{de^{-C/[RT(T_E-T)^2]}}{dT} = 0$$

$$e^{-C/[RT(T_E-T)^2]} \left\{ \frac{C}{RT^2(T_E - T)^2} - \frac{2C}{RT(T_E - T)^3} \right\} = 0$$

This expression has a real root when the portion inside in the curly brackets equals zero:

$$\frac{C}{RT^2(T_E - T)^2} - \frac{2C}{RT(T_E - T)^3} = 0 \qquad T = \frac{T_E}{3}$$

Thus, if the nucleation explanation is correct, it would only be "too cold to snow" when T decreases well below $T_E/3$. This would imply temperatures below 91 K, or $-182\,°C$! This is far colder than any recorded temperature on Earth, indicating that the nucleation factor is not an adequate explanation for snowfall suppression at frigid temperatures. The better explanation considers the water vapor capacity of the atmosphere, which decreases exponentially with decreasing temperature. Cold air holds exponentially less water vapor than warm air, and so the precipitation potential decreases dramatically as temperatures become more frigid. This is the main reason why warm snowstorms, where the air temperature is just slightly below the freezing point, typically produce the greatest snowfall rates.

6.6 GROWTH

Once a nucleus is successfully established, its continued growth is determined by the rate at which additional atoms can be added. Sometimes, when the transformation produces (or consumes) a large amount of heat, the growth rate can also depend on how quickly heat can be removed from (or conveyed to) the interface.

For the time being, however, we will consider growth under the scenario where the attachment of mass to the interface limits the rate. During the initial stages of growth, the growth rate is typically linear with time. However, as growth proceeds, the rate generally begins to slow. As with many diffusion-controlled processes (recall Chapters 4 and 5), the growth rate often becomes parabolic (i.e., the new phase advances with the *square root* of time). During the latest stages of a transformation, either the driving force for the transformation or the supply of the reacting (parent) phase will be exhausted, at which time the growth rate will fall to zero.

The linear growth rate $\dot{G}$ typically observed during the initial stages of attachment-limited growth depends on the size of the driving force for the transformation process and the frequency at which atoms can successfully transfer themselves from the reactant phase to the product phase. Expressed mathematically,

$$\dot{G} = -\Delta G_{rxn} v \phi \tag{6.47}$$

where $\dot{G}$ is the growth rate (m/s), ΔG_{rxn} is the molar Gibbs free energy for the transformation reaction (J/mol), v is the frequency (1/s) with which fresh atoms are added to the growing phase, and ϕ (m $\cdot$ mol/J) can be considered as a combined geometric and energy factor term that takes into account the size/shape of the growing phase and the effectiveness of the transformation driving force for causing growth. The negative sign is needed to ensure that the growth rate is a positive value since a thermodynamically favorable transformation process will have a negative ΔG_{rxn}. You should already be familiar with the frequency factor v, as it also appeared in the expression for the nucleation rate $\dot{N}$. This makes sense, because both the activation of a viable nucleus and its subsequent growth depends on the same fundamental (mass-transport-controlled) rate at which atoms are added to the nucleus. Expressing, as before, the driving force for the transformation process (ΔG_{rxn}) in terms of ΔH_{rxn} and the degree of undercooling (or superheating) from the equilibrium temperature T_E, we can therefore write for a condensed phase transformation process:

$$\dot{G}_{(condensed\ phase)} = -\frac{\Delta H_{rxn} \Delta T}{T_E} \phi v_0 e^{(-\Delta G_m/RT)} \tag{6.48}$$

For a condensed-matter transformation that occurs on cooling below an equilibrium temperature T_E, the overall growth rate as a function of temperature is shown in Figure 6.23. The overall growth rate is a result of two factors: the driving force for transformation, which increases linearly with decreasing temperature below T_E, and the mass transfer term $v = v_0 e^{(-\Delta G_m/RT)}$, which decreases exponentially with decreasing temperature. As a result, the overall growth rate $\dot{G}$ obtains a peak at intermediate temperatures. As T approaches T_E, although mass transfer is fast, the driving force for transformation falls to zero (and thus so does $\dot{G}$). At very low temperatures, although the driving force for transformation is large, mass transfer is slow and thus $\dot{G}$ again approaches zero. Thus, an *intermediate* undercooling temperature produces the greatest growth rate.

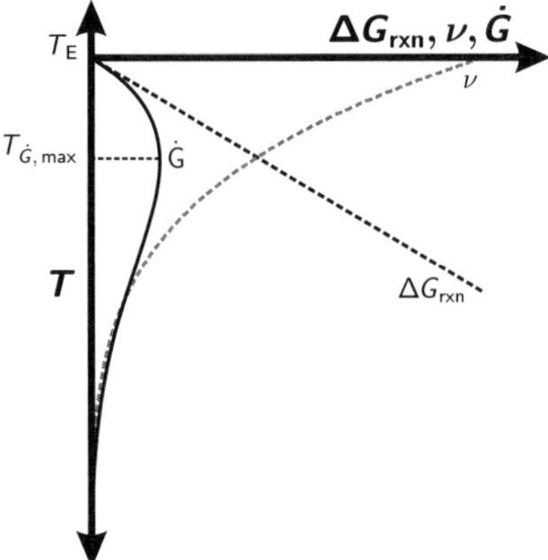

FIGURE 6.23 For a phase transformation that occurs on cooling a system below T_E, the overall growth rate $\dot{G}$ is a result of two factors: the driving force for transformation, which increases linearly with decreasing temperature below T_E, and the mass transfer term $v = v_0 e^{(-\Delta G_m/RT)}$, which decreases exponentially with decreasing temperature. As a result, the overall growth rate $\dot{G}$ obtains a peak at intermediate temperatures.

Example 6.7

Question: Calculate the mass-transport-limited growth rate for the solidification of Fe(s) at $T = 1200$ K given the following information:

- $\Delta H^{\circ}_{solidification} = -15.2$ kJ/mol
- $T_m = 1812$ K
- $\Delta G_m = -120$ kJ/mol
- $v_0 = 10^7/s$
- $\phi = 10^{-8}$ m · mol/J

Approximately how long would it take for a 20-cm-long ingot with a 3-cm × 6-cm square cross section to solidify at 1200 K assuming nucleation only occurs along the mold walls followed by linear mass-transport-limited growth inward to the center?

Solution: To answer this question, we apply the information provided to Equation 6.48, yielding

$$\dot{G} = -\frac{\Delta H_{rxn} \Delta T}{T_E} \phi v_0 e^{(-\Delta G_m/RT)}$$

$$= -\frac{-15200 \text{ J/mol}(1812 \text{ K} - 1200 \text{ K})}{1812 \text{ K}} \cdot 10^{-8} \text{ m} \cdot \text{mol/J}$$

$$(10^7/\text{s})\exp\left(-\frac{120{,}000\ \text{J/mol}}{8.314\ \text{J/(mol}\cdot\text{K)}\cdot 1200\ \text{K}}\right)$$

$$= 3.1 \times 10^{-3}\ \text{m/s} = 3.1\ \text{mm/s}$$

Growth will proceed inward from the mold walls toward the center. The shortest dimension of the ingot cross section determines the time required. The time required for the growth to proceed 1.5 cm (from the closest wall to the center of the ingot) under the linear mass-transport-limited growth regime is therefore

$$t = \frac{\Delta x}{\dot{G}} = \frac{15\ \text{mm}}{3.1\ \text{mm/s}} = 4.8\ \text{s}$$

Thus, we predict that the ingot will solidify in less than 5 s! This is extremely rapid solidification. In all likelihood, the actual solidification rate under these conditions might instead be controlled by the rate at which the heat of fusion can be conducted away from the solidifying ingot. Thus a solidification growth rate based on heat transport should be calculated and the slower of the two limiting growth rates should be adopted to provide a best estimate of the solidification time required. This is discussed in the dialog box below.

Estimating Heat-Transport-Limited Growth Rates

Sometimes, growth rates are controlled by the rate of heat transport rather than the rate of mass transport. This is particularly true during melting or solidification processes, which involve a large absorption or release of heat. Under such circumstances, accurate treatment of the growth rate requires coupling the heat produced/absorbed during growth to the transient heat conduction equation (similar to the transient diffusion equation) within the confines of a moving interface boundary condition. This treatment typically results in growth that proceeds with the square root of time. This detailed treatment is beyond the scope of this textbook. However, for the purposes of obtaining a rough estimate for heat-transport-limited growth, here we will employ a simple approach that ignores the moving interface issue and assumes a linear steady-state heat transport rate.

Consider a planar growth front as illustrated in Figure 6.24. The growth of a thin slice of new phase of thickness dx in a time dt generates a heat flux Q (W/m^2) given by

$$Q = \Delta H_{\text{rxn}}\frac{\rho}{M}\frac{dx}{dt} \tag{6.49}$$

where ΔH_{rxn} is the molar heat of the reaction, ρ is the density of the material, and M is the molecular weight. Under steady-state growth, this heat flux must be exactly balanced by the heat transported away (or toward) the growth front by thermal conduction. Ignoring the movement of the growth front and assuming steady-state heat conduction due to a uniform temperature gradient across the critical

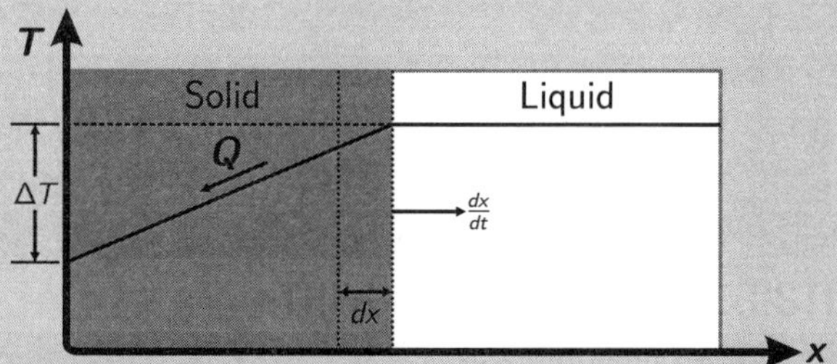

FIGURE 6.24 Schematic illustration of heat-transport-limited growth. The rate of growth of a slice of transformed phase of thickness dx is determined by the rate at which the heat absorbed (or released) by this transformed volume can be conducted to (or away from) the interface.

(usually shortest) dimension associated with the transforming volume, the heat flux due to conduction can be estimated as

$$Q = -\kappa \frac{dT}{dx} \approx -\kappa \frac{\Delta T}{L} \tag{6.50}$$

where κ is the thermal conductivity,[a] L is the critical dimension along which the transformation is proceeding (e.g., for a solidification process proceeding from the walls to the center, L would be the shortest distance from the wall to the center of the transforming volume), and ΔT is the temperature drop along L from the center to the edge of the solidifying volume.

Equating expressions 6.49 and 6.50 and solving for the growth rate $(dx/dt = \dot{G})$ yields

$$\dot{G} = \frac{dx}{dt} = -\frac{\kappa M \, \Delta T}{\Delta H_{rxn} \rho L} \tag{6.51}$$

For the case of the solidification of Fe(s) discussed in Example 6.7, if $\kappa = 25$ W/(m · K), we could estimate the heat-transport-limited solidification rate as

$$\dot{G} = -\frac{25 \text{ W/(m · K)} \cdot 55.85 \times 10^{-3} \text{ kg/mol}(1812 \text{ K} - 1200 \text{ K})}{-15200 \text{ J/mol} \cdot 7870 \text{ kg/m}^3 \cdot 0.015 \text{ m}}$$

$$= 4.8 \times 10^{-4} \text{ m/s} = 0.48 \text{ mm/s}$$

The heat-transport-limited solidification rate estimated by this approach is almost one order of magnitude slower than the mass-transport-limited growth rate that was calculated in Example 6.7. Thus, it is likely that heat transport will limit the solidification of this ingot under these conditions.

[a]For simplicity, here it is assumed that the thermal conductivity of the parent and transformed phase are identical.

6.7 NUCLEATION AND GROWTH COMBINED

During a phase transformation, nucleation and growth typically occur *concurrently*. That is, as soon as a few viable nuclei begin to form, they will start growing, even as additional new nuclei continue to be formed. To understand the overall rate of transformation, it is therefore necessary to consider the rates of nucleation and growth together. Figure 6.25 schematically compares the rate of nucleation and growth as a function of temperature for a typical condensed-matter transformation that occurs on cooling below an equilibrium temperature T_E. This figure is a combination of Figures 6.22 and 6.23. The overall transformation rate, given on the figure by $\dot{F}$, is a nonlinear product of the nucleation ($\dot{N}$) and growth ($\dot{G}$) rates. Because both $\dot{N}$ and $\dot{G}$ obtain a peak at intermediate temperatures, so does $\dot{F}$. However, the peaks for $\dot{N}$ and $\dot{G}$ typically do not occur at the same temperature ($\dot{G}$ typically peaks at higher temperatures than $\dot{N}$). As will be discussed below, this has important implications on the *type of microstructure* that develops depending on the degree of undercooling during transformation.

6.7.1 Effect of Nucleation Rate versus Growth Rate on Microstructure

Figure 6.26 illustrates how the choice of the undercooling temperature can impact the typical microstructure that is obtained during a phase transformation that occurs on cooling a system below T_E. Representative microstructures corresponding to four different isothermal undercooling temperatures (T_1, T_2, T_3, and T_4, respectively) are shown. At temperature T_1, which represents a small degree of undercooling (just slightly below T_E), the nucleation rate is low, but the growth rate is high. Thus, only

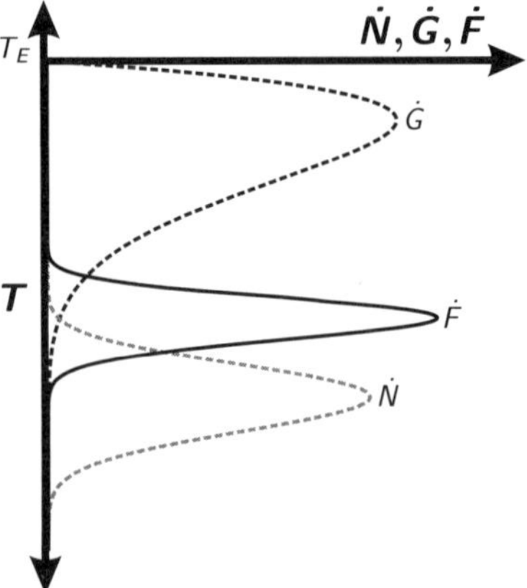

FIGURE 6.25   For a phase transformation that occurs on cooling a system below T_E, the overall transformation rate $\dot{F}$ is a nonlinear product of the nucleation rate $\dot{N}$ and the growth rate $\dot{G}$. Because both $\dot{N}$ and $\dot{G}$ obtain a peak at intermediate temperatures, so does $\dot{F}$.

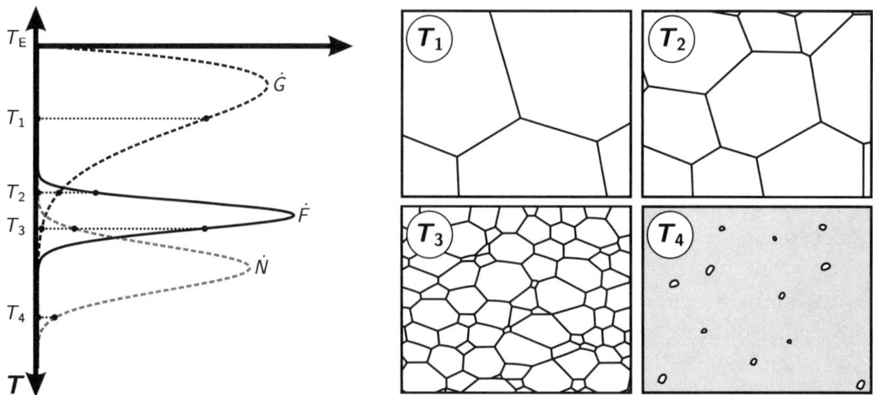

FIGURE 6.26 Influence of transformation temperature on microstructure. Carrying out a liquid–solid phase transformation isothermally at a temperature just slightly below the equilibrium transformation temperature (T_1) tends to produce a coarse-grained microstructure since the growth rate is high but the nucleation rate is low. Transforming at lower temperatures (greater amount of undercooling) results in microstructural refinement (e.g., T_2, T_3). In certain systems a rapid quench to low temperature (T_4) may produce an amorphous or glasslike microstructure as the crystalline phase transition can be kinetically impeded.

a small number of nuclei are formed, but they have the opportunity to grow quickly. As a result, a coarse (e.g., large-grained) microstructure is obtained. At temperature T_2, which represents a moderate degree of undercooling, both the nucleation rate and the growth rate are high. Thus, the microstructural transformation occurs quickly and a relatively equiaxed, moderately refined microstructure is typically obtained. At temperature T_3, which represents a large degree of undercooling, the nucleation rate is still significant, but the growth rate is low. As a result, a very fine microstructure is obtained. Finally, at temperature T_4, which represents a rapid quench to very low temperatures, both the nucleation rate and the growth rate are near zero. If the original transforming phase was a liquid, an amorphous or glasslike microstructure may therefore be obtained (for certain materials systems). If the original transforming phase was a solid, the phase transformation may be kinetically prevented by this rapid quenching to T_4, and the original phase may therefore persist as a metastable phase.

Geology of Ingenous Rocks: Natural Case Study in Nucleation and Growth

The next time you go hiking in the mountains, take a closer look at the rocks around you—especially if they are igneous in nature. Igneous rocks are formed when molten magma from beneath Earth's surface solidifies. Depending on the rate of solidification, the resulting rock can take on a variety of different appearances due to differences in the relative rates of nucleation and growth during solidification. For example, you may be familiar with granite, rhyolite, and obsidian (see Figure 6.27). Chemically, these three rocks are all quite similar. However, microstructurally they are quite different, and this is due to how quickly they solidified from molten magma. Granite typically forms underground and solidifies very slowly at relatively high temperatures. As a result, nucleation is slow while the growth rate is fast, leading to a beautiful, highly textured, coarse-grained microstructure. The individual grains are often visible to the naked eye and can be centimeters in size. In contrast, rhyolite typically forms on Earth's surface (e.g., in lava flows during a volcanic eruption) and thus cools much more quickly. The lower temperatures and faster cooling rate during solidification lead to a much finer (although still fully crystalline) microstructure. Obsidian—also known as *volcanic glass*—is typically formed during violent volcanic eruptions. During such eruptions, molten fragments are ejected high into the air where they solidify rapidly before crystallization can occur. As a result, the amorphous structure of the melt is frozen in place, leading to the characteristic glasslike properties of obsidian.

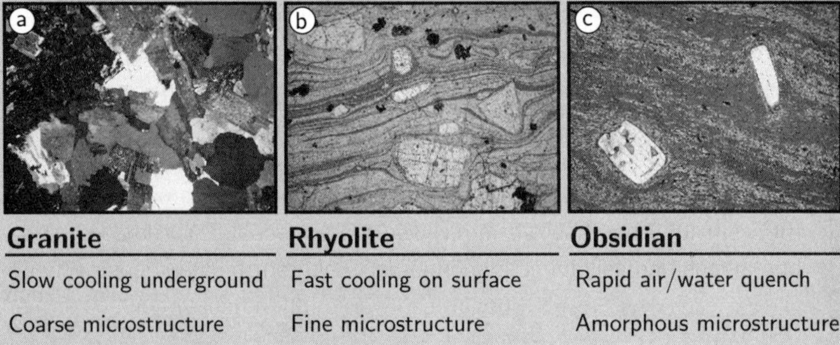

FIGURE 6.27 Examples of (*a*) granite, (*b*) rhyolite, and (*c*) obsidian. All three are igneous rocks of similar chemical composition. The marked microstructual differences between the three are due to dramatic differences in the rate of solidification during their formation. Granite typically forms underground and solidifies very slowly at relatively high temperatures. As a result, nucleation is slow while the growth rate is fast, leading to a highly textured, coarse-grained microstructure. Rhyolite typically forms on the Earth's surface (e.g., in lava flows during a volcanic eruption) and thus cools much more quickly. The lower temperatures and faster cooling rate during solidification lead to a much finer microstructure. Obsidian is typically formed during violent volcanic eruptions where molten fragments ejected high into the air solidify rapidly before crystallization can occur. As a result, the amorphous structure of the melt is frozen in place, leading to the characteristic glasslike properties of obsidian. Micrographs courtesy of the Oxford Earth Sciences Image Store (http://www.earth.ox.ac.uk/ oesis/index.html)

6.7.2 Overall Rate of Transformation: Johnson–Mehl and Avrami Equations

The overall fraction of material transformed as a function of time $[F(t)]$ can be derived assuming nucleation and growth of spherical particles where both the nucleation rate $(\dot{N})$ and the growth rate $(\dot{G})$ are constant as a function of time. The derivation is beyond the scope of the present text but results in a formula for $F(t)$ with the following form:

$$F(t) = 1 - e^{-(\pi/3)\dot{G}^3 \dot{N} t^4} \tag{6.52}$$

This equation, which relates the fraction transformed to the nucleation rate, the growth rate, and the time elapsed since the start of the transformation (at constant temperature), is known as the *Johnson–Mehl* equation. The fact that the exponential term depends on $\dot{G}^3$ can be understood on the basis that growth is assumed to proceed spherically, and thus the volume transformed increases with the cube power of the linear growth rate.

Figure 6.28 plots Equation 6.52 as a function of time for the phase transformation process previously illustrated in Figure 6.26 at undercooling temperatures $T_1, T_2, T_3,$ and T_4. In all cases, the transformation is *sigmoidal*, meaning that the fraction transformed first increases exponentially in time before slowing down and asymptotically approaching complete transformation. Sigmoidal behavior is observed in a wide variety of natural systems, including the propagation of species into new ecological niches, the rate of alcohol production by yeast in beer, and the spread of infectious diseases through a population. The sigmoidal shape of $F(t)$ can be understood as shown in Figure 6.29. As the new phase nucleates and grows, at first it can grow rapidly

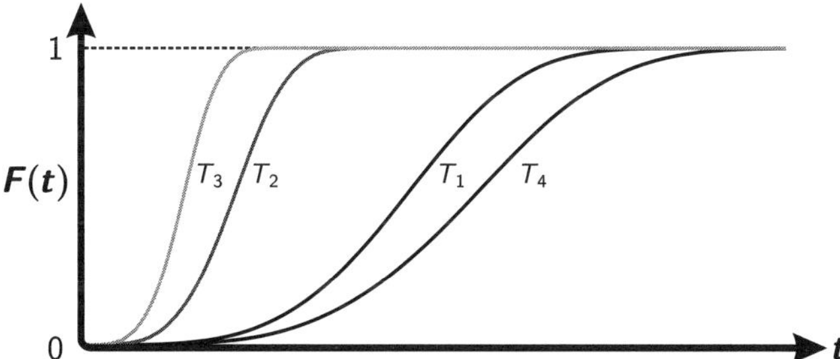

FIGURE 6.28   Fraction transformed $F(t)$ during a phase transformation as a function of the time elapsed at four different undercooling temperatures. The fraction transformed starts at 0 and increases *sigmoidally* with time until full transformation is obtained ($F = 1$). In this example, the overall transformation process is fastest for $T = T_2$, corresponding to the temperature where the growth rate was maximized (refer to Figure 6.26). At both $T = T_1$ and $T = T_4$, the transformation rate is slow, although for different reasons in each case. At $T = T_1$, transformation is slow because the driving force for transformation is small, while at $T = T_4$, transformation is slow because atomic mobility (i.e., diffusion) is slow.

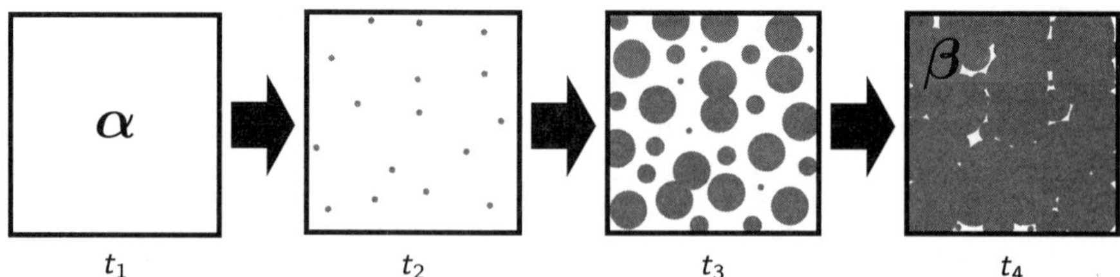

FIGURE 6.29 Schematic illustration of the progress of a phase transformation ($\alpha \to \beta$). During the initial stages of transformation, the new phase can nucleate and grow rapidly without restriction. However, as the transformation process continues, growing particles begin to impinge on one another and less parent phase is available for continued nucleation, thus slowing the rate of transformation.

without restriction. However, as the transformation process continues, growing particles begin to impinge on one another and less parent phase is available for continued nucleation, thus slowing the rate of transformation.

In this example, the overall transformation process is fastest for $T = T_2$, corresponding to the temperature where the growth rate is maximized. At both $T = T_1$ and $T = T_4$, the transformation rate is slow, although for different reasons in each case. At $T = T_1$ transformation is slow because the driving force for transformation is small, while at $T = T_4$, transformation is slow because atomic mobility (i.e., diffusion) is slow.

A simplified version of the Johnson–Mehl equation, known as the Avrami equation, is often employed. The Avrami equation is typically expressed as

$$F(t) = 1 - e^{-kt^n} \tag{6.53}$$

where n is known as the "Avrami exponent" and k is an empirical constant that accounts for both the nucleation and growth rates associated with the phase transformation. The larger the value of k and the higher the value of n, the faster the rate of transformation. The value of n in the Avrami equation often contains dimensional information about the transformation process. For 3D phase transformations, n typically varies between 4 and 3. For 2D transformations, such as may occur in a thin film or sheet, n typically varies between 3 and 2. For 1D transformations, such as may occur in a wire or rod, n typically varies between 2 and 1. Variations in n may also reflect other nonidealities in the nucleation and growth process, such as highly nonspherical growth or nucleation and growth rates that decrease as a function of time.

6.7.3 Time–Temperature–Transformation Diagrams

The kinetics of solid-state phase transformations are often summarized in time–temperature–transformation (TTT) diagrams. Figure 6.30 illustrates the construction of a TTT diagram using the transformation kinetics for the system in Figures 6.26 and 6.28. The TTT diagram shows the time required to achieve certain amounts of transformation as a function of the temperature of the transformation.

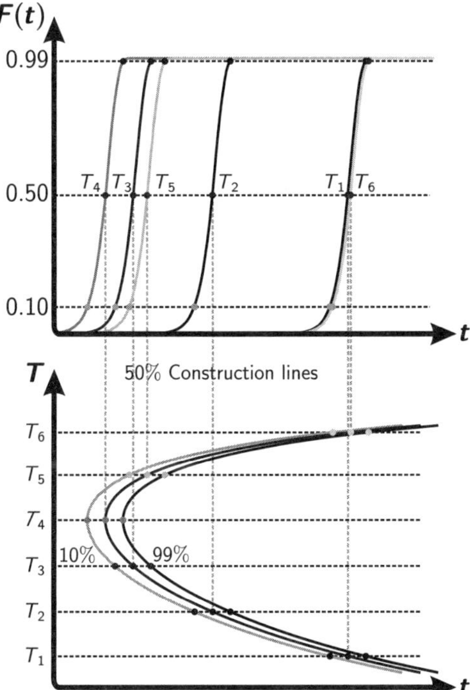

FIGURE 6.30   Constructing a TTT diagram from $F(t)$ curves at several different temperatures, where the $F(t)$ curves are constructed from the $\dot{G}(T)$ and $\dot{N}(T)$ curves for the transformation process. The TTT diagram provides the time required to achieve certain amounts of transformation as a function of the temperature of the transformation. TTT curves are also known as "C curves," because of their characteristic C-like shape. The C-like shape is due to the fact that both nucleation and growth (and hence the overall transformation rate) peak at intermediate temperatures.

The transformation contours obtained in TTT diagrams are also known as "C curves," because of their characteristic C-like shape. The C-like shape is due to the fact that both nucleation and growth (and hence the overall transformation rate) peak at intermediate temperatures. The TTT diagram illustrates how it is possible to form metastable structures such as glassy phases by cooling a material rapidly enough to slide in front of the "nose" of the C-curve contours, thereby avoiding the transformation process and instead freezing in a nonequilibrium structure.

The homework provides an opportunity to develop a TTT diagram for a solidification process. The process involves three principal steps:

1. Calculate $\dot{N}(T)$ and $\dot{G}(T)$.
2. Calculate $F(t)$ at 5–10 different temperatures (using values calculated for $\dot{N}$ and $\dot{G}$ at these 5–10 different temperatures).
3. Generate a TTT diagram with 3–4 'C curves' corresponding to 3–4 different values of F (e.g., $F = 0.01, 0.1, 0.5, 0.99$).

Differences between Liquid–Solid and Solid–Solid Phase Transformations

The kinetics of solid–solid phase transformations tend to be much more sluggish than liquid–solid transformations. This is due to a number of factors:

- Solid–solid interfacial energies tend to be larger than liquid–solid interfacial energies, which leads to an increase in ΔG^*.
- Solid–solid transformations tend to have smaller ΔG_V than liquid–solid phase transformations, which also leads to an increase in ΔG^*.
- The volume changes associated with solid–solid phase transformations produce a large strain energy term, which also increases ΔG^*.
- Solid-state diffusion is much slower than diffusion in liquids, reducing both $\dot{N}$ and $\dot{G}$.

Because of these factors, solid–solid phase transformations rarely reach equilibrium. The retention of metastable phases is nearly guaranteed. In fact, the retention of metastable phases is purposely exploited in many material processes, for example, in steelmaking, to create intricate composite microstructures with exceptional properties. In the discussions of phase transformations that follow, you should therefore keep in mind the fact that incomplete phase transformation and metastable phase retention is the rule, not the exception. Complete conversion to the equilibrium phase composition is rare in solid–solid phase transformations (and even in many liquid–solid phase transformations).

6.8 SOLIDIFICATION

The nucleation rate, growth rate, and transformation rate equations that we developed in the preceding sections are sufficient to provide a general, semiquantitative understanding of nucleation- and growth-based phase transformations. However, it is important to understand that the kinetic models developed in this introductory text are generally not sufficient to provide a microstructurally predictive description of phase transformation for a specific materials system. It is also important to understand that real phase transformation processes often do not reach completion or do not attain complete "equilibrium." In fact, extended defects such as grain boundaries or pores should not exist in a true equilibrium solid, so nearly all materials exist in some sort of metastable condition. Many phase transformation processes produce microstructures that depart wildly from our equilibrium expectation. The limited atomic mobilities associated with solid-state diffusion can frequently cause (and preserve) such nonequilibrium structures. In this section, we will focus more deeply on solidification (a liquid–solid phase transformation) as a way to discuss some of these issues. In particular, we will examine a few kinetic concepts/models

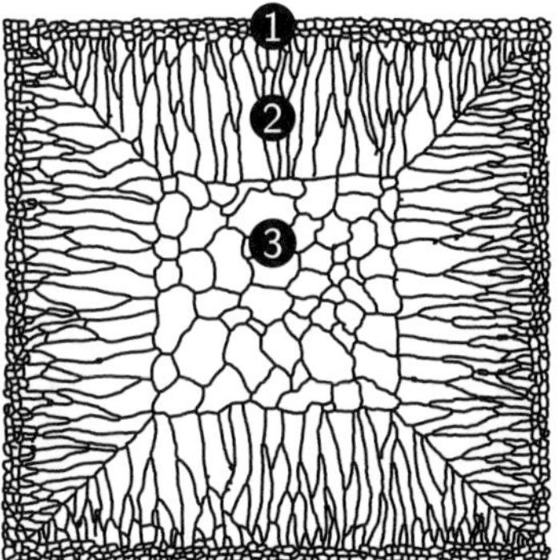

FIGURE 6.31 Typical microstructure obtained during ingot casting: (1) chill zone, (2) columnar zone, (3) equaixed grain zone. Adapted from Bower and Flemings 1967 [12].

that can help explain a number of interesting nonequlibrium phenomena encountered in condensed-matter phase transformations such as coring and dendritic growth.

6.8.1 Casting Microstructures

We first consider the typical overall microstructure that is obtained when casting (solidifying) a material from its melt in a mold (Figure 6.31). To induce solidification, the walls of the casting mold are typically held at a temperature well below the solidification temperature. The cold wall temperatures combined with a high heterogeneous nucleation rate at the mold surface typically results in a region of very small crystal grains in the contact zone near the mold wall (referred to as the *chill zone*). Solidification then proceeds inward parallel to the flow of heat out from the melt. In this stage, solidification typically proceeds more slowly and is highly directional, with less heterogeneous nucleation. As a result, *columnar zone* regions are formed with elongated or columnar grains extending toward the central region of the ingot. If the growth of these columnar grains is sufficiently rapid, they may extend all the way to the center of the ingot. However, the center of the ingot typically begins to solidify in a relatively slow but uniform fashion prior to encroachment of the columnar zones. In this case, a central zone of relatively isotropic, equiaxed grains is obtained.

Factors such as slow, uniform cooling without mixing lead to a coarse-grained, highly textured ingot. On the other hand, fast cooling and/or significant melt turbulence during solidification tends to yield a fine-grained structure that is often more desirable for good mechanical properties.

6.8.2 Plane Front Solidification (Scheil Equation)

We now consider the details of the solid–liquid growth front in more depth. As a solidification process proceeds, the boundary between the solid and liquid phases progressively advances through the volume of the material until the whole material is solidified. Consider, for example, the linear growth propagation mode that gives rise to the columnar zones during the ingot solidification process described above. Two main modes of solid–liquid growth front propagation are typically observed. The solid–liquid boundary can exhibit either stable propagation, also known as *plane front solidification*, or instabilities, thereby leading to cellular or dendritic growth. We first consider the case of stable plane front solidification.

For simplicity, consider a binary (two-component) alloy composed of A and B atoms that exhibits complete solid solubility. The alloy mixture is initially liquid and possesses a uniform composition c_0. We now consider the stable, planar, directional solidification (from left to right) of an ingot of this alloy as shown schematically in Figure 6.32a. As the alloy begins to solidify, a *partitioning* of A and B atoms between the solid and liquid phases takes place. Partitioning occurs because, as shown by the phase diagram for the alloy in Figure 6.32b, the initial solid that forms is enriched in A atoms compared to the starting alloy composition, while the remaining liquid will be somewhat depleted of A atoms. As solidification proceeds, this partitioning continues, although the composition of both the solid and liquid slowly becomes more and more depleted of A atoms as the remaining liquid composition becomes more and more enriched in B atoms. The last part of the ingot to solidify will be greatly enriched in B atoms compared to the initial composition. This process results in a nonuniform compositional profile across the ingot at the end of the solidification process, as shown in the figure. The severity of the compositional nonuniformity depends on the degree of solute partitioning between the solid and the

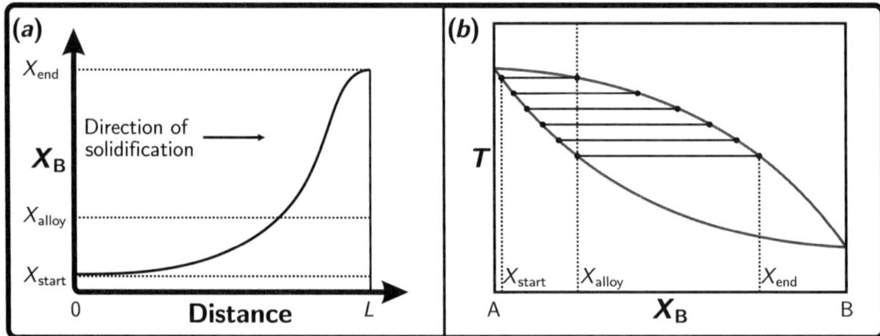

FIGURE 6.32 Plane front solidification of an isomorphous binary alloy system. (*a*) As the alloy begins to solidify (from left to right), *partitioning* of A and B atoms between the solid and liquid phases takes place. This process results in a nonuniform compositional profile across the ingot as solidification progresses from left to right. (*b*) The origin of the partitioning is explained by the phase diagram for the alloy. At the beginning of the solidification process, the solid that forms is greatly enriched in A atoms compared to the initial composition, while the remaining liquid is somewhat depleted of A atoms as a consequence. As solidification proceeds, this partitioning continues, although the composition of both the solid and liquid slowly becomes more and more depleted of A atoms as the remaining liquid composition becomes more and more enriched in B atoms. The last part of the ingot to solidify will be greatly enriched in B atoms compared to the initial composition.

liquid as well as the rate of solidification relative to the rate of solid-state diffusion. If solid-state diffusion is fast, then the composition gradient in the solid can be quickly erased, leading to reequilibration and a more uniform composition profile. However, in most cases, solid-state diffusion is slow compared to solidification, and thus these nonuniform composition gradients tend to be frozen in. The *Scheil equation* provides a way to model the nonuniform concentration profile that arises in such a plane front solidification process:

$$c(x) = k'c_0 \left[1 - \frac{x}{L}\right]^{(k'-1)} \tag{6.54}$$

In this equation, c_0 is the starting (uniform) composition of the liquid alloy, L is the length of the solidifying bar (ingot), and k' is the effective solute partitioning ratio. Under the limiting case where mixing in the liquid is rapid and solid-state diffusion is slow relative to the rate of solidification, k' can be approximated from the phase diagram as the ratio of the compositions of the solidus-versus-liquidus lines at the temperature where the solidification begins (i.e., $X_{\text{start}}/X_{\text{alloy}}$ in Figure 6.32b).

Nonuniform composition occurs not only in a linear plane front solidification process, but can occur in almost any nucleation-and-growth processes (solid/liquid or solid/solid). In such cases, compositional partitioning between the growing nuclei and the parent matrix phase leads to a characteristic "cored" microstructure, as shown in Figure 6.33. Coring can have undesirable effects on materials properties, and so it is often removed by heat treating the microstructure at a temperature just below the transformation temperature to allow solid-state diffusion to homogenize the composition.

6.8.3 Cellular or Dendritic Growth

While the phase front during a transformation process often grows in a stable manner, there are a number of factors that can lead to unstable growth. Figure 6.34 illustrates the difference between stable planar (or spherical) growth (*a*) versus unstable cellular (*b*) or dendritic (*c*) growth. Growth front instabilities lead to both cellular and dendritic growth; the difference is in the degree of instability. Minor instability leads to

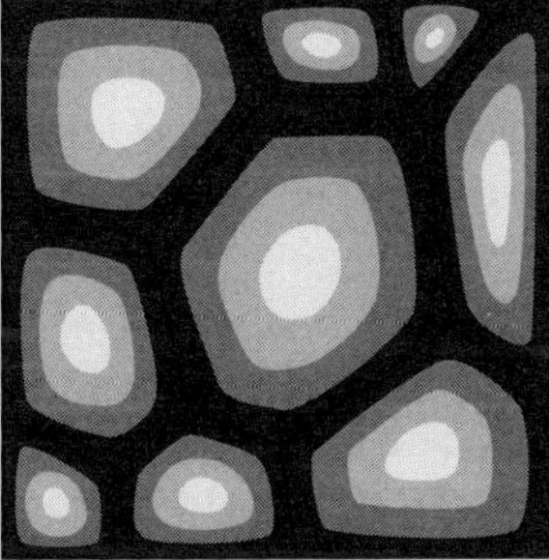

FIGURE 6.33 Schematic illustration of the characteristic "cored" microstructure that can arise due to compositional partitioning effects during a nucleation-and-growth process.

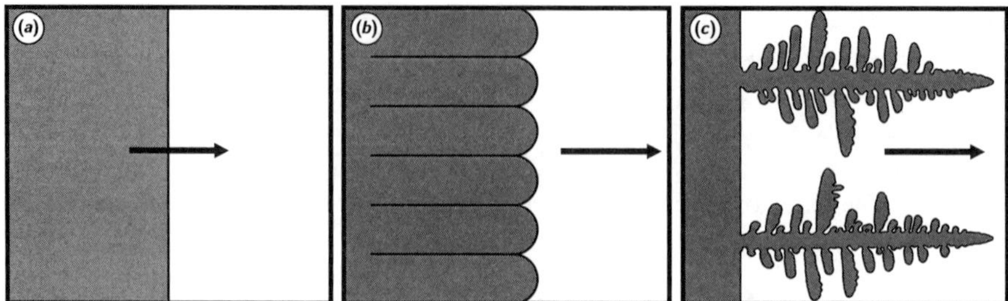

FIGURE 6.34   Schematic illustration of (*a*) stable planar (or spherical) growth, (*b*) unstable cellular growth, and (*c*) dendritic growth.

the formation of primary protuberances, called *cells*, which advance perpendicular to the interface. If the instability increases, these primary protuberances can themselves spawn secondary protuberances perpendicular to the primary protuberances, and a dendritic microstructure develops. Cellular and dendritic microstructures are most commonly observed in vapor–solid or liquid–solid phase transformations, although they can also be formed in solid–solid phase transformations. In fact, when pure metallic Li is used in batteries, it tends to form needlelike dendrites that can short circuit the battery. This problem has prevented the use of pure Li metal in Li batteries despite its excellent energy density. Carbon-based Li intercalation anodes, where the Li atoms are chemically (and safely) incorporated into a carbon superstructure, are typically used instead.

Figure 6.35 illustrates one common cause for growth front instability during solidi-fication. Because solidification is exothermic, heat must be continually removed from the interface to allow the solidification to proceed. Any minor protuberances that form along the growth front will project into the liquid, and under the common case where the liquid is at least somewhat supercooled relative to the equilibrium transformation temperature, heat will tend to be removed from these protuberances more quickly than the surrounding areas (Figure 6.35*a*). As a result, the local temperature at the protu-berance will tend to be slightly lower than along the planar growth front. As shown in Figure 6.35*b*, depending on the temperature, this can lead to an increase or a decrease in the local growth rate ($\dot{G}$) for the protuberance relative to the planar front. At temper-atures above the maximum growth rate peak, a slight decrease in the local temperature at the protuberance will lead to an increase in $\dot{G}$, and hence the protuberance will continue to grow faster than the planar front. At temperatures below the maximum growth rate peak, a slight decrease in the local temperature at the protuberance will lead to a decrease in $\dot{G}$, and hence the growth of the protuberance will be suppressed. In general, then, a low degree of undercooling during transformation tends to favor cellular/dendritic instability, while a larger degree of undercooling tends to suppress cellular/dendritic instability. Compositional gradients can produce similar instabil-ity effects on growing interfaces. Certain actions, for example, actively mixing the

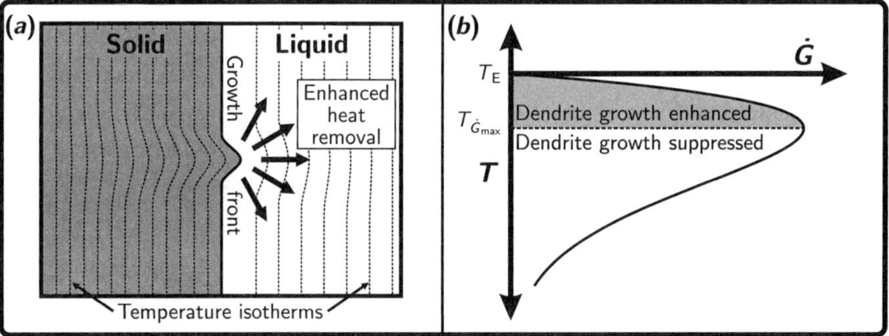

FIGURE 6.35 Inhomogeneous heat removal from the growth front is a common source of growth front instability. (*a*) Under supercooling conditions, heat will tend to be removed from growth front protuberances more quickly than the surrounding areas. As a result, the local temperature at the protuberance will tend to be slightly lower than along the planar growth front. (*b*) Due to the variation of $\dot{G}$ with the degree of undercooling, the protuberance will therefore experience either an increase or a decrease in the local growth rate relative to the planar front depending on the temperature. At temperatures above the maximum growth rate peak, a slight decrease in the local temperature at the protuberance will lead to an increase in $\dot{G}$, and hence the protuberance will continue to grow faster than the planar front. At temperatures below the maximum growth rate peak, further slight decreases in the local temperature at the protuberance will lead to decreases in $\dot{G}$, and hence the growth of the protuberance will be suppressed. In general, then, a low degree of undercooling during transformation tends to favor cellular/dendritic instability, while a larger degree of undercooling tends to suppress cellular/dendritic instability.

liquid, can help suppress growth front instability by eliminating these thermal and/or compositional gradients.

6.8.4 Eutectic Lamellae

During a eutectic transformation, as illustrated in the phase diagram in Figure 6.36*a*, solidification produces two distinct solid phases:

$$L \rightarrow \alpha + \beta \tag{6.55}$$

If the interfacial energy between the α and β phases is low compared to the liquid–α and liquid–β interfacial energies, then the α and β phases will prefer to wet each rather than the liquid. As shown in Figure 6.36*b*, this can lead to the coordinated growth of long alternating *lamellar* or plateletlike formations of the α and β phases. Such lamellar microstructures commonly occur for rapidly solidified eutectic phase transformations where insufficient time is provided to approach a more equilibrium microstructure.

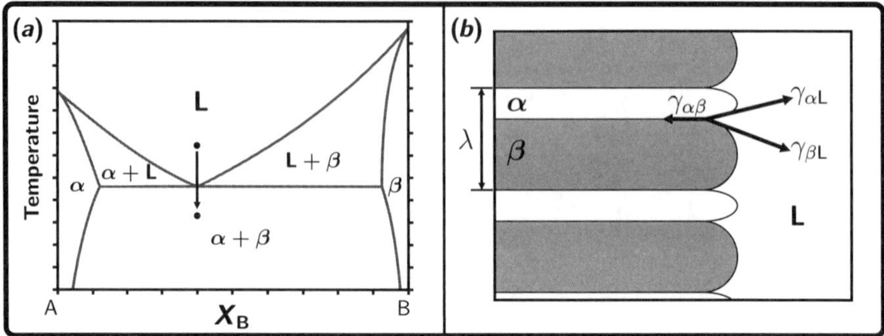

FIGURE 6.36 (*a*) Binary eutectic phase diagram. (*b*) During a eutectic solidification L → α + β, if the interfacial energy between the α and β phases is low compared to the liquid–α and liquid–β interfacial energies, then the α and β phases will prefer to wet each rather than the liquid. Under rapid, directional solidification, this can lead to the coordinated growth of long alternating *lamellar* or plateletlike formations of the α and β phases. Such lamellar microstructures are common features of eutectic phase transformations.

As with the growth of spherical nuclei, the growth of eutectic lamellae is governed by the balance between the volume free energy released by the transformation and the interfacial free energy consumed by the creation of the α–β lamellae. Considering a cube of the eutectic lamellar structure with sides of dimension λ (where λ is the lamellar spacing), this balance can be captured by the following simple equation:

$$\Delta G_{\text{tot}} = \lambda^3 \Delta G_V + 2\lambda^2 \gamma_{\alpha\beta} \tag{6.56}$$

The transformation is only favorable if $\Delta G_{\text{tot}} < 0$, which establishes a minimum threshold for the lamellar spacing, λ[5]:

$$\lambda^3 \Delta G_V + 2\lambda^2 \gamma_{\alpha\beta} < 0$$

$$\lambda > -\frac{2\gamma_{\alpha\beta}}{\Delta G_V} \tag{6.57}$$

This expression indicates that a finer lamellar spacing is enabled by a smaller interfacial energy between the α and β phases and/or a larger driving force for transformation. Since the driving force for transformation is directly proportional to the amount of undercooling, greater undercooling therefore favors a finer lamellar spacing:

$$\Delta G_V = \Delta H_V \frac{\Delta T}{T_{\text{E}}}$$

$$\lambda > -\frac{2T_{\text{E}}\gamma_{\alpha\beta}}{\Delta H_V \, \Delta T} \tag{6.58}$$

This spacing dependence is illustrated in Figure 6.37.

---

[5]Note the change in direction of the inequality that occurs in the second step of Equation 6.57 because we divide through by a negative number (ΔG_V must be negative for the phase transformation to occur).

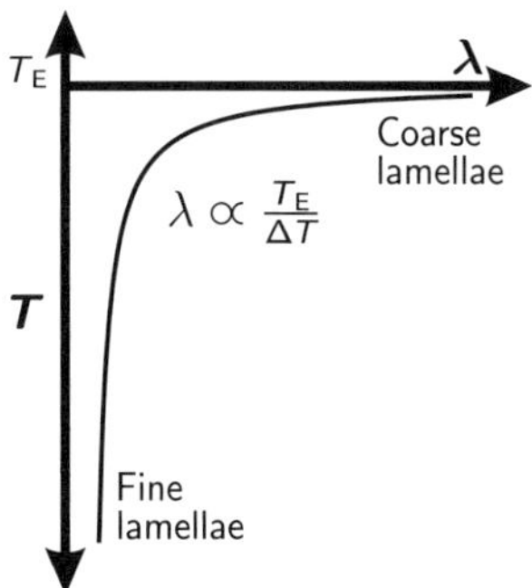

FIGURE 6.37 Minimum lamellar spacing versus temperature as predicted from Equation 6.58 for a eutectic solidification reaction that occurs on cooling below T_E. Greater undercooling favors a finer lamellar spacing.

Example 6.8

Question: Consider the eutectic solidification of a liquid to $\alpha + \beta$ where the volume fraction of the β phase is relatively low. In this case, rather than solidifying in a lamellar structure, the β phase may solidify as a series of rods embedded in a matrix of the α phase. Assuming that the β rods are evenly spaced in a square array in the α matrix, calculate the critical β-phase volume fraction (ϕ_β) below which the rodlike morphology is more energetically stable than the lamellar morphology.

Solution: Figure 6.38 compares the lamellar-versus-rodlike eutectic microstructures. To determine the β volume fraction where the rodlike microstructure is more energetically stable than the lamellar microstructure, we need to compare the change in free energy associated with forming each. Based on the geometry of the two structures, we can write equations describing their respective total free energies of transformation as

$$\Delta G_{\text{tot,lamellar}} = \lambda^3 \Delta G_V + 2\lambda^2 \gamma_{\alpha\beta}$$

$$\Delta G_{\text{tot,rod}} = \lambda^3 \Delta G_V + 2\pi r \lambda \gamma_{\alpha\beta} \tag{6.59}$$

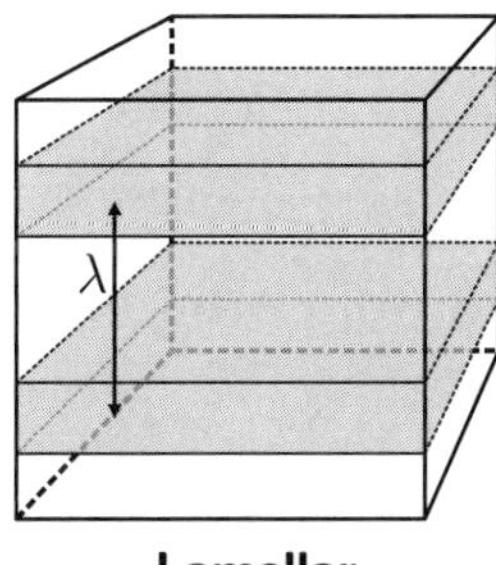

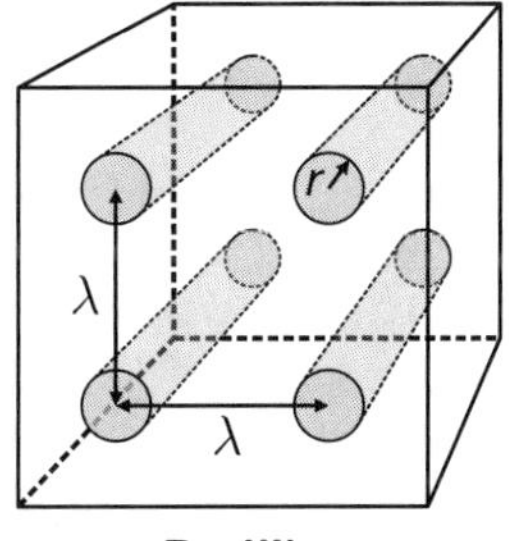

FIGURE 6.38 Schematic illustration of the lamellar–rodlike eutectic microstructures discussed in Example 6.38.

The free energy of the lamellar structure does not depend on the volume fraction of the β phase, only on the lamellar spacing λ. The free energy of the rodlike structure depends both on the rod spacing λ and the volume fraction of the β phase. We can explicitly incorporate this volume fraction dependence by expressing how ϕ_β depends on λ and r:

$$\phi_\beta = \frac{\pi r^2}{\lambda^2}$$

Thus

$$r = \sqrt{\frac{\phi_\beta}{\pi}}\,\lambda \tag{6.60}$$

Inserting Equation 6.60 into the expression for the total free energy of the rodlike microstructure (Equation 6.59) yields

$$\Delta G_{\text{tot,rod}} = \lambda^3 \Delta G_V + 2\pi \sqrt{\frac{\phi_\beta}{\pi}} \lambda^2 \gamma_{\alpha\beta} \tag{6.61}$$

We can now construct an inequality involving the free energies of these two microstructures and solve for the critical volume fraction below which the rodlike microstructure has the lower free energy:

$$\text{When } \phi_\beta < \phi_\beta^{\text{crit}}: \quad \Delta G_{\text{tot,rod}} < \Delta G_{\text{tot,lamellar}}$$

$$\lambda^3 \Delta G_V + 2\pi \sqrt{\frac{\phi_\beta}{\pi}} \lambda^2 \gamma_{\alpha\beta} < \lambda^3 \Delta G_V + 2\lambda^2 \gamma_{\alpha\beta}$$

$$\sqrt{\phi_\beta \pi} < 1 \tag{6.62}$$

$$\phi_\beta < \frac{1}{\pi}$$

$$\phi_\beta^{\text{crit}} = \frac{1}{\pi}$$

Thus, when the β volume fraction is below $1/\pi$ (about 0.318), the rodlike microstructure is energetically favored.

6.8.5 Peritectic Solidification

A peritectic solidification process is illustrated in Figure 6.39 and can be exemplified by the following reaction:

$$L + \beta \rightarrow \alpha \tag{6.63}$$

Because a peritectic transformation requires solid-state diffusion as β reacts with the liquid phase to convert to α, it tends to be more sluggish than a eutectic reaction. As the α phase builds up as a layer between the β and liquid phases, solid-state

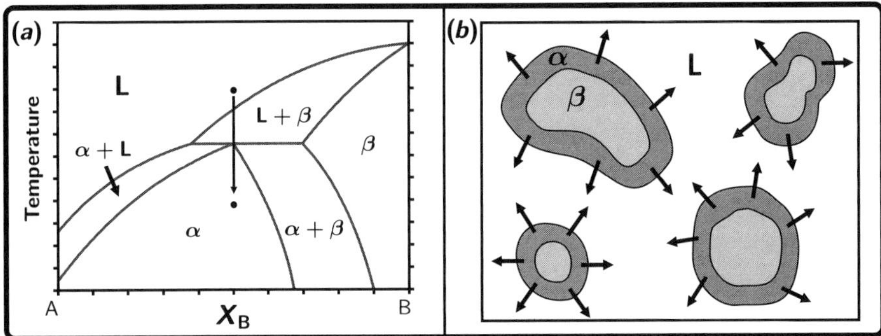

FIGURE 6.39 (*a*) Peritectic phase diagram. (*b*) Schematic illustration of phase formation during a peritectic solidification.

diffusion becomes increasingly rate limiting, and so it is typical for some nonequilibrium β phase to be retained after cooling below the peritectic temperature. Just as with the growth of passive oxide coatings discussed in the last chapter (Section 5.5), during a peritectic transformation the thickness of the α phase will frequently increase proportionally to $\sqrt{Dt}$.

6.9 MARTENSITIC TRANSFORMATIONS

Martensitic transformations are *displacive* rather than diffusional transformations. In a martensitic transformation, the atoms transform from one crystal structure to a new crystal structure via a highly coordinated shearinglike movement that propagates through a domain. A classic example of the martensitic transformation is the FCC to body-centered-tetragonal (BCT) conversion in Fe–Ni–C steels. Because the atoms simply transform their crystallographic orientations but do not participate in long-range diffusion, the original and transformed phases must have the same chemical composition. The change in structure, orientation, and lattice volume associated with a martensitic transformation can introduce macroscopic shape and volume changes, as shown in Figure 6.40. When confined by a matrix of the parent phase, a partial martensitic transformation can induce significant internal stresses and strains, which can act to harden (but also potentially embrittle) the material. Careful quenching and tempering protocols to control the nucleation, growth, and partial decomposition of martensitic phases enable the creation of composite microstructures with excellent mechanical properties. This approach has been used for thousands of years by sword-smiths and is used to this day in the fabrication of high-hardness steels.

Although the martensitic transformation does not involve long-range diffusion, it does require nucleation. Large internal stresses and strains are usually induced when a local region transforms martensitically. These additional terms must therefore be included in calculating the free energy of transformation, ΔG_V:

$$\Delta G_{V,\text{tot}} = \Delta G_V + \sigma_s \epsilon_s + \sigma_{\text{nn}} \epsilon_{\text{nn}} \tag{6.64}$$

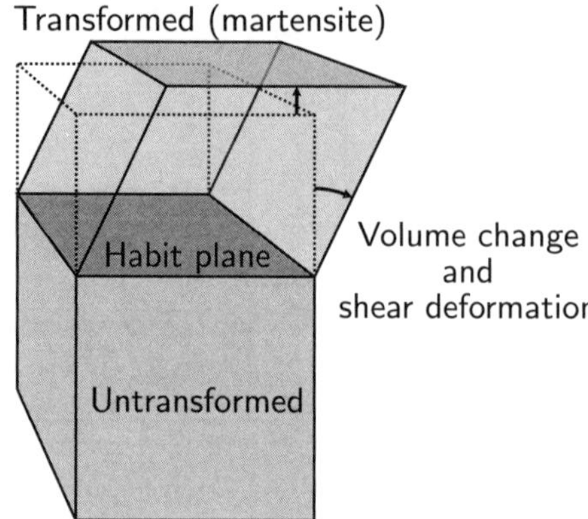

FIGURE 6.40   Macroscopic shape changes induced during a martensitic transformation.

Where $\sigma_s \epsilon_s$ accounts for the energy due to the shear induced by the martensitic transformation and $\sigma_{nn}\epsilon_{nn}$ accounts for the energy due to the dilation (volume change) associated with the martensitic transformation. Because both $\sigma_s \epsilon_s$ and $\sigma_{nn}\epsilon_{nn}$ are positive, significant undercooling is typically required to trigger martensite nucleation. Homogeneous nucleation of martensite is almost never observed. Instead, martensite nucleates heterogeneously at sparse locations, such as at tilt or grain boundaries. Rapid quenching or deep undercooling is thus often needed to induce martenstite nucleation. Once a martensite domain has nucleated, however, propagation can occur quite rapidly. Because diffusion is not required, the martensite domain grows by the propagation of the habit plane, often assisted by the internal stresses arising from the transformation itself. Propagation speeds can approach the speed of sound and can even be audible to the human ear. Unlike diffusional-based nucleation and growth transformations, nucleation and growth processes in matertensitic transformations are not significantly affected by low temperatures. Upon quenching a material to induce a partial martensitic transformation, further reduction in temperature will usually induce further martensitic nucleation and growth. Thus, greater transformation is accomplished by deeper and more rapid quneching. This is in contrast to diffusion-based nucleation and growth transformations, where rapid and deep undercooling can prevent phase transformation.

6.10 CHAPTER SUMMARY

This chapter discussed the kinetics of liquid–solid and solid–solid phase transformations. The main points introduced in this chapter include:

- A phase transformation can occur when a region of matter can *lower its total free energy* by changing its composition, structure, symmetry, density, or any other phase-defining aspect.

- Condensed-matter phase transformations can be broadly divided into two main categories: *diffusional* transformations and *diffusionless* (or "fluxless") transformations.

- Diffusionless phase transformations do not require the net transport of atoms across a phase boundary. For example, phase transformations involving a change in spin or magnetic moment or certain changes in crystal structure or symmetry do not require diffusional fluxes. Examples of such processes include the martensitic transformation in steel or certain cubic-to-tetragonal phase transformations.

- Diffusional transformations require the net transport/rearrangement of atoms. Diffusional can be further subdivided into two main types: *continuous* and *discontinuous*.

- Spinodal decomposition is an example of a *continuous* phase transformation. In a spinodal transformation, a single phase separates into two phases via gradual changes in local composition. The spinodal decomposition process gradually occurs everywhere (small in degree, large in extent).

- The nucleation-and-growth process is an example of a *discontinuous* phase transformation. The new phase nucleates in highly localized regions with properties that are abruptly and distinctly different from the parent phase (large in degree, small in extent).

- A driving force must be present for a phase transformation to occur. The most common driving forces for condensed-matter phase transformation include temperature and composition, although pressure-induced phase transformations are also possible.

- Phase diagrams can be used to predict phase transformations as a function of temperature or compositional changes.

- For an arbitrary phase transformation that occurs at an equilibrium temperature T_E, the Gibbs free-energy change per unit volume of the material, ΔG_V, may be calculated as

$$\Delta G_V = \Delta H_V \frac{\Delta T}{T_E}$$

  where $\Delta T = T_E - T$ is known as the amount of supercooling below (or superheating above) the equilibrium transformation temperature. Thus, the size of the driving force depends linearly on the deviation from the equilibrium temperature.

- Volumetric free energy is related to molar free energy by the molar volume of the material:

$$\Delta G_V = \frac{\Delta G}{V_m} = \frac{\Delta G}{M/\rho}$$

  where the molar volume V_m (m^3/mol) is given by the molecular weight of the material M (kg/mol) divided by the density of the material ρ (kg/m^3).

- Surfaces and interfaces generally have higher energy than the bulk of a material because atoms at the surface/interface have missing or dangling bonds and/or possess altered structural arrangement. The "extra" energy associated with a surface or interface compared to the bulk is quantified by a surface energy term γ:

$$\gamma = \left(\frac{\partial G}{\partial A}\right)_{T,P}$$

Typical units for γ are J/m^2 or J/cm^2.

- Kinetic implications of surface/interface energy include the following: (1) Since surfaces and interfaces carry "excess" free energy, systems will seek to minimize the amount of surface/interface area per unit volume—this leads to phenomena like coarsening, grain growth, and sintering (discussed in Chapter 7). (2) Smaller particles have higher surface/volume ratios and thus tend to be less stable, more reactive, have lower melting point (relative to the bulk), etc. (3) The energy cost to create surfaces/interfaces leads to a *nucleation barrier* in condensed-matter phase transformations. Therefore, nucleation-based phase transformations can only occur if *the energy released by creating the new volume of the second phase sufficiently offsets the energy expended in creating the new interfacial area*. This leads to a minimum viable nucleation size and thus helps determine the rate at which nucleation can proceed.

- *Homogeneous nucleation* occurs without specific locational preference anywhere inside the volume of a material. *Heterogeneous nucleation* preferentially occurs at specific favored locations such as along surfaces, walls, or grain boundaries. The activation barrier for heterogeneous nucleation is usually lower than the barrier for an otherwise equivalent homogeneous nucleation process due to a reduction in the interface energy costs required to nucleate a new phase heterogeneously.

- The energetic cost associated with creating the new interfacial area during nucleation leads to a minimum viable nucleation size (r^*) and an activation barrier for nucleation (ΔG^*). For homogeneous nucleation

$$r^* = -\frac{2\gamma}{\Delta G_V} \qquad \Delta G^* = \frac{16\pi\gamma^3}{3(\Delta G_V)^2}$$

For heterogeneous nucleation

$$r^*_{\text{het}} = -\frac{2\gamma}{\Delta G_V} \qquad \Delta G^*_{\text{het}} = \frac{16\pi\gamma^3}{3(\Delta G_V)^2}f(\theta)$$

where $f(\theta)$ accounts for the decreased interfacial energy cost associated with the geometry of the heterogeneous nucleation process compared to a homogeneous nucleation process.

- Heterogeneous nucleation lowers ΔG^* but does not affect r^*.

- The overall rate of nucleation ($\dot{N}$) is determined by two factors: the concentration of viable nuclei n^* and the rate of atomic diffusion to the nuclei. For transformations that occur on cooling, n^* obtains a peak at intermediate temperatures below T_E while diffusion decreases exponentially with cooling. As a result, an *intermediate* undercooling temperature results in the greatest rate of nucleation.

- The peak in $\dot{N}$ at intermediate temperatures explains how glasses can be formed, even from materials systems that do not tend to form into glasses (e.g., metallic glasses). The key to forming glasses in such systems is to *rapidly quench them* from the melt, so that they solidify without having time for crystalline nuclei to form. In this way, the random amorphous structure of the liquid melt is "frozen" in place before the material has time to nucleate and grow organized crystalline domains.

- Like the nucleation rate ($\dot{N}$), the growth rate ($\dot{G}$) for a new phase also depends on two factors: the driving forces for the transformation and the rate of atomic diffusion (or sometimes the rate of heat transfer for heat-transport-limited growth). For transformations that occur on cooling, the driving force increases linearly with undercooling below T_E, while diffusion decreases exponentially with undercooling. As a result, an *intermediate* undercooling temperature results in the greatest rate of growth.

- The *overall transformation rate* for a nucleation and growth process, $\dot{F}$, is a nonlinear product of the nucleation ($\dot{N}$) and growth ($\dot{G}$) rates. Because both $\dot{N}$ and $\dot{G}$ obtain a peak at intermediate temperatures, so does $\dot{F}$. However, the peaks for $\dot{N}$ and $\dot{G}$ typically do not occur at the same temperature, and this has important implications on the type of microstructure that develops depending on the degree of undercooling during transformation.

- As an example of the microstructural repercussion of differing nucleation and growth rates, carrying out a liquid–solid phase transformation isothermally at a temperature just slightly below the equilibrium transformation temperature tends to produce a coarse-grained microstructure since the growth rate is high but the nucleation rate is low. Transforming at lower temperatures (greater amount of undercooling) results in microstructural refinement. A rapid quench to low temperature may produce an amorphous or glasslike microstructure as the crystalline phase transition can be kinetically impeded.

- The Johnson–Mehl equation describes the overall fraction of material transformed as a function of time [$F(t)$] assuming spherically growing nuclei and constant nucleation and growth rates:

$$F(t) = 1 - e^{-(\pi/3)\dot{G}^3 \dot{N} t^4}$$

This equation relates the fraction transformed to the nucleation rate, the growth rate, and the time elapsed since the start of the transformation (at constant temperature) and predicts sigmoidal transformation behavior, meaning that the fraction transformed first increases exponentially in time before slowing down and asymptotically approaching complete transformation.

- The kinetics of solid-state phase transformations are often summarized in time-temperature-transformation (TTT) diagrams. The TTT diagram shows the time required to achieve certain amounts of transformation as a function of the temperature of the transformation. The transformation contours obtained in TTT diagrams are also known as "C curves" because of their characteristic C-like shape. The C-like shape is due to the fact that both nucleation and growth (and hence the overall transformation rate) peak at intermediate temperatures.

- Real phase transformation processes rarely reach completion or attain complete "equilibrium." Many phase transformation processes produce microstructures that depart wildly from our equilibrium expectation. The limited atomic mobilities associated with solid-state diffusion can frequently cause (and preserve) such nonequilibrium structures. Common examples include coring, dendritic growth, and lamellar-type microstructures.

- During a plane front solidification process, *partitioning* of species between the solid and liquid phases can take place, which results in a nonuniform compositional profile at the end of the solidification process. The severity of the compositional nonuniformity depends on the degree of solute partitioning between the solid and the liquid as well as the rate of solidification relative to the rate of solid-state diffusion. If solid-state diffusion is fast, then the composition gradient in the solid can be quickly erased, leading to a more uniform composition profile. However, in most cases, solid-state diffusion is slow compared to solidification, and thus these nonuniform composition gradients tend to be frozen in. The *Scheil equation* provides a way to model the nonuniform concentration profile that arises in such a plane front solidification process:

$$c(x) = k' c_0 \left[1 - \frac{x}{L}\right]^{(k'-1)}$$

where c_0 is the starting (uniform) composition of the liquid alloy, L is the length of the solidifying bar (ingot), and k' is the effective solute partitioning ratio.

- Nonuniform composition occurs not only in a linear plane front solidification process, but can occur in almost any nucleation and growth processes (solid–liquid or solid–solid). In such cases, compositional partitioning between the growing nuclei and the parent matrix phase leads to a characteristic "cored" microstructure that is typically undesirable for most applications.

- Growth front instability during transformation can lead to cellular or dendritic microstructures, depending on the severity of the instability. Minor instability leads to the formation of primary protuberances, called *cells*, which advance perpendicular to the interface. If the instability increases, these primary protuberances can themselves spawn secondary protuberances perpendicular to the primary protuberances, and a dendritic microstructure develops. Cellular and dendritic microstructures are most commonly observed in vapor–solid or liquid–solid phase transformations, although they can also be formed in solid–solid phase transformations.

- Local variations in temperature (and hence growth rate) along the growth front can locally accelerate (or suppress) growth, thereby destabilizing (or stabilizing) the growth process. In general, a low degree of undercooling during transformation tends to favor cellular/dendritic instability, while a larger degree of undercooling tends to suppress cellular/dendritic instability. Compositional gradients can have similar instability effects on growing interfaces. Certain actions, for example, actively mixing the liquid, can help suppress growth front instability by eliminating these thermal and/or compositional gradients.

- During a eutectic solidification L $\rightarrow \alpha + \beta$, if the interfacial energy between the α and β phases is low compared to the liquid–α and liquid–β interfacial energies, then the α and β phases will prefer to wet each rather than the liquid. Under rapid, directional solidification, this can lead to the coordinated growth of long alternating *lamellar* or plateletlike formations of the α and β phases. Such lamellar microstructures are common feature of eutectic phase transformations

- Martensitic transformations are *displacive* rather than diffusional transformations. In a martensitic transformation, the atoms transform from one crystal structure to a new crystal structure via a highly coordinated shearinglike movement that propagates through a domain. A classic example of the martensitic transformation is the FCC-to-BCT conversion in Fe–Ni–C steels.

- Although the martensitic transformation does not involve long-range diffusion, it does require nucleation. Large internal stresses and strains are usually induced when a local region transforms martensitically. Because of this issue, rapid quenching or deep undercooling is often needed to induce martenstite nucleation. Once a martensite domain has nucleated, however, propagation can occur quite rapidly.

- Upon quenching a material to induce a partial martensitic transformation, further reduction in temperature will usually induce further martensitic nucleation and growth. Thus, greater transformation is accomplished by deeper and more rapid quenching. This is in contrast to diffusion-based nucleation and growth transformations, where rapid and deep undercooling can prevent phase transformation.

6.11 CHAPTER EXERCISES

Review Questions

Problem 6.1. Define/explain the following. Use schematic illustrations/diagrams as appropriate.

(a) Continuous phase transformation

(b) Discontinuous phase transformation

(c) Spinodal decomposition

(d) Critical nucleus size r^*

(e) Heterogeneous nucleation

(f) Coring

(g) Eutectic lamellae

(h) Martensitic transformation

Problem 6.2. You are tasked to design a new mold material for iron casting that ensures the best possible wetting of the mold surface with molten iron. To ensure that iron wetting on the mold surface is energetically favorable, should the surface energy between the mold and the molten iron be much greater or much smaller than the surface energy between the mold and air? Fully explain your answer using diagrams/equations as appropriate.

Problem 6.3. Explain the difference between stable growth and dendritic growth. Use diagrams in your explanation.

Calculation Questions

Problem 6.4. At 1800 K, the surface energy of alumina (Al_2O_3) is 8.5×10^{-5} J/cm^2. The surface energy for liquid nickel against its own vapor at 1800 K is 1.7×10^{-4} J/cm^2. At the same temperature, the interfacial energy between liquid nickel and alumina is 1.8×10^{-4} J/cm^2. From these data, calculate the contact angle of a droplet of liquid nickel on an alumina plate at 1800 K.

Problem 6.5. A small droplet of molten silver is cooled until it solidifies via a homogeneous nucleation-and-growth process. Calculate the amount of undercooling required to initiate the onset of solidification (i.e., the temperature at which at least one viable homogeneous nuclei forms in the volume of the droplet). Given: Droplet volume $(V_{drop}) = 0.10$ cm^3, $T_m = 962\,°C$, $\gamma_{sl} = 1.3 \times 10^{-5}$ J/cm^2, $\rho_{Ag} = 9.3$ g/cm^3, $M_{Ag} = 107.9$ g/mol, and $\Delta H^°_{solidification} = -11.3$ kJ/mol.

Problem 6.6. Figure 6.41 is a TTT diagram for 1054 steel. The heavy line to the left is for 1% transformed and the heavy line to the right is for 99% transformed. The dashed line in the middle is for 50% transformed. Calculate and plot (using the computer application of your choice) the rate of transformation for 300–750 °C at 50 °C intervals by taking the times for 50% transformed and dividing 50% transformed by the time. Plot the temperature on the vertical axis and the rate of transformation on the horizontal axis.

Problem 6.7. Figure 6.42 gives a set of rate of transformation $[F(t)]$ curves for the crystallization of a glass. Using this information, calculate and plot (using a computer program of your choice) a TTT (time–temperature–transformation) diagram for this crystallization that includes curves for 10, 50, and 90% transformed with temperature on the vertical axis versus log (time) on the horizontal axis.

Problem 6.8. Part of the aluminum–copper (Al–Cu) phase diagram is given in Figure 6.43. Aluminum is strengthened by precipitating $CuAl_2$ particles from the κ solid solution. Consider an alloy composition that contains 3 wt % Cu and that the $CuAl_2$ precipitates form by a nucleation-and-growth process.

(a) Estimate the equilibrium temperature for the precipitation of $CuAl_2$ particles from the 3 wt % Cu alloy.

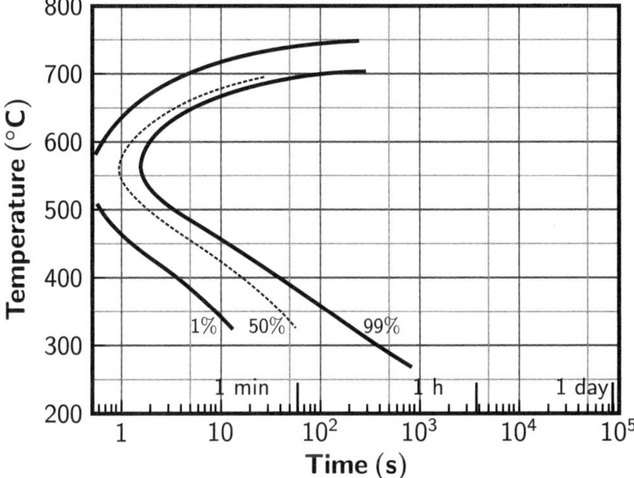

FIGURE 6.41 TTT diagram for 1054 steel.

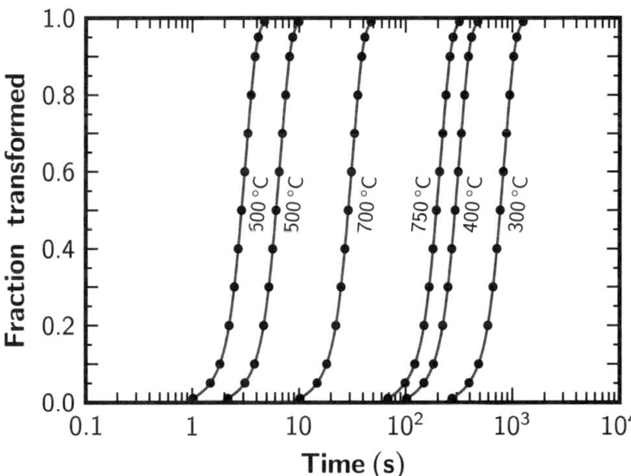

FIGURE 6.42 Crystallization of a glass as a function of time for various transformation temperatures.

(b) Sketch schematic time–temperature–transformation (TTT) curves for the transformation of κ to $CuAl_2$. Provide curves for both 10 and 90% extent of transformation and note the equilibrium temperature for the transformation.

(c) For a 3.0 wt% Cu solid solution at 300 °C, assume that the Gibbs free energy of transformation for $\kappa \rightarrow CuAl_2$ is -5 kJ/mol. Assuming that $CuAl_2$ has a density of 4 g/cm^3 and that the surface energy of the $\kappa/CuAl_2$ interface is 1.7×10^{-5} J/cm^2, calculate r^* and ΔG^* for this transformation.

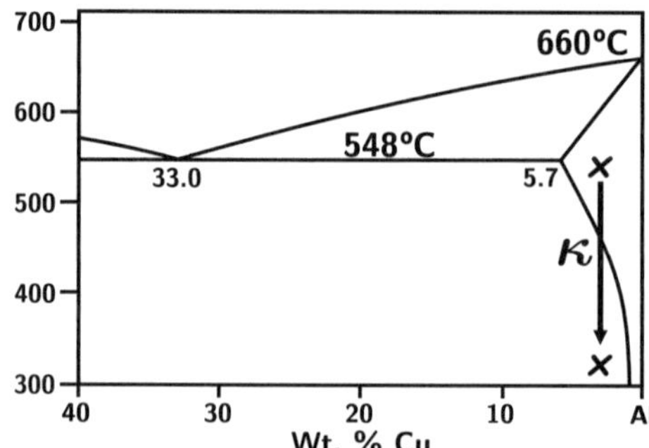

FIGURE 6.43 Partial Al–Cu phase diagram highlighting the κ-CuAl$_2$ precipitate composition.

(d) If the temperature is lowered from 300 to 200 °C, will r^* increase or decrease? Why?

(e) If the temperature is lowered from 300 to 200 °C, will ΔG^* increase or decrease? Why?

(f) If $\dot{G} = 2 \times 10^{-6}$ cm/s and $\dot{N} = 50/(\text{cm}^3 \cdot \text{s})$ at $T = 300$ °C, plot (using a computer application of your choice) $F(t)$ as a function of time for this transformation at 300 °C.

CHAPTER 7

MICROSTRUCTURAL EVOLUTION

The previous chapter discussed the kinetics of condensed-matter phase transformations. We learned that a phase transformation occurs when it is energetically favorable to change the structural and/or chemical arrangement of a body of matter. Phase transformations are driven by the ability to decrease the overall volumetric free energy of a system by creating a new phase. In this chapter, we turn our attention to the kinetics of morphological or microstructural evolution. Here, we are concerned with changes in the shape or microstructure of a material in the absence of a phase change. Such changes are generally driven by the energetic benefit provided by decreasing the surface or interfacial energy of the system. In other words, microstructural evolution is powered by *geometric* driving forces while phase transformations are powered by *chemical* driving forces. Examples of microstructrual evolution include changes in surface morphology, coarsening, grain growth, and sintering.

7.1 CAPILLARY FORCES

The primary internal driving force for morphological or microstructural evolution is the *capillary force*. Capillary forces arise when changes to the area or morphology of a surface or interface will lower its total free energy. As an example, local differences in surface curvature, especially at the micrometer scale and below, lead to strong capillary forces that tend to favor the transport of matter from regions of high covexity to regions of high concavity (see Figure 7.1). This process can lead to a gradual smoothening or flattening of a surface over time.

In the solid state, the capillary force can be expressed as an increase or decrease in chemical potential of a species in the vicinity of a curved interface relative to a

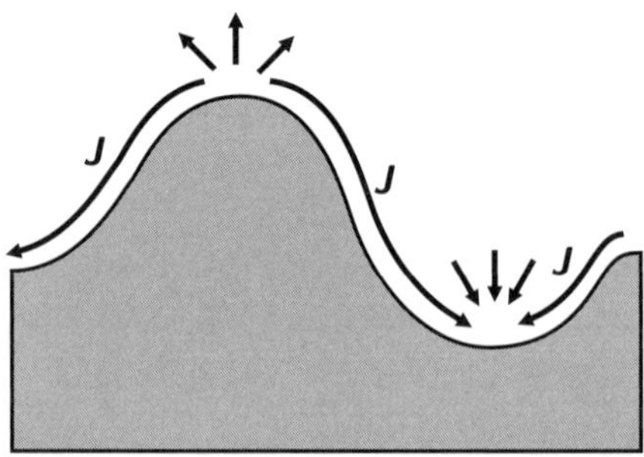

FIGURE 7.1   Capillary forces arising due to differences in surface curvature lead to net transport of matter from the convex to concave regions, resulting in smoothing of the surface over time.

planar interface. The change in chemical potential is directly equal to the work done in going from a flat interface to a curved interface. For an ideal solution,

$$\text{Work} = \gamma \, dA = \Delta\mu = RT \, \ln c_i - RT \, \ln c_{i,0} \tag{7.1}$$

where γ is the surface energy and dA is the change in area associated with going from a flat interface to a curved interface; c_i is the solubility of species i in the vicinity of the curved interface while $c_{i,0}$ is the equilibrium solubility of species i at a flat interface.

Consider the increase in solubility that arises at a spherical interface relative to a planar interface. For 1 mol of material transfered from a flat to a spherical interface, the change in surface area, dA, is given by $8\pi r \, dr = 8\pi r(V_\mathrm{m}/4\pi r^2)$.[1] Thus:

$$\gamma \, dA = RT \, \ln \frac{c_i}{c_{i,0}}$$

$$\gamma 8\pi r \left(\frac{V_\mathrm{m}}{4\pi r^2} \right) = RT \, \ln \frac{c_i}{c_{i,0}} \tag{7.2}$$

$$c_i = c_{i,0} \exp \left(\frac{2\gamma V_\mathrm{m}}{RT} \frac{1}{r} \right)$$

The more general expression for solubility in the vicinity of a curved (but nonspherical) interface is

$$c_i = c_{i,0} \exp \left[\frac{\gamma V_\mathrm{m}}{RT} \left(\frac{1}{r_1} + \frac{1}{r_2} \right) \right] \tag{7.3}$$

where r_1 and r_2 are the principal radii of curvature of the interface at the point of interest.

Equations 7.2 and 7.3 indicate that highly convex surfaces ($r =$ small, positive) lead to enhanced solubility relative to a planar interface, which drives the dissolution or removal of atoms from such areas. In contrast, highly concave surfaces ($r =$ small, negative) lead to suppressed solubility relative to a planar interface,

---

[1] $V_\mathrm{m} = 4\pi r^2 dr$ where V_m is the molar volume.

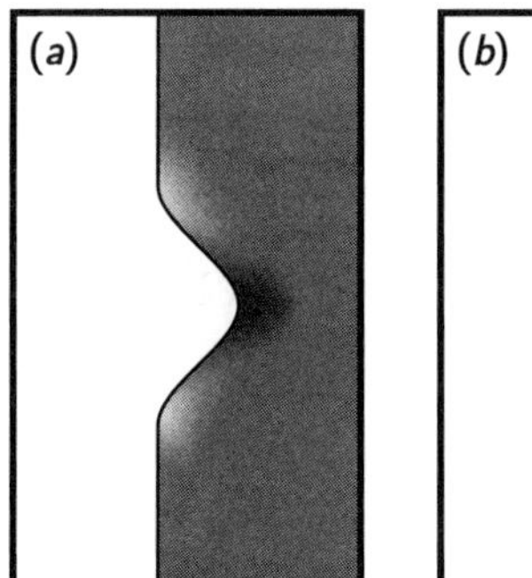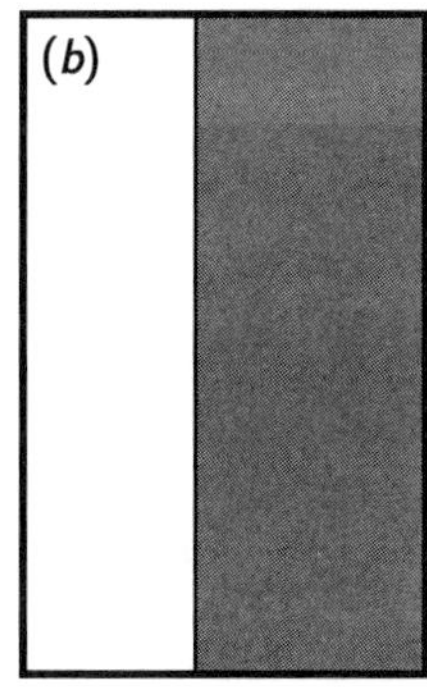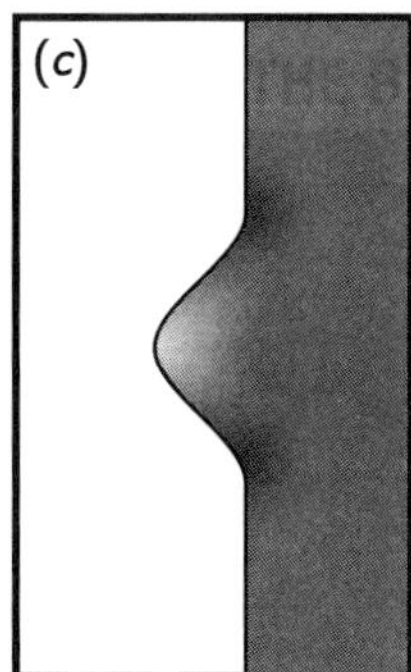

FIGURE 7.2   The equilibrium solubility over a positive curvature surface (*a*) is enhanced compared to a planar surface (*b*). The equilibrium solubility over a negative curvature surface (*c*) is suppressed (regions of dark shading indicate increased local solubility while regions of lighter shading indicate decreased local solubility).

which drives the precipitation or addition of atoms to such areas (see Figure 7.2). The enhanced solubility of small spherical particles or precipitates explains their instability during annealing/coarsening processes but also highlights the utility of such fine-particle-size materials in sintering or other powder-processing applications where reaction/densification is desired.

Example 7.1

Question: (a) A precipitate phase has a density of 5.0 g/cm^3 and a molecular weight of 200 g/mol. Spherical particles of this precipitate are embedded in a matrix. If decreasing the radius of the spherical precipitates from 50 to 25 nm increases the precipitate atom solubility by 1% at 500°C, what is the interfacial energy associated with the precipitate phase? (b) If the spherical precipitate radius is reduced by another factor of 2, what is the additional percent increase in solubility?

Solution: (a) Setting up a ratio using Equation 7.2 and solving for γ yields:

$$\frac{c_1}{c_2} = \frac{\exp\left(\frac{2\gamma V_{\mathrm{m}}}{RT}\frac{1}{r_1}\right)}{\exp\left(\frac{2\gamma V_{\mathrm{m}}}{RT}\frac{1}{r_2}\right)}$$

$$\gamma = \frac{RT}{2V_{\mathrm{m}}}\left(\frac{r_2 r_1}{r_2 - r_1}\right)\ln\left[\frac{c_1}{c_2}\right]$$

Inserting values from the problem statement (being careful to use SI units) yields

$$\gamma = \frac{8.314\ \mathrm{J/(mol \cdot K)} \cdot 773\ \mathrm{K}}{2 \cdot 4.0 \times 10^{-5}\ \mathrm{m^3/mol}}\left(\frac{25 \times 10^{-9}\ \mathrm{m} \cdot 50 \times 10^{-9}\ \mathrm{m}}{25 \times 10^{-9}\ \mathrm{m} - 50 \times 10^{-9}\ \mathrm{m}}\right)\ln\left[\frac{1}{1.01}\right]$$

$$= 4.0 \times 10^{-2}\ \mathrm{J/m^2} = 4.0 \times 10^{-6}\ \mathrm{J/cm^2}$$

(b) Now that we have γ, we can again make use of the ratio approach to calculate the additional percent increase in solubility with a further factor-of-2 reduction in precipitate radius:

$$\frac{c_1}{c_2} = \frac{\exp\left(\frac{2\gamma V_{\mathrm{m}}}{RT}\frac{1}{r_1}\right)}{\exp\left(\frac{2\gamma V_{\mathrm{m}}}{RT}\frac{1}{r_2}\right)} = \exp\left[\frac{2\gamma V_{\mathrm{m}}}{RT}\left(\frac{1}{r_1} - \frac{1}{r_2}\right)\right]$$

$$= \exp\left[\frac{2 \cdot 4.0 \times 10^{-2}\ \mathrm{J/m^2} \cdot 4.0 \times 10^{-5}\ \mathrm{m^3/mol}}{8.314\ \mathrm{J/(mol \cdot K)} \cdot 773\ \mathrm{K}} \right.$$

$$\left. \times \left(\frac{1}{25 \times 10^{-9}\ \mathrm{m}} - \frac{1}{12.5 \times 10^{-9}\ \mathrm{m}}\right)\right]$$

$$= 0.98$$

Thus, with a second 2× reduction in the precipitate radius, the solubility increases by $\approx 2\%$ (e.g., twice as much). With further reductions in radius, the solubility effect becomes more and more pronounced. Thus, for particles >100 nm in size, this curvature effect is typically negligible, while for particles <10 nm in size the effect can become significant.

Melting Point Depression in Nanoparticles

In addition to enhancing solubility and driving morphological changes, the increased chemical potential associated with curved interfaces also gives rise to a number of other important effects. Among the most notable is a marked depression in melting point often observed for nanoscale particles. This effect can be understood from basic thermodynamic principles. As previously discussed in Chapter 2, a material will melt at temperatures above T_{m}, where the Gibbs free energy of the liquid phase (G_1) is lower than the Gibbs free energy of the solid phase (G_s). The additional chemical potential associated with the curvature of a small particle increases G_s, which pushes T_{m} to lower temperatures. This effect is illustrated in Figure 7.3.

Mathematically, for a melting process $\mathrm{s} \to \mathrm{l}$, G_1 and G_s can be described by

$$G_s(r) = H_s - TS_s + \frac{2\gamma V_{\mathrm{m}}}{r} \tag{7.4}$$

$$G_1 = H_1 - TS_1 \tag{7.5}$$

where the term $2\gamma V_{\mathrm{m}}/r$ in the expression for $G_s(r)$ accounts for the chemical potential contribution of a curved particle. As G_1 represents the liquid-phase Gibbs free energy, it does not contain this solid-phase curvature contribution.

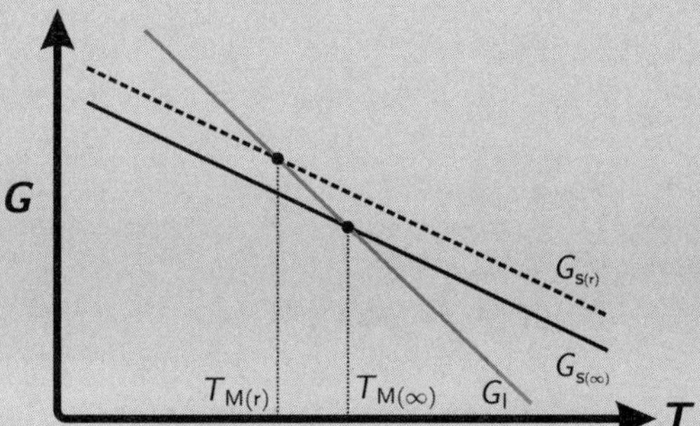

FIGURE 7.3   Thermodynamic explanation for the melting point depression observed in nanoscale particles. Compared to $G_s(\infty)$ for the bulk solid, $G_s(r)$ for a finite-sized particle is slightly higher due to the excess chemical potential associated with the curved interface of the particle. G_l is not affected. This results in a measurable decrease in T_m for small-radius particles compared to the bulk solid.

The Gibbs free energy of melting for a particle of radius r, $\Delta G_m(r)$, can be calculated as

$$\Delta G_m(r) = G_l - G_s(r) = (H_l - TS_l) - \left(H_s - TS_s + \frac{2\gamma V_m}{r} \right)$$

$$= \Delta H_m - T\Delta S_m - \frac{2\gamma V_m}{r} \tag{7.6}$$

where ΔH_m and ΔS_m are the bulk enthalpy and entropy of melting, respectively. For a bulk solid ($r \to \infty$), the "bulk" melting point $T_m(\infty)$ can therefore be calculated as

$$\Delta G_m(\infty) = \Delta H_m - T\Delta S_m$$

$$0 = \Delta H_m - T_m(\infty)\Delta S_m \tag{7.7}$$

$$T_m(\infty) = \frac{\Delta H_m}{\Delta S_m}$$

For a particle of radius r, the melting point $T_m(r)$ can likewise be calculated as

$$\Delta G_m(r) = \Delta H_m - T\Delta S_m - \frac{2\gamma V_m}{r}$$

$$0 = \Delta H_m - T_m(r)\Delta S_m - \frac{2\gamma V_m}{r} \tag{7.8}$$

$$T_m(r) = \frac{\Delta H_m - 2\gamma V_m/r}{\Delta S_m}$$

From Equation 7.7 , we have $\Delta S_{\mathrm{m}} = \Delta H_{\mathrm{m}}/T_{\mathrm{m}}(\infty)$ and thus we can write $T_{\mathrm{m}}(r)$ in terms of $T_{\mathrm{m}}(\infty)$ as

$$T_{\mathrm{m}}(r) = \frac{\Delta H_{\mathrm{m}} - 2\gamma V_{\mathrm{m}}/r}{\Delta H_{\mathrm{m}}/[T_{\mathrm{m}}(\infty)}]$$

$$T_{\mathrm{m}}(r) = T_{\mathrm{m}}(\infty)\left[1 - \frac{2\gamma V_{\mathrm{m}}}{r\Delta H_{\mathrm{m}}}\right]$$

(7.9)

Since melting is an endothermic process (ΔH_{m} is positive), Equation 7.9 indicates that the melting point will decrease as the particle radius r decreases. Since typical values for γ are on the order of 10^{-1} J/m^2, typical values for V_{m} are on the order of 10^{-5} m^3/mol, and typical values for ΔH_{m} are on the order of 10–100 kJ/mol, the melting point depression effect typically only becomes appreciable when r is on the order of nanometers.

7.2 SURFACE EVOLUTION

One important effect of capillary forces is that they tend to cause rough surfaces or interfaces to smoothen over time. Consider the surface profile $h(x)$ depicted in Figure 7.1. Assuming isotropic surface properties, capillary forces will drive the transport of atoms from areas of high convexity to areas of high concavity, resulting in the gradual smoothing of this surface over time. This smoothing process can occur by a number of different mechanisms. The two most common are (1) solid-state diffusion of atoms along the surface or (2) vapor-phase transport of atoms evaporated from the surface. We will briefly consider both mechanisms.

7.2.1 Surface Evolution by Solid-State Diffusion

If the smoothing process occurs by solid-state diffusion of atoms along the surface, it can be shown [13] that the height will change in time according to

$$\frac{\partial h}{\partial t} = -B^{\mathrm{S}}\frac{\partial^4 h}{\partial x^4}$$

(7.10)

where B^{S} is a constant that depends on various material-specific parameters, including the surface energy and the surface diffusivity. This partial differential equation illustrates that the height of the surface changes most rapidly in the regions of highest surface curvature. A Fourier series solution to Equation 7.10 yields

$$h(x, t) = A(t)\sin\frac{2\pi x}{\lambda}$$

(7.11)

where the roughness of the surface is broken down into a series of Fourier wavelengths (values of λ) whose time-dependent magnitudes [$A(t)$] are given by

$$A(t) = A(0)e^{-B^{\mathrm{S}}(2\pi/\lambda)^4 t}$$

(7.12)

Thus, the shortest wavelength surface roughness features decay most rapidly while the longer wavelength roughness features decay more slowly. This feature-dependent decay speed is analogous to other transient diffusion processes (recall, e.g., the discussion of transient diffusion in Chapter 4, where we saw that the sharpest features in a concentration profile decay most rapidly).

7.2.2 Surface Evolution by Vapor-Phase Transport

Another common mechanism for surface evolution involves vapor-phase transport. If a material is exposed to temperatures or atmospheres where significant condensation/evaporation of atoms to/from the surface can occur, capillary forces will favor evaporation from the areas of highest positive curvature and will favor condensation to the areas of highest negative curvature. Therefore, just like surface diffusion, this vapor-phase-mediated evaporation-and-condensation process can result in surface smoothing over time. When the vapor-phase mechanism controls the surface evolution, it can be shown [13] that the surface height will change in time according to

$$\frac{\partial h}{\partial t} = B^{\mathrm{V}} \frac{\partial^2 h}{\partial x^2} \tag{7.13}$$

where B^{V} is a constant that depends on various material-specific parameters including the surface energy and the equilibrium vapor pressure above the surface (for a planar reference surface). Just as with the surface diffusion mechanism, the most rapid height changes occur in the regions of highest surface curvature, although for vapor-phase transport, this sensitivity is based on the second derivative of the surface profile rather than the fourth derivative.

Surface Faceting: Anisotropic Surface Energy Effects

Isotropic surfaces possess a surface energy γ that is independent of surface orientation or inclination. Surface evolution for isotropic surfaces is therefore simply governed by the drive to reduce overall surface area. However, many materials show *anisotropic* surface properties, where γ varies with surface orientation or inclination. In such cases, the overall surface energy can sometimes be decreased by the introduction of surface *facets*. Surface faceting occurs when a planar or smoothly varying curved crystal surface can instead be replaced by a series of angled facets that correspond to overall lower energy orientations of the surface. Although the total area of the faceted surface may be higher than that of an analogous smooth surface, the total surface energy of the faceted surface can be lower since it is composed entirely of favored, low-energy surface orientations. Surface faceting can occur at both small and large length scales. For example, surface faceting can drive submicrometer-scale rearrangement of crystal surfaces. It also underlies the beautiful geometric surface terminations of many natural crystals and gemstones. Figure 7.4 illustrates the microscale faceting of an inclined crystal surface (*a*) and the macroscale geometric faceting of a natural single-crystal gemstone (*b*).

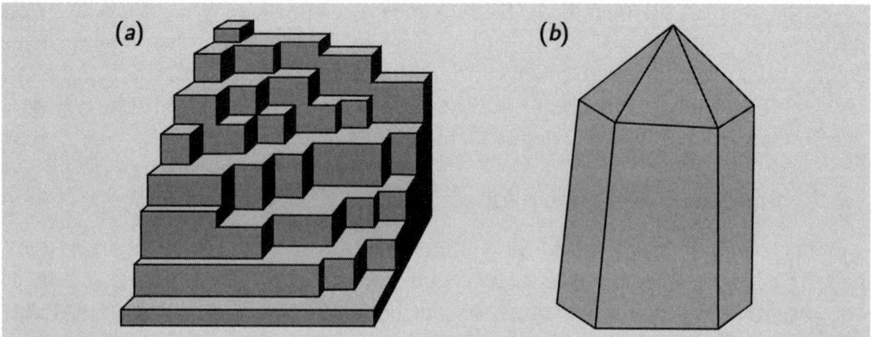

FIGURE 7.4 Examples of anisotropic surface-faceting phenomena at the micro- and macroscale; (*a*) microscale faceting of an inclined crystal surface; (*b*) macroscale geometric faceting of a natural single-crystal gemstone.

7.3 COARSENING

As we have previously discussed, capillary forces lead to an increase in solubility in the vicinity of highly curved surfaces. For two-phase systems consisting of a distribution of particles embedded in a matrix as illustrated in Figure 7.5, this effect can lead to gradual dissolution of the smallest particles compensated by growth of the largest particles. This phenomenon is known as *coarsening*. Coarsening has important implications (usually negative) for many materials applications. For example, coarsening leads to degradation in high-surface-area catalysts and also leads to degradation in structural materials where fine particles are used as a strengthening mechanism.

Figure 7.5 illustrates the basic steps involved in coarsening. Because atomic solubility increases with decreasing particle radius, particles smaller than the mean particle tend size to lose atoms over time, acting as "sources." The atoms lost from the source particles diffuse through the matrix and are eventually captured by particles larger than the mean particle size. These large particles act as "sinks," because they attract the atoms lost by the smaller particles and thereby grow over time. Based on this picture, two main processes can limit the coarsening kinetics: (1) the rate at which atoms diffuse from the small to large particles or (2) the rate at which atoms are added or subtracted from the sink/source particles. We will briefly consider both mechanisms.

7.3.1 Diffusion-Limited Coarsening

If the coarsening process is limited by the rate at which atoms diffuse through the matrix from the source particles to the sink particles, it can be shown [13] that the mean particle size of the distribution will increase with time according to

$$\langle R(t)\rangle^3 - \langle R(0)\rangle^3 = K_{\mathrm{D}}t \tag{7.14}$$

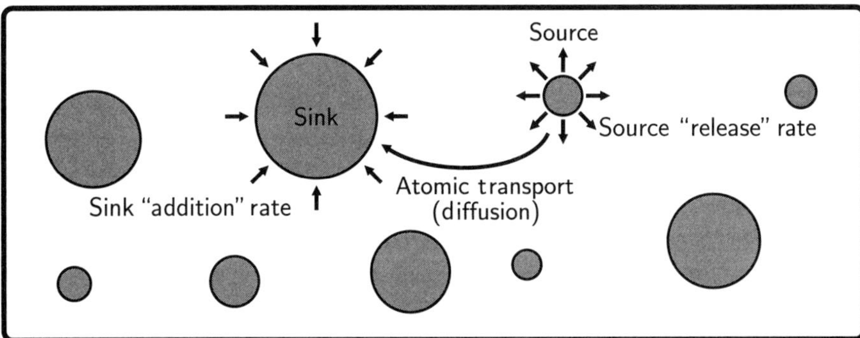

FIGURE 7.5 Coarsening of a distribution of embedded particles in a matrix. Coarsening involves several steps. Because atomic solubility increases with decreasing particle radius, particles smaller than the mean particle size tend to lose atoms over time, acting as "sources." The atoms lost from the source particles diffuse through the matrix and are eventually captured by particles larger than the mean particle size. These large particles act as "sinks," because they attract the atoms lost by the smaller particles and thereby grow over time. Two main processes can limit the coarsening kinetics: (1) the rate at which atoms diffuse from the small to large particles or (2) the rate at which atoms are added or subtracted from the sink/source particles.

where $\langle R(t)\rangle$ is the mean particle size at time t, $\langle R(0)\rangle$ is the initial mean particle size at time $t = 0$, and K_D is a kinetic constant that depends on various material-specific parameters, including the diffusivity, the surface energy, and the equilibrium solubility. This equation shows that diffusion-limited coarsening leads to an increase in the mean particle size with the *cube root of time* ($t^{1/3}$). Measurements of mean particle size in experimental systems where coarsening is limited by volume diffusion are generally consistent with this prediction.

7.3.2 Source/Sink-Limited Coarsening

If the coarsening process is limited by the rate at which atoms can be removed from (or attached to) source/sink particles, it can be shown [13] that the mean particle size of the distribution will increase with time according to

$$\langle R(t)\rangle^2 - \langle R(0)\rangle^2 = K_S t \tag{7.15}$$

where $\langle R(t)\rangle$ is the mean particle size at time t, $\langle R(0)\rangle$ is the initial mean particle size at time $t = 0$, and K_S is a kinetic constant that depends on various material-specific parameters, including the rate constant for the interfacial attachment/detachment process, the surface energy, and the equilibrium solubility. This equation shows that source/sink-limited coarsening leads to an increase in the mean particle size with the *square root of time* ($t^{1/2}$).

Example 7.2

Question: The following mean particle size–time data were obtained for a distribution of particles embedded in a matrix during a coarsening experiment:

t (h)	$\langle R \rangle$ (nm)
0	7
11	11.6
50	22
90	27
150	35
300	50
500	64

Determine the likely mechanism regulating coarsening in this system.

Solution: Figure 7.6 plots the provided data in two alternative fashions: (a) $\langle R(t) \rangle^2 - \langle R(0) \rangle^2$ versus t and (b) $\langle R(t) \rangle^3 - \langle R(0) \rangle^3$ versus t. In both cases, a linear fit to the data is also provided. A comparison of the two plots clearly reveals that the $\langle R(t) \rangle^2 - \langle R(0) \rangle^2$ versus t coarsening law does a much better job of describing the data. Based on this analysis, it is likely that this system is regulated by source/sink-limited coarsening.

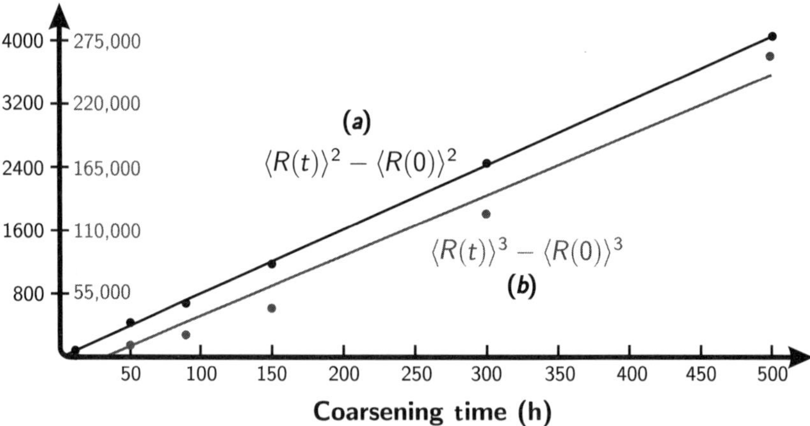

FIGURE 7.6 Plots of the experimental coarsening data provided in Example 7.2: (a) $\langle R(t) \rangle^2 - \langle R(0) \rangle^2$ vs. t; (b) $\langle R(t) \rangle^3 - \langle R(0) \rangle^3$ vs. t. The $\langle R(t) \rangle^2 - \langle R(0) \rangle^2$ vs. t coarsening law does a much better job of fitting the data, indicating that this system is likely regulated by source/sink-limited coarsening.

7.4 GRAIN GROWTH

The drive to reduce interfacial area and curvature leads to an increase in the average grain size of polycrystalline materials when annealed at elevated temperatures. As with coarsening, capillary forces cause larger grains to grow at the expense of smaller grains, which shrink and eventually disappear. This process causes the total number of grains to decrease over time and the average grain size to increase (see Figure 7.7).

The kinetics of grain growth in two dimensions (2D grain growth) are well established. One important conclusion derived from 2D grain growth models is the *rule of six*. This rule holds that grains with more than six sides tend to grow while grains with fewer than six sides tend to shrink. Grains with exactly six sides tend to be stable and relatively static.

For 2D grain growth, kinetic models generally predict a *parabolic growth* law of the form [13]

$$R_{\text{rms}}^2(t) - R_{\text{rms}}^2(0) = \frac{C}{\pi}t \qquad (7.16)$$

where $R_{\text{rms}}(t)$ is the equivalent root-mean-square radius of the average grain[2] at time t, $R_{\text{rms}}(0)$ is the initial equivalent root-mean-square radius of the average grain at time $t = 0$, and C is a kinetic constant that depends on various material-specific parameters, including the grain boundary mobility and the grain boundary interfacial energy.

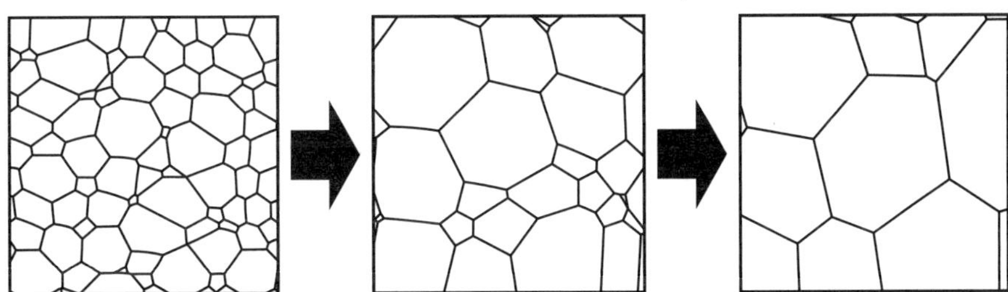

FIGURE 7.7   Grain growth in a polycrystalline material. The drive to reduce interfacial area and curvature causes larger grains to grow at the expense of the smaller grains, which shrink and eventually disappear. This process causes the total number of grains to decrease over time and the average grain size to increase. Grain growth is a virtually universal kinetic process that occurs in many polycrystalline materials when heated to temperatures close to their melting point (i.e., typically above $\frac{2}{3}T_{\text{m}}$ to $\frac{3}{4}T_{\text{m}}$).

---

[2]Real grains are, of course, not circular in shape. Grain "size" is treated in this model by assigning an effective radius to each grain that quantifies its effective area irrespective of its actual shape. In other words, an arbitrarily shaped grain of area A would have an equivalent radius of $\sqrt{(A/\pi)}$.

Factors Affecting Grain Growth

The factor C in Equation 7.16 is a material-dependent kinetic constant that depends on a number of parameters, including the grain boundary mobility and the grain boundary interfacial energy. High grain boundary mobility and high grain boundary energy yield more rapid grain growth. Grain boundary mobility itself depends on a number of more fundamental parameters—in particular the rate of solid-state diffusion in the material. This dependence makes sense because grain boundary motion requires the transfer (diffusion) of atoms across grain interfaces. Since solid-state diffusion is an exponentially temperature activated process, grain boundary mobility also generally increases exponentially with increasing temperature.

Grain growth is often an undesirable process, as small-grained materials tend to possess better mechanical properties than large-grained materials. Materials are therefore engineered to limit grain boundary mobility—usually by introducing chemical and/or microstructural inhomogeneities. Examples include solute impurity atoms, small-scale precipitates, and pores—all of which tend to impede grain boundary movement and thereby reduce mobility.

Attempts to develop complete kinetic models of 3D grain growth have yielded only limited success. Rules similar to the *rule of six* for 2D grain growth have been suggested for 3D grain growth. The most common approaches classify grains according to the number of edges or faces they possess. Grains with edges or faces larger than a critical value tend to grow, while grains with edges or faces smaller than a critical value tend to shrink. Unlike 2D growth, these critical edge/face values do not appear to be universal constants but depend on the ensemble properties of the entire collection of grains and can change as the grain structure evolves.

Example 7.3

Question: A polycrystalline material with an initial equivalent root-mean-square grain radius of 1 μm is annealed at 700°C for 5 h. After annealing, the equivalent root-mean-square grain radius is determined to be 3 μm. The annealed sample is then annealed a second time at 800°C for 2 h. After this second anneal, the equivalent root-mean-square grain radius has increased to 5 μm. Assuming the kinetics of the grain growth process follow an Arrhenius relationship, determine the activation energy for this process.

Solution: Starting with the parabolic grain growth law (Equation 7.16), we can explicitly incorporate the Arrhenius-type temperature dependence of the growth kinetics into the growth constant C, giving

$$R_{\text{rms}}^2(t) - R_{\text{rms}}^2(0) = \frac{C_0 \exp\left(-\frac{\Delta G_{\text{act}}}{RT}\right)}{\pi} t$$

where ΔG_{act} is the activation energy for the grain growth process. The problem statement provides two data points for annealing temperature, annealing time, and grain size. By setting up a ratio based on these two sets of data, we can eliminate C_0:

$$\frac{R^2_{1,rms}(t) - R^2_{1,rms}(0)}{R^2_{2,rms}(t) - R^2_{2,rms}(0)} = \frac{\left[C_0 \exp\left(-\frac{\Delta G_{act}}{RT_1}\right)\Big/\pi\right]t_1}{\left[C_0 \exp\left(-\frac{\Delta G_{act}}{RT_2}\right)\Big/\pi\right]t_2}$$

$$= \frac{t_1}{t_2}\exp\left[-\frac{\Delta G_{act}}{R}\left(\frac{1}{T_1} - \frac{1}{T_2}\right)\right]$$

Solving this expression for the unknown quantity ΔG_{act} yields

$$\Delta G_{act} = -R\frac{T_1 T_2}{T_2 - T_1}\ln\left[\left(\frac{t_2}{t_1}\right)\left(\frac{R^2_{1,rms}(t) - R^2_{1,rms}(0)}{R^2_{2,rms}(t) - R^2_{2,rms}(0)}\right)\right]$$

$$= -8.314 \text{ J/(mol} \cdot \text{K)}\frac{973 \text{ K} \cdot 1073 \text{ K}}{1073 \text{ K} - 973 \text{ K}}\ln\left[\frac{5 \text{ h}}{2 \text{ h}} \cdot \frac{(3 \text{ μm})^2 - (1 \text{ μm})^2}{(5 \text{ μm})^2 - (3 \text{ μm})^2}\right]$$

$$= -190000 \text{ J/mol}$$

Note that because this expression involves a ratio, any units can be used for the R_{rms} and t values as long as the same units are used in the numerator and denominator.

7.5 SINTERING

The drive to reduce interfacial area and curvature can cause the particles in a powder compact to consolidate together at sufficiently elevated temperatures (but below the melting point). This process is known as sintering. Because sintering enables the solid-state consolidation of a material well below the melting point, it is an important fabrication process for materials with high melting points such as ceramics, which can be formed into complex shapes from a powder and then sintered to produce near-net-shape final parts.

The initial stages of sintering are driven by the large reductions in particle surface area and surface curvature that can be achieved by the formation and growth of *necks* between particles, as illustrated in Figure 7.8. Neck formation and growth requires the transport of mass to the neck region from other areas of the compact. A number of different mass transport pathways are possible, as indicated in Figure 7.9. These pathways include bulk, grain boundary, and surface diffusion pathways, as well as vapor-phase transport and viscous flow. The major mass transport sintering mechanisms are summarized in Table 7.1.

Sintering can occur with or without densification. Sintering without densification, as illustrated in Figure 7.8a, occurs by surface-based mass transport mechanisms that

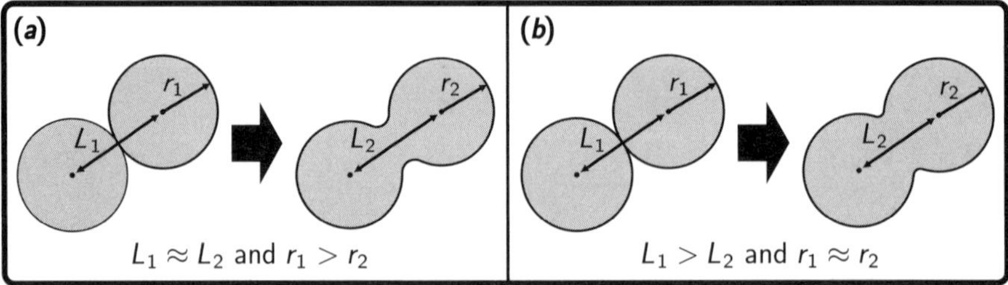

FIGURE 7.8   (*a*) Nondensifying sintering and (*b*) sintering accompanied by densification.

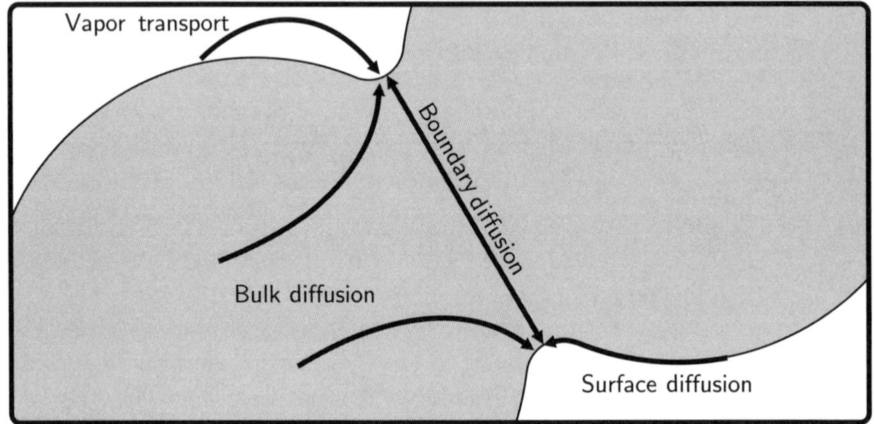

FIGURE 7.9   Sintering mechanisms.

TABLE 7.1   Summary of Common Sintering Mechanisms

Mechanism	Densifying or Nondensifying
Bulk diffusion to neck region	Densifying
Boundary diffusion to neck region	Densifying
Surface diffusion to neck region	Nondensifying
Vapor transport to neck region	Nondensifying
Viscous flow	Either

simply redistribute surface material and contribute to neck growth between particles without a consequent reduction in porosity. In this case, the particle centers of mass remain fixed. Densification during sintering, as illustrated in Figure 7.8*b*, requires bulk transport of material from the interior volume of particles to the pore zones, which causes a net shrinkage of the compact and a reduction in the porosity. In this case, the particle centers of mass gradually move closer to together as shrinkage and densification take place.

Major factors that determine sintering behavior include material properties such as the relative rates of bulk, grain boundary, and surface diffusion as well as the material's vapor pressure and surface energy. External factors such as the starting particle size and sintering temperature are also crucial. Sintering rates generally increase significantly with increasing temperature and increase with decreasing particle size. Although small particle sizes can greatly enhance sintering, the challenges associated with synthesizing and processing extremely fine powders frequently render this approach uneconomic.

The relationship between starting particle size and sintering rate can be captured by a number of sintering scaling laws. These laws are often known as the *Herring scaling laws*. The Herring scaling laws enable estimation of the comparative sintering rates for sintering systems that are identical in all respects except for their characteristic particle/feature size. For a sintering system possessing a characteristic particle or feature size of r, the sintering rate will scale as

$$\text{Sintering rate} \; \propto \frac{1}{r^n} \tag{7.17}$$

where n is a scaling exponent that depends on the dominant sintering mechanism. If sintering is controlled by surface or grain boundary diffusion, $n = 4$. If sintering is controlled by bulk diffusion, $n = 3$. If sintering is controlled by vapor-phase transport, $n = 2$. Finally, if sintering is controlled by viscous flow (generally only applicable to amorphous materials near T_m), $n = 1$.

> **Example 7.4**
>
> **Question:** Changing the starting particle size from 2 μm to 500 nm results in a 16-fold increase in the sintering rate. Based on this observation, what is the likely dominant sintering mechanism in this system and why? What additional observations could be conducted to further verify this conclusion?
>
> **Solution:** The particle size is reduced by a factor of 4 and the sintering rate increases by a factor of 16. This represents a $1/r^2$ dependence ($n = 2$), which suggests that the sintering rate is controlled by vapor-phase transport. Sintering controlled by vapor-phase transport is nondensifying, so examining the sintered compact to verify lack of densification would help to further confirm this conclusion. In addition, examining the sintering furnace walls, the setter, or other sacrificial parts used during sintering for evidence of film deposition due to vapor-phase transport would also help confirm this conclusion.

7.6 CHAPTER SUMMARY

This chapter examined the kinetics of microstructural evolution. Microstructural evolution processes include surface smoothing/faceting, coarsening, grain growth, and sintering. Fundamentally, microstructural evolution is powered by the drive to decrease surface energy. Unlike the phase transformation processes discussed in the previous chapter, these processes can occur in a single-phase material and simply involve changes in the microstructure/morphology of the material, rather than its phase composition or chemistry. The main points introduced in this chapter include:

- The *capillary force* is the primary internal driving force for most morphological or microstructural evolution processes. Capillary forces arise when changes to the area or morphology of a surface or interface will lower its free energy.
- Surface/interface curvature gives rise to capillary forces. The higher the curvature, the greater the effect. In general, capillary forces act to reduce surface curvature over time. This drives many kinetic processes, including surface smoothing, coarsening, grain growth, and sintering.
- Surface curvature also leads to important thermodynamic effects such as solubility enhancement and melting point depression in very small (typically nanometer-scale) particles.
- Capillary forces drive the transport of matter from surfaces/interfaces of high convexity to areas of high concavity. This leads to the gradual smoothing or flattening of a surface or interface with time. This surface evolution process can occur either by solid-state diffusion of matter across the surface of the material itself or by a vapor-phase process where matter is evaporated/condensed from/to the surface. As with other transient mass transport processes, the sharpest (or shortest wavelength) surface features tend to be smoothed out first, followed by the longer wavelength features.
- In a two-phase system consisting of a distribution of fine particles embedded in a matrix, capillary forces can promote the gradual dissolution of the smallest (most soluble) particles compensated by growth of the largest particles. This phenomenon is known as *coarsening* and leads to a gradual decrease in the number of particles and an increase in the average particle size over time.
- If the coarsening process is limited by the rate at which atoms diffuse through the matrix from the source particles to the sink particles, the mean particle size of the distribution will increase with time according to

$$\langle R(t) \rangle^3 - \langle R(0) \rangle^3 = K_\mathrm{D} t \tag{7.18}$$

  Diffusion-limited coarsening therefore leads to an increase in the mean particle size with the cube root of time ($t^{1/3}$).
- If the coarsening process is limited by the rate at which atoms can be removed from (or attached to) source/sink particles, the mean particle size of the distribution will increase with time according to

$$\langle R(t) \rangle^2 - \langle R(0) \rangle^2 = K_\mathrm{S} t \tag{7.19}$$

  Source/sink-limited coarsening therefore leads to an increase in the mean particle size with the square root of time ($t^{1/2}$).
- Polycrystalline materials often experience grain growth when annealed at elevated temperatures. As with coarsening, capillary forces cause larger grains to grow at the expense of smaller grains, which shrink and eventually disappear. This causes the total number of grains to decrease over time and the average grain size to increase.

- The *rule of six*, derived for 2D grain growth, holds that grains with more than six sides tend to grow while grains with fewer than six sides tend to shrink. There is no equivalent rule for 3D grain growth, although in general larger grains possessing more faces/edges tend to grow while smaller grains possessing fewer faces/edges tend to shrink.

- Two-dimensional grain growth often follows a *parabolic growth* law of the form

$$R_{\text{rms}}^2(t) - R_{\text{rms}}^2(0) = \frac{C}{\pi}t \tag{7.20}$$

An analogous law is not available for 3D grain growth. As with 2D grain growth, however, 3D grain growth usually proceeds sublinearly with time.

- In a process known as *sintering*, the drive to reduce interfacial area and curvature can cause the particles in a powder compact to consolidate together at elevated temperatures (but below the melting point).

- The initial stage of sintering involves the formation and growth of necks as particles grow together. This process requires the transport of mass to the neck region from other areas of the compact. If mass is moved to the neck region via surface or vapor-phase transport processes from other surface regions of the sample, pore volume is simply redistributed and the sintering process occurs without densification of the compact. If mass is moved to the neck region via bulk processes from the interior of the particles, volumetric consolidation can occur and the sintering process may be accompanied by densification (and shrinkage) of the compact.

- Major factors that determine sintering behavior include material properties such as the relative rates of bulk, grain boundary, and surface diffusion as well as the material's vapor pressure and surface energy. External factors such as the starting particle size and sintering temperature are also crucial. Sintering rates generally increase significantly with increasing temperature and also increase with decreasing particle size.

- The relationship between starting particle size and sintering rate can be captured by a number of sintering scaling laws. These laws are often known as the *Herring scaling laws*. The Herring scaling laws enable estimation of the comparative sintering rates for sintering systems that are identical in all respects except for their characteristic particle/feature size. For a sintering system possessing a characteristic particle or feature size r, the sintering rate will scale as

$$\text{Sintering rate} \propto \frac{1}{r^n} \tag{7.21}$$

where n is a scaling exponent that depends on the dominant sintering mechanism. If sintering is controlled by surface or grain boundary diffusion, $n = 4$. If sintering is controlled by bulk diffusion, $n = 3$. If sintering is controlled by vapor-phase transport, $n = 2$. Finally, if sintering is controlled by viscous flow (generally only applicable to amorphous materials near T_{m}), $n = 1$.

- Because surface smoothing, coarsening, grain growth, and sintering all generally involve solid-state mass transport, they tend to be temperature-activated processes that obey an exponential Arrhenius-type rate law. Thus, the kinetic constants appearing in the equations for these processes (e.g., B^S, B^V, K_D, K_S, C) can all typically be modeled with a temperature dependence of the type (using K_D as an example) $K_D = K_{D,0}e^{-\Delta G_{act}/RT}$.

7.7 CHAPTER EXERCISES

Review Questions

Problem 7.1. Define the following:

(a) Capillary force

(b) Convex

(c) Concave

(d) Curvature-induced melting point depression

(e) Coarsening

(f) Equivalent grain radius

(g) Sintering

(h) Rule of six

Problem 7.2. Changing the starting particle size from 1 μm to 500 nm results in a 16-fold increase in the sintering rate, although the density of the sintered compact is not affected. Based on these observations, what is the likely dominant sintering mechanism in this system and why?

Calculation Questions

Problem 7.3. Solid silver has a density of 9.3 g/cm^3, a liquid–solid surface energy of 1.8×10^{-4} J/cm^2, and an enthalpy of melting of -11.3 kJ/mol. Given that silver has a bulk melting point of 962 °C, calculate the estimated melting point of a silver nanoparticle with a radius of 7.0 nm.

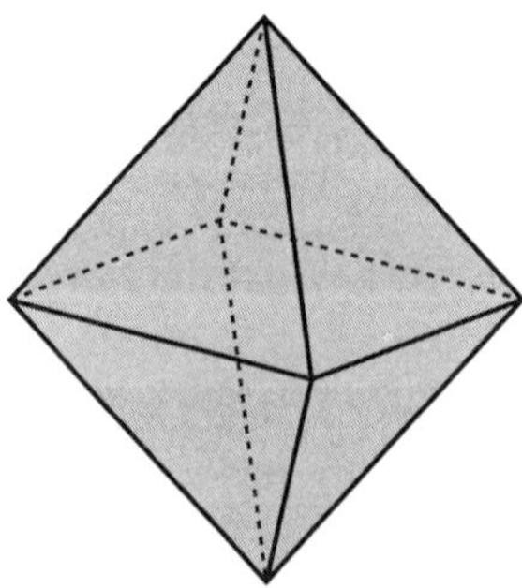

FIGURE 7.10   Schematic illustration of the octahedral single crystal in problem 7.4.

Problem 7.4. Consider the octahedral single crystal shown in Figure 7.10 composed entirely of (111) surface terminations. Compared to a spherical single crystal of the same total volume with an isotropic surface energy γ, what is the maximum allowed surface energy of the (111) surface terminations if the octahedral crystal is to be favored compared to the spherical crystal?

Problem 7.5. A polycrystalline material with an unknown initial grain size is subjected to an annealing study. After 3 h of annealing at 600°C, the grain size is measured to be 2 μm. The material is then annealed for an additional 4 h at 600°C and the grain size is measured to be 3 μm. Assuming grain growth in this material can be modeled using the 2D grain growth law. Determine the initial grain size of the material prior to annealing.

REFERENCES

1. K. J. Laidler, *Chemical Kinetics*, 3rd ed, Prentice Hall, Upper Saddle River, NJ, 1987.
2. J. E. House, *Principles of Chemical Kinetics*, Academic, Burlington, MA. 1996.
3. J. Espenson, *Chemical Kinetics and Reaction Mechanisms*, McGraw Hill, New York, 2002.
4. M. E. C. Robert, J. Kee, and P. Glarborg, *Chemically Reacting Flow: Theory and Practice.* Wiley, Hoboken, NJ, 2003.
5. J. Crank, *The Mathematicsof Diffusion*, Clarendon, Oxford, 1975.
6. L. Onsager, "Reciprocal relations in irreversible processes. I," *Physical Review*, vol. 37, no. 4, p. 405, 1931.
7. L. Onsager, "Reciprocal relations in irreversible processes. II," *Physical Review*, vol. 38, no. 12, p. 2265, 1931.
8. M. Hillert, *Phase Equilibria, Phase Diagrams and Phase Transformations: Their Thermodynamic Basis*, Cambridge University Press, New York 2007.
9. D. V. Ragone, *Thermodynamics of Materials*, Wiley, Hoboken, NJ, 1994.
10. F. Hummel, *Introduction to Phase Equilibria in Ceramic Systems*, CRC Press, Boca Raton, FL, 1984.
11. J.W. Cahn, "Spinodal decomposition," *The 1967 Institute of Metals Lecture, TMS AIME*, vol. 242, pp. 166–180, 1968.
12. T. Bower and M. Flemings, "Formation of the chill zone in ingot solidification," *AIME Met. Soc Trans.*, vol. 239, no. 2, pp. 216–219, 1967.
13. R. W. Balluffi, S. M. Allen, and W. C. Carter, *Kinetics of Materials*, Wiley, Hoboken, NJ, 2005.

PART III

APPENDIXES

APPENDIX A

UNITS

Issues relating to units inevitably crop up in any technical undertaking. For those of us in the United States, the challenge is further heightened by our continued insistence on imperial-based units such as inches, pounds, and degrees Fahrenheit. (Besides the United States, only two other countries in the world have not adopted the metric system: Liberia and Myanmar.) Units issues doomed the \$328 million Mars Climate Observer spacecraft in 1999, which crashed into the red planet thanks to a thruster impulse instruction that was erroneously provided in imperial lbs $\cdot$ s units instead of metric N $\cdot$ s units.

While a nice solution to units issues would be to use SI units for everything (see Table A.1 for the list of base SI units and common derived SI units), such units are not always convenient. In this textbook, a concerted effort has been made to use metric rather than imperial units for mass and length, so at least we do not have to deal with imperial-to-metric conversions. Nevertheless, we will always run into situations where other unit conversions are required. The goal of this section is to review the most common units and unit conversion we will likely be confronted with in the treatment of materials kinetics.

Among the most common non-SI units to appear in many kinetics problems are the calorie, the electron-volt, and the watt-hour—which are all units of energy—and the atmosphere, torr, and bar—which are all units of pressure. Their relationships to the SI units of energy (J) and pressure (Pa) are documented in Table A.2.

Most equations that we must evaluate involve a mixture of variables and fundamental constants (i.e., consider $PV = nRT$, where P, V, n, and T are variables and R is a fundamental constant). When plugging numbers into such equations, the units that are used for the variables should be consistent with the units used for the fundamental constants. The simplest way to ensure this consistency is to use SI units for

TABLE A.1   Selection of Commonly Used SI Units

Quantity	Unit Name	Unit Symbol
	SI Base Units	
Length	meter (or metre)	m
Mass	kilogram	kg
Time	second	s
Electric current	ampere	A
Temperature	kelvin	K
Amount of substance	mole	mol
Luminous intensity	candela	cd
	Common SI Derived Units	
Force	newton	$N = kg \cdot m/s^2$
Pressure, stress	pascal	$Pa = N/m^2$
Energy	joule	$J = N \cdot m = C/V = W/s$
Power	watt	$W = J/s = V/A$
Charge	coulomb	$C = A \cdot s$
Voltage, electric potential	volt	$V = J/C = W/A$
Capacitance	farad	$F = C/V$
Resistance	ohm	$\Omega = V/A$

TABLE A.2   Common Alternative (Non-SI) Units of Energy and Pressure

Quantity	Unit Name	Unit Symbol	Relationship to SI Unit
Energy	calorie	cal	1 cal = 4.184 J
	electron-volt	eV	$1 \text{ eV} = 1.602 \times 10^{-19} \text{ J}$
	watt-hour	Wh	1 Wh = 3600 J
Pressure	atmospheres	atm	1 atm = 101,325 Pa = 760 torr
	torr(mmHg)	torr	1 torr = 133.3 Pa
	bar	bar	$1 \text{ bar} = 10^5 \text{ Pa}$

Note: A human "calorie" (i.e., the calorie unit used on food package labels) is actually 1 kcal = 1000 cal.

all variables and apply the SI-based value for the fundamental constant. Table A.3 provides a helpful listing of the most common fundamental constants used in this textbook.

The bullets below provide a few more comments and recommendations about best practices for dealing with units.

- Equations involving pressure are a common source for units errors. Be careful! The SI unit for pressure is the pascal (Pa), but pressure is almost always given in atmospheres (atm). The safest policy is to always use SI units when dealing with expressions involving pressure. Thus, anytime you see a pressure, make sure to convert it to Pa (1 atm = 101,325 Pa).

TABLE A.3 Fundamental Constants

Constant Name	Constant Symbol	Value
Avogadro's number	N_A	6.022×10^{23}/mol
Gas constant	R	8.314 J/(mol $\cdot$ K)
		82.05 atm $\cdot$ cm^3/(mol $\cdot$ K) (non-SI)
Boltzmann's constant	$k = R/N_A$	1.381×10^{-23} J/K
		8.62×10^{-5} eV/K (non-SI)
Planck's constant	h	6.626×10^{-34} J $\cdot$ s
		3.76×10^{-15} eV $\cdot$ s (non-SI)
Elementary charge	e	1.602×10^{-19} C
Faraday's constant	$F = e \cdot N_A$	96,485 C/mol
Electron rest mass	m_e	9.11×10^{-31} kg
Vacuum permittivity	ϵ_0	8.85×10^{-12} F/m

Note: For R, k, and h, both the SI value of the constant and a commonly used non-SI value of the constant are provided. The SI value of the constant is listed first in each case.

- Another common confusion involves activation energies, which are often given in eV units rather than J/mol units. Activation energies given in eV units are typically denoted with the symbol E_a, while activation energies given in J/mol units are typically denoted with the symbol ΔG_a. Different symbols are used for these two activation energies because they do not have the same basis of comparison! ΔG_a represents a molar-normalized activation energy—it is given on the basis of *energy per mol*. (You can perhaps think of ΔG_a as the amount of activation energy required to process an entire mole of material.) On the other hand, E_a is not molar normalized. It represents a direct energy, not an energy per mole. (You can perhaps think of E_a as a direct measure of the height of the energy barrier, or the activation energy required per activation event.) Thus, when these activation energies are deployed in Arrhenius-type activation equations, they must be properly coupled with either the Boltzmann constant or the gas constant: E_a (eV) goes with k (eV/K), while ΔG_a (J/mol) goes with R (J/(mol $\cdot$ K)). Don't mix and match!

$$\text{Arrhenius activation expression equivalence:} \quad e^{-E_a/(kT)} = e^{-\Delta G_a/(RT)}$$

To convert an activation energy that is provided in eV units to an activation energy that is provided in J/mol units, it is necessary to multiply by Faraday's constant, which accounts for both the eV-to-J conversion and the conversion to a per-molar basis:

$$E_a \cdot F = \Delta G_a$$

Thus, activation energy $E_a = 1$ eV is the same as $\Delta G_a \approx 96.5$ kJ/mol.

PERIODIC TABLE

Periodic Table of the Elements

MAIN-GROUP ELEMENTS · TRANSITION ELEMENTS · INNER TRANSITION ELEMENTS

KEY

20	Atomic number
Ca	Symbol
Calcium	Element
40.078	Atomic mass

Color key: Metals (main group); Metals (transition); Metals (inner transition); Nonmetals; Metalloids

Main-Group and Transition Elements

Period	IA(1)	IIA(2)	IIIB(3)	IVB(4)	VB(5)	VIB(6)	VIIB(7)	VIII(8)	(9)	(10)	IB(11)	IIB(12)	IIIA(13)	IVA(14)	VA(15)	VIA(16)	VIIA(17)	VIII(18)
1	1 H 1.00794 Hydrogen																	2 He 4.002602 Helium
2	3 Li 6.941 Lithium	4 Be 9.01218 Beryllium											5 B 10.811 Boron	6 C 12.0107 Carbon	7 N 14.00674 Nitrogen	8 O 15.9994 Oxygen	9 F 18.99840 Fluorine	10 Ne 20.1797 Neon
3	11 Na 22.98977 Sodium	12 Mg 24.3050 Magnesium											13 Al 26.98153 Aluminum	14 Si 28.0855 Silicon	15 P 30.973761 Phosphorus	16 S 32.066 Sulfur	17 Cl 35.4527 Chlorine	18 Ar 39.948 Argon
4	19 K 39.0983 Potassium	20 Ca 40.078 Calcium	21 Sc 44.95591 Scandium	22 Ti 47.867 Titanium	23 V 50.9415 Vanadium	24 Cr 51.9961 Chromium	25 Mn 54.93809 Manganese	26 Fe 55.845 Iron	27 Co 58.93320 Cobalt	28 Ni 58.6934 Nickel	29 Cu 63.546 Copper	30 Zn 65.39 Zinc	31 Ga 69.723 Gallium	32 Ge 72.61 Germanium	33 As 74.92160 Arsenic	34 Se 78.96 Selenium	35 Br 79.904 Bromine	36 Kr 83.80 Krypton
5	37 Rb 85.4678 Rubidium	38 Sr 87.62 Strontium	39 Y 88.90585 Yttrium	40 Zr 91.224 Zirconium	41 Nb 92.90638 Niobium	42 Mo 95.94 Molybdenum	43 Tc (98) Technetium	44 Ru 101.07 Ruthenium	45 Rh 102.9055 Rhodium	46 Pd 106.42 Palladium	47 Ag 107.8682 Silver	48 Cd 112.411 Cadmium	49 In 114.818 Indium	50 Sn 118.710 Tin	51 Sb 121.760 Antimony	52 Te 127.60 Tellurium	53 I 126.90447 Iodine	54 Xe 131.29 Xenon
6	55 Cs 132.9054 Cesium	56 Ba 137.327 Barium	57 La 138.9055 Lanthanum	72 Hf 178.49 Hafnium	73 Ta 180.9479 Tantalum	74 W 183.84 Tungsten	75 Re 186.207 Rhenium	76 Os 190.23 Osmium	77 Ir 192.217 Iridium	78 Pt 195.078 Platinum	79 Au 196.96655 Gold	80 Hg 200.59 Mercury	81 Tl 204.3833 Thallium	82 Pb 207.2 Lead	83 Bi 208.98038 Bismuth	84 Po (209) Polonium	85 At (210) Astatine	86 Rn (222) Radon
7	87 Fr [223.0197] Francium	88 Ra [226.0254] Radium	89 Ac [227.0278] Actinium	104 Unq [261.11] Unnilquadium	105 Unp [262.114] Unnilpentium	106 Unh [263.118] Unnilhexium	107 Uns [262.12] Unnilseptium	108 Uno (265) Unniloctium	109 Une (265) Unnilennium									

Inner Transition Elements

6 Lanthanides

58	59	60	61	62	63	64	65	66	67	68	69	70	71
Ce Cerium 140.116	Pr Praseodymium 140.90765	Nd Neodymium 144.24	Pm Promethium (145)	Sm Samarium 150.36	Eu Europium 151.964	Gd Gadolinium 157.25	Tb Terbium 158.92534	Dy Dysprosium 162.50	Ho Holmium 164.93032	Er Erbium 167.26	Tm Thulium 168.93421	Yb Ytterbium 173.04	Lu Lutetium 174.967

7 Actinides

90	91	92	93	94	95	96	97	98	99	100	101	102	103
Th Thorium 232.0381	Pa Protactinium 231.03588	U Uranium 238.0289	Np Neptunium 237.0482	Pu Plutonium (244)	Am Americium [243.0614]	Cm Curium (247)	Bk Berkelium (247)	Cf Californium (251)	Es Einsteinium [252.083]	Fm Fermium [257.0951]	Md Mendelevium (258)	No Nobelium 259.1009	Lr Lawrencium [262.11]

ANSWERS TO SELECTED CALCULATION QUESTIONS

Chapter 1

1.8 (a) 1.987 cal/(mol · K), (b) 0.08205 L · atm/(mol · K), (c) 82.05 cm^3 · atm/(mol · K)

1.10 (a) 1.602×10^{-19} J, (b) 96,470 J, (c) 23.06 kcal/mol, (d) 8.618×10^{-5} eV/K, (e) 2.59×10^{-2} eV and 1.097×10^{-1} eV

Chapter 2

2.5 $P_{HCl} = 0.446$ atm, $P_{FeCl_2} = 0.277$ atm, and $P_{H_2} = 0.277$ atm

2.7 $P_{HI} = 0.926$ atm, $P_{I_2} = 0.0372$ atm, and $P_{H_2} = 0.0372$ atm

2.9 6.23×10^{-4} g/cm^3

2.10 1.59 g/cm^3 and 5.67×10^{-2} mol/cm^3

Chapter 3

3.5 27.7 kJ/mol

3.6 (a) 1.21×10^{-4}/yr, (b) 16,000 yr, (c) 38,000 yr

Chapter 4

4.7 (a) 2.46×10^{-1} cm^3/(cm^2 · s), (b) 2.02×10^{-5} g/(cm^2 · s), (c) 6.02×10^{18} molecules/(cm^2 · s), (d) 1.93 A/cm^2

4.8 260 s = 4.3 min

4.12 (a) 1.1×10^3 m/s, (b) 5.76×10^{18} molecules/cm^3, (c) 0.38 μm, (d) 1.4×10^{-2} cm^2/s

4.14 $D_{N_2,O_2} = D_{O_2,N_2} = 1.58 \times 10^{-5}$ m^2/s vs. $D_{O_2} = 1.65 \times 10^{-5}$ m^2/s

Chapter 5

5.5 4.8×10^{-9} g/s

5.7 (a) 2.4 cm/s, (b) 52 kJ/mol, (c) 29 h, (d) Same

Chapter 6

6.4 120°

6.5 660 °C

Chapter 7

7.3 450 °C

7.5 0.5 μm

INDEX

F

G